INTERNATIONAL UNION OF CRYSTALLOGRAPHY
MONOGRAPHS ON CRYSTALLOGRAPHY

IUCr Monographs on Crystallography

1 *Accurate molecular structures: their determination and importance*
   A. Domenicano and I. Hargittai, editors
2 *P.P. Ewald and his dynamical theory of Xray diffraction*
   D. W. J. Cruickshank, H. J. Juretschke and N. Kato, editors
3 *Electron diffraction techniques, Vol. 1*
   J. M. Cowley, editor
4 *Electron diffraction techniques, Vol. 2*
   J. M. Cowley, editor
5 *The Rietveld method*
   R. A. Young, editor
6 *Introduction to crystallographic statistics*
   U. Shmueli, G. H. Weiss
7 Crystallographic instrumentation
   L. A. Aslanov, G. V. Fetisov, G. A. K. Howard
8 *Direct phasing in crystallography: fundamentals and applications*
   C. Giacovazzo
9 *The weak hydrogen bond in structural chemistry and biology*
   G. R. Desiraju and T. Steiner
10 *Defect and microstructure analysis by diffraction*
   R. L. Snyder, J. Fiala and H. J. Bunge
11 *Dynamical theory of X-ray diffraction*
   A. Authier
12 *The chemical bond in inorganic chemistry*
   I. D. Brown
13 *Structure determination from powder diffraction data*
   W. I. F. David, K. Shankland, L. B. McCusker and Ch. Baerlocher, *editors*

Diffuse X-ray Scattering and Models of Disorder

T. R. WELBERRY

Institute of Advanced Studies
Australian National University

OXFORD
UNIVERSITY PRESS

Great Clarendon Street, Oxford OX2 6DP

Oxford University Press is a department of the University of Oxford.
It furthers the University's objective of excellence in research, scholarship,
and education by publishing worldwide in

Oxford New York

Auckland Bangkok Buenos Aires Cape Town Chennai
Dar es Salaam Delhi Hong Kong Istanbul Karachi Kolkata
Kuala Lumpur Madrid Melbourne Mexico City Mumbai Nairobi
São Paulo Shanghai Taipei Tokyo Toronto

Oxford is a registered trade mark of Oxford University Press
in the UK and in certain other countries

Published in the United States
by Oxford University Press Inc., New York

© Oxford University Press, 2004

A catalogue record for this title is available from the British Library

Library of Congress Cataloging in Publication Data
(Data available)
ISBN 0 19 852858 2 (Hbk)
10 9 8 7 6 5 4 3 2 1

Printed in India
on acid-free paper by
Thomson Press (I) Ltd.

PREFACE

Over the ninety years or so since the first discovery of the diffraction of X-rays by crystals, crystallography has grown into a very precise, widely applicable, and definitive tool. The undoubted success of this conventional crystallography (crystal structure determination) is due, to a very large extent, to the fact that the same basic method may be applied to materials as diverse as, on the one hand, a simple salt which contains only a few atoms per cell to, on the other, macromolecular crystals which may contain thousands of atoms per cell. The assumption in all cases is that the crystal consists of a three dimensional array of identical units, and this gives rise to a diffraction pattern consisting of discrete diffraction peaks (called Bragg reflections). The measurement and analysis of such discrete diffraction data has become routine for all but the most complex of examples.

Real materials, however, only approximate this ideal and the diffraction patterns of most materials contain, in addition to sharp Bragg peaks, a weak continuous background known as *diffuse scattering*. This scattering necessarily arises whenever there are departures of any kind from the ideal of a perfectly regular array of identical units. Such departures from ideality may arise in a whole variety of different ways and to different extents, but all of these effects may be brought together under a common name; *disorder*. Although much of present day knowledge of the solid-state has been derived from crystallographic studies using Bragg diffraction, the properties of many important materials are dependent not simply on the average crystal structure but are often crucially dependent on the departures from ideality that are present. For example, the useful mechanical properties of many alloys and ceramics, the opto-electronic properties of many materials, many electrical properties of semiconductors, high-temperature superconductivity, etc. depend upon the presence of various types of disorder.

For X-rays, the basic scattering event of a photon interacting with a crystal lattice occurs on a time scale of $\sim 10^{-16}$ s. This is several orders of magnitude faster than typical thermal vibration frequencies, so to a good approximation the X-ray diffraction experiment sees atoms or molecules statically displaced from their average positions as a result of such thermal motion. This may be termed thermal disorder and the scattering associated with it as thermal diffuse scattering or TDS. Disorder may also arise as a result of mixing different atomic or molecular species (solid solutions) or where a molecular species can pack into the basic crystal lattice in two (or more) quite different orientations. Such defects may occur as isolated point substitutional defects, clusters of such defects, or as interstitials where the defect atoms occur on sites not normally occupied in the average lattice. The scattering associated with this so-called occupational or chemical disorder is often termed short-range order (SRO) diffuse scattering. Occupational disorder will normally be accompanied by (static) atomic distortions that help alleviate local stresses and these distortions also give rise to diffuse scattering. In most real materials thermal, substitutional, and static displacement diffuse scattering will be present to varying degrees and instances can be found where any one of these dominates the diffraction pattern.

Whereas the conventional analysis of Bragg peaks provides information about the average crystal structure (information such as atomic coordinates, site-occupancies, or mean-square atomic displacements — all properties of single atom sites), diffuse scattering contains information about how pairs of atoms behave and is thus potentially a rich source of information on how atoms and molecules interact. Crystallographers have been aware of such scattering since the earliest times, but development of techniques for recording and analysing it have lagged well behind the advances made in conventional crystallography. Typical diffuse intensities are several orders of magnitude below Bragg peak intensities, and this was clearly a major impediment to earlier generations of researchers. However, with advances in X-ray sources (rotating anodes, synchrotron radiation) and in methods of detection (linear-detectors, area detectors, CCDs, image plates, etc.) good quality diffuse diffraction data are now much more readily accessible.

Obtaining good quality data is no longer the main impediment to obtaining information from diffuse scattering, and is not the main focus of this book. Nevertheless in order to set the scene for what follows, in Part I a brief description is given of the experimental methods that may be used to obtain diffuse scattering data.

The one major impediment to the utilisation of the rich source of information that diffuse scattering contains is the diversity and often complex combination of disorder effects that arise in nature and the fact that until recently no one simple method of analysis has been available which can usefully deal with all of them. The main aim in this book is to show how computer simulation of a model crystal *does* provide such a general method by which diffuse scattering of all kinds and from all types of materials can be interpreted and analysed. Such methods have only been feasible since the advent of powerful and relatively inexpensive computers. Part III of the book describes example studies of a wide variety of real systems which contain disorder of various kinds. These examples are intended not only to document the development of computer simulation methods for investigating and analysing disorder problems but also to provide a resource to help future researchers recognise the kinds of effects that can occur and to point the way to how new problems that are encountered might be tackled.

Of crucial importance in the study of disorder problems is an appreciation of the theory of disordered systems. This not only means an understanding of how different types of disorder give rise to particular diffraction effects but more importantly an appreciation of what is meant by *correlation*, since the diffracted intensity is the Fourier transform of the *pair correlation* function. What does it mean for occupancies and for displacements? How does it occur in a lattice? How does it vary with distance? What constraints are there on values that correlation can take? What is meant by *pair* correlations and *multi-site* correlations and what is the relationship between them? In order to address these questions Part II of the book describes a number of simple stochastic models of disorder which allow various concepts to be established and enable simple examples to be generated to illustrate the key principles. This provides the necessary background to the methods that are utilised in the analyses of real disordered systems that are discussed in Part III.

This book developed out of a set of notes that were prepared for a course of grad-uate lectures that were given under the auspices of the Troisième Cycle de la Physique en Suisse Romande in Lausanne in 1998. I am particularly grateful to Gervais Chapuis for the invitation to present these lectures. I am also greatly endebted to the numerous colleagues and collaborators, too numerous to mention, with whom I have had the priv-ilege to work over the years. All contributed enormously both to the studies described and to my understanding of disordered systems generally. In the latter context I might say that my early collaborations with Rex Galbraith, Ian Enting, Dave Pickard and Clark Carroll were crucial and provided me with the grounding which allowed many subse-quently encountered disorder problems to be solved. I would also like to acknowledge the contribution made by Brent Butler who was responsible (among other things) for writing the program *DIFFUSE*, which was used to calculate the diffraction patterns in most of the studies described in Part III of the book and which continues to be an invaluable research tool. I also wish to give special thanks to two of my current col-leagues: Ray Withers with whom I have had many years of fruitful collaboration and with whom I have been able to discuss in depth the most complex of disorder problems and Aidan Heerdegen for his tireless and expert help in preparing the manuscript of this volume. Finally I would like to thank my wife, Linda, for her constant encouragement and support particularly during the time the book was being written.

Canberra   T.R.W.
May 2004.

CONTENTS

Part I

Experiment

1

MEASUREMENT OF DIFFUSE SCATTERING

1.1 Introduction

Because typical diffuse scattering intensities are $\sim 10^3$–10^4 lower than those of Bragg peaks, in order to be able to record substantial regions of a full reciprocal-space intensity distribution some kind of multi-detection is required. If a single point counter were to be used to record data at increments of say 0.1 of each of the three reciprocal lattice vectors this already represents a 10^3 increase in the time required for an experiment compared to the Bragg experiment. Coupled with the 10^3–10^4 lower intensity the combined factor is more like 10^6–10^7. Although the advent of synchrotrons has meant that available incident X-ray intensities may be more intense than conventional laboratory sources by factors which exceed these kinds of magnitudes this is not a cost-effective way of using these expensive facilities. The real gain is obtained by recording many reciprocal points simultaneously.

Early measurements of diffuse scattering used X-ray film to record the scattered patterns. Commonly used was the so-called Laue method in which a narrow collimated beam of X-rays from a tube source was directed onto a stationary crystal. Even without the use of a monochromator the resulting scattering pattern largely resulted from the intense characteristic radiation (e.g. $CuK\alpha$). The Ewald sphere corresponding to the characteristic radiation intersected the diffuse features in reciprocal-space, producing a continuous distribution of intensity in the scattered pattern. This was either recorded on a flat plate placed behind the crystal or (to obtain more extensive coverage) on a cylindrical film surrounding the crystal as shown in Fig. 1.1. Note that in the diagram the Ewald sphere does not pass through any of the Bragg peaks so the scattered intensity shows only the diffuse peaks indicated by the broad dark line segments. If the incident X-ray beam also contains the 'bremstrahlung' or white radiation the scattered pattern also contains sharp peaks resulting from Bragg reflection from the reciprocal lattice points using the appropriate wavelength selected from the white radiation spectrum.

X-ray film is in many ways a very good two-dimensional detector since it has good sensitivity and excellent spatial resolution. It's main limitations are the fact that it needs to be chemically developed making it inconvenient for any automated or quantitative measurement. In addition it has fairly limited dynamic range. Film has therefore largely been superceded by other forms of detection. Nevertheless the stationary crystal or Laue method developed for use with film forms the basis of many of the currently used methods using other detectors.

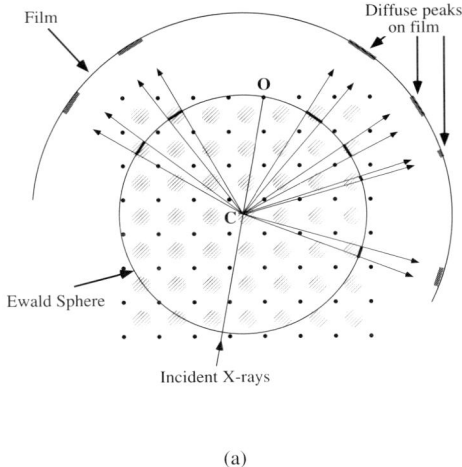

(a)

Fig. 1.1   The stationary crystal or 'Laue' method of recording diffuse scattering on film.

1.2   2D Data collection

1.2.1   *Using a linear position-sensitive detector*

Many of the observed diffraction patterns shown in this book were recorded on the diffractometer system which is shown in Fig. 1.2 (see Osborn and Welberry, 1990). This was designed specifically for the purpose of measuring complete 2D planar sections of diffuse scattering data from aligned single crystals and is based on a linear curved position-sensitive wire detector (PSD) manufactured by STOE, and described in detail by Wölfel (1983). The PSD essentially simultaneously records scattered X-rays over the range of the diffraction angle, 2θ, from $\sim 3°$ to $54°$. It may be repositioned to cover a higher-angle range if desired. The resolution along the detector is about $0.1°$ in 2θ and the data is binned accordingly into an array of 512 pixels, each of which corresponds to a $0.1°$ increment in 2θ and may be considered to be an independent measurement. Compared to a single point detector, therefore, the PSD provides an increase in detection speed of about this factor of 512. The entrance window to the detector may be chosen in the range 1–8 mm, the former value providing a resolution normal to the length of the detector comparable to that along it.

For use with laboratory sources it is still necessary to use crystals that are larger than would normally be considered suitable for Bragg reflection measurement. An optimum size is one with all three dimensions ~ 0.6 mm, giving a volume which is a factor of perhaps 3^3 greater than a typical specimen used for Bragg data collection. Combining this with the 512 factor gives a total enhancement relative to a single point detector and a Bragg data sized crystal of 1.4×10^4. Even with this enhancement factor measuring diffuse scattering is still relatively time-consuming and data collection times of ~ 1–3 days per section are typically required.

A number of features of the design of the instrument are worth mentioning since they contribute greatly to the quality of data that may be recorded. First, the collimation and beam-stop arrangement is designed to eliminate air-scattering as far as possible. The collimator extends to within about 2 mm of the sample and the beam stop is placed only about 8 mm after the sample. The main X-ray beam therefore only passes through ~ 1 cm of air, thus virtually eliminating air scattering. A second feature is the use of the curved detector with the crystal positioned at the centre of curvature. Scattered X-rays are thus incident normal to the detector window for all 2θ angles and this avoids parallax problems that occur for flat linear PSDs. Thirdly, the instrument is used in a mode where the detector is stationary during the data collection for a whole reciprocal layer. This avoids the possibility that the sensitivity characteristics of the detector may be altered by unwanted movements of the detection wire. The curved wire is held in place by a magnetic field and may be susceptible to mechanical shocks. Finally the open geometry of the system makes it easy to use ancillary equipment such as gas-flow cooling systems.

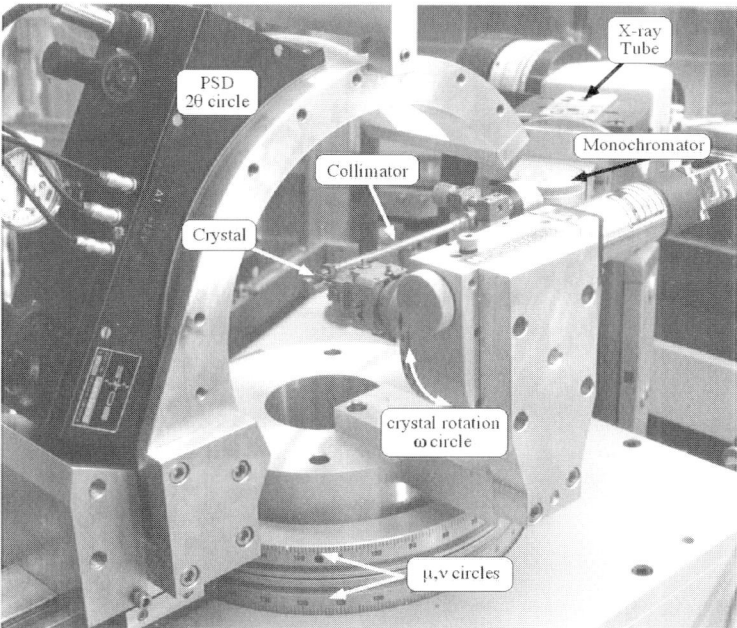

Fig. 1.2   The PSD diffractometer system.

Data is collected using a stationary position of the PSD by recording PSD scans as the crystal is rotated incrementally through the angle ω. Typically 1000 scans at increments in ω of $0.36°$ are recorded for each section. Figure 1.3 shows a set of data recorded for a single reciprocal section of the compound 1,3-dibromo-2,5-diethyl-4,6-dimethylbenzene (Bemb2) (see Chapter 8). Figure 1.3(a) shows a plot of the raw data rendered as a greyscale map. Each vertical line in the plot corresponds to a single PSD scan and successive lines in the horizontal direction are obtained as ω is incremented. Figure 1.3(a) is therefore analogous to a Weissenberg photograph. A simple transformation is required to rebin this data into undistorted reciprocal space coordinates and the resulting intensity distribution is shown in Fig. 1.3(c).

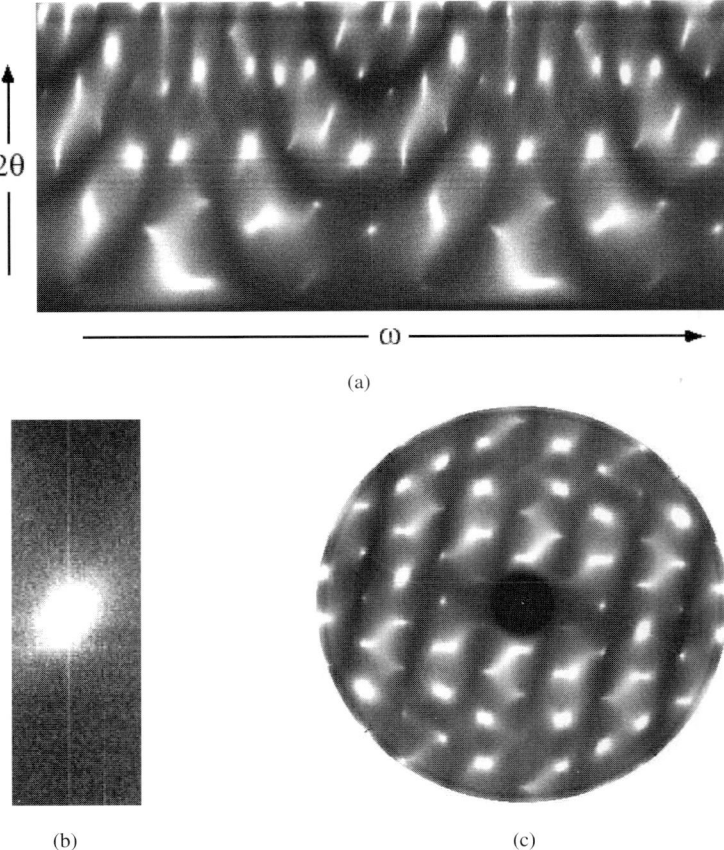

(a)

(b)   (c)

Fig. 1.3   A typical section of diffuse scattering data recorded on the linear PSD system. (a) Raw data in Weissenberg form. (b) An enlargement of a small region of (a) showing streaking through a Bragg peak position. (c) Undistorted reciprocal-space plot obtained from (a) after correction for non-uniform PSD response.

Figure 1.3 also serves to illustrate some limitations of the PSD. The counting efficiency of the PSD is not completely uniform along its entire length. In particular there are a number ripples in the sensitivity as a function of angle, which are most pronounced near the middle of the 2θ range. These are seen as narrow horizontal bands of lighter and darker intensity in the Weissenberg pattern Fig. 1.3(a). It is relatively straightforward to correct for this non-uniform response since for a given PSD gas pressure and HT supply the ripples are quite stable. The response is shown in Fig. 1.4(a) and a calibration curve is obtained by taking the ratio of the observed response curve to a smooth polynomial fitted to it. The resulting calibration curve has been used n the production of the undistorted pattern of Fig. 1.3(c) and no sign of the ripples can be seen in this figure.

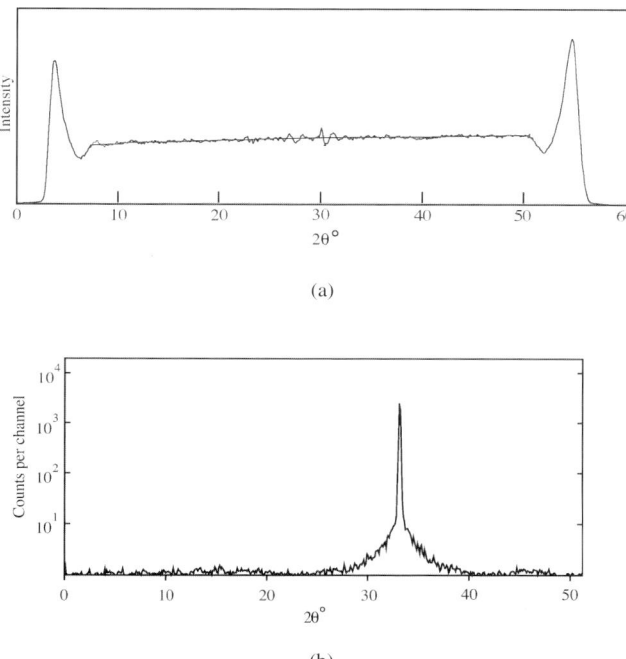

(a)

(b)

Fig. 1.4 (a) The counting efficiency of the PSD as a function of angle. The useful range lies between about $7°$ and $50°$ of 2θ. The smooth line is a sixth order polynomial fitted to this region of the response curve and provides the means of calibrating the response. (b) Logarithmic intensity plot along the PSD when a Bragg peak is incident. Note that the tails of the Bragg peak with intensity in the range of ~ 0.5–0.05% of the peak intensity covers about a $15°$ range of 2θ. Units are counts per channel in a 10 s interval.

A second problematic feature that may be seen in Fig. 1.3(a) is a narrow streak that extends some considerable distance in a vertical direction from some of the strongest Bragg peak positions. Figure 1.3(b) shows an enlargement of one of these. Since ω is stepped $0.36°$ between exposures most Bragg peaks will not fall directly onto the detector but when one does the intensity is so high that the detector is completely swamped and anomalous counts are recorded over a considerable range of 2θ. This is shown in Fig. 1.4(b) where the counts recorded along the detector when a Bragg peak is incident are plotted on a log-scale. The Bragg peak has large wings extending over a range of

$15°$ of 2θ. The magnitude of intensity in these wings is only $\sim 0.5\%$ of the Bragg peak intensity, which would not be too much of a problem if Bragg peak intensities were being measured, but is clearly detrimental if accurate diffuse scattering measurements are required. In fact the patterns shown in Fig. 1.3 were obtained from a quite small crystal sample with a small mosaic spread. In such a situation it is relatively straightforward if a Bragg peak is encountered to step away from the exact Bragg position by a fraction of the ω increment. The small gap in the data may be replaced by interpolation using neighbouring ω scans.

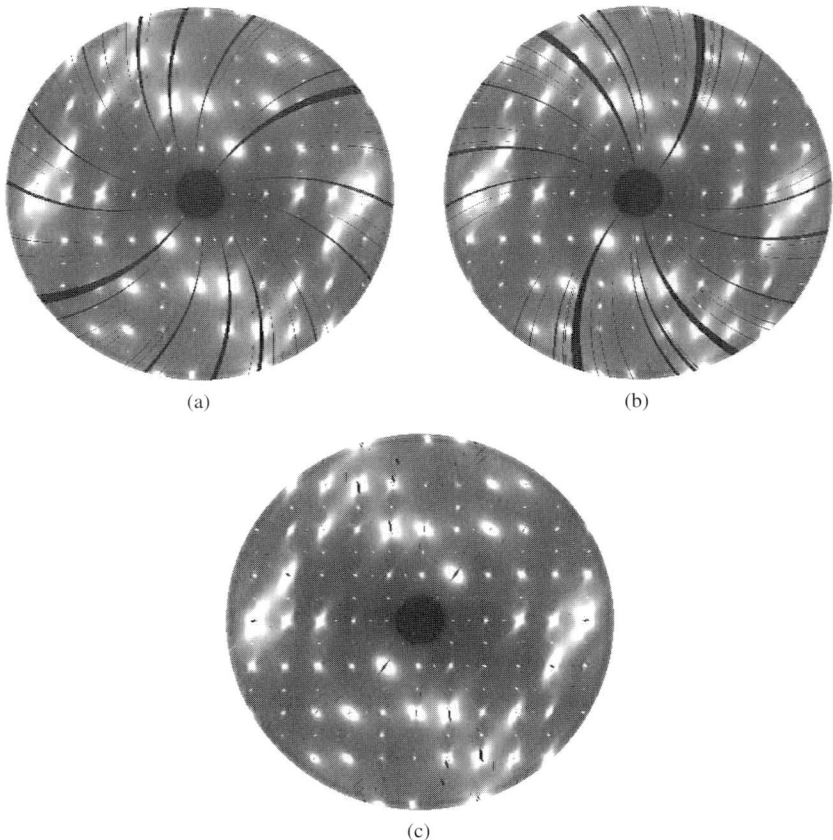

(a)   (b)

(c)

Fig. 1.5 Showing how data lost through the stepping away from (mainly low-order) Bragg peaks may be minimised by combining data collected from the two sides of the Ewald sphere. Combination of (a) and (b), each of which show quite large regions of data loss, yields (c) in which only some small regions of data are missing.

If the mosaic spread of the crystal is greater than the $0.36°$ ω increment then the problem is much worse and it may be necessary to step away from the Bragg position by a much larger amount. The worst cases occur when there are large Bragg peaks occurring at low 2θ values. In some cases the thermal diffuse scattering close to such strong Bragg peaks may be sufficient to swamp the detector and data from a number of consecutive ω scans may be lost. Such an example is shown in Fig. 1.5(a) where the black arcs indicate regions that have been lost in this way. Such lost data can often be replaced if the diffraction pattern possess symmetry (e.g. a mirror plane) but this is not possible for the monoclinic example shown. In this case an alternative strategy can be employed. This is to record a second set of data in which the ω-goniometer is placed on the other side of the PSD (i.e. the direction of the ω-axis is reversed). In this case the Ewald sphere passes through the reciprocal section in the opposite (clockwise) direction and the lost regions are quite different (see Fig. 1.5(b)). By combining (a) and (b) the pattern shown in Fig. 1.5(c) can be obtained. Now there are only a few very small regions where no data have been recorded, at the points where the black arcs in (a) and (b) intersect.

1.2.2 *Using image-plates at a synchrotron*

Attempts have been made to record comparable data from single 2D sections using high-energy synchrotron radiation (see Butler *et al.* (2000), Welberry *et al.* (2003*b*)). The use of high-energy X-rays has the advantage that diffraction angles are much smaller than for the laboratory based PSD system and the scattered X-rays may be recorded on a simple flat image plate (IP). The use of high-energies also overcomes problems encountered for the PSD system when the sample is highly absorbing. The experimental arrangement used is shown in Fig. 1.6. The X-ray beam is incident on the crystal which is prealigned about the desired crystal axis. As the crystal is rotated incrementally about this axis an IP is translated behind the stationary layer-screen slit to record successive vertical lines of scattering. The translation corresponding to the basic rotation increment $\Delta\omega$ is the same as the slit width so that each vertical line on the IP is an independent measurement corresponding to a particular value of ω. The exposed IP thus contains a Weissenberg image similar to those obtained for the PSD sytem.

Figure 1.7 shows data recorded by this method. For a single section of data 1440 individual exposures at 0.25° intervals of ω were recording using a 0.5 mm layer-screen slit and a corresponding 0.5 mm IP translation per step. This required four separate IPs for the complete section of data and a total measurement time of ~ 2 h. The final data obtained has much better counting statistics (much smoother pattern), somewhat better resolution in ω and much better resolution in 2θ than achievable with the laboratory based PSD system described in Section 1.2.1. However there were considerable technical problems encountered in processing the raw IP data to produce a reciprocal lattice image. Problems arose because of the fact that for a given plate the exposure to the X-rays had occurred progressively over a period of about 30 min. Consequently when the IP was subsequently read the most recently recorded parts of the image had only a few minutes in which the latent image could decay while the earlier parts had had in excess of 30 min. This sometimes resulted in the scale factor required at opposite ends of the plate differing by as much as a factor of two. This made it very difficult to place the data on a consistent intensity scale. Added to this was the further complication that the ex-

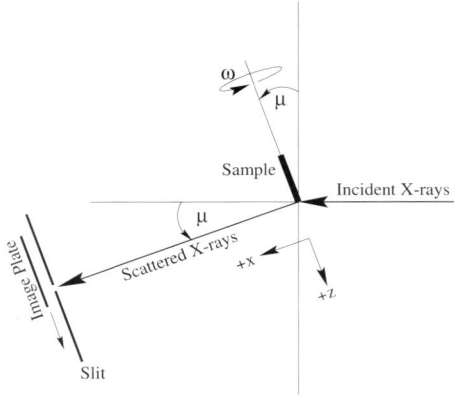

Fig. 1.6 The geometric arrangement (viewed from above) for recording a single reciprocal layer using the Weissenberg method on a high-energy synchrotron beam-line. As the crystal is rotated through the angle ω the IP is translated horizontally behind the layer-screen slit. The angle μ is altered to select different reciprocal layers normal to the rotation axis.

posures were made using incident photon count monitoring so the actual elapsed times for the exposures varied from plate to plate because of the decay of the synchrotron ring current. For the example shown it was possible to use the four-fold symmetry of the pattern to derive suitable internal scaling strategies, but for a sample with no such symmetry these scaling problems would have been difficult to solve.

1.3   3D data collection

1.3.1   *Using an automatic IP detector*

Both the laboratory based linear PSD method (Section 1.2.1) and the synchrotron layer-screen slit method (Section 1.2.2) provide the means of recording high quality data from particular 2D sections of a diffraction pattern. In selecting a single section, however, both methods waste a large fraction of the X-rays that are incident on the sample. X-rays that are scattered into other regions of the diffraction pattern are simply lost. Since no additional time would be required to record this scattering it would seem appropriate to do this if the means were available. This could be achieved simply by removing the layer-screen from the arrangement shown in Fig. 1.6 and recording a whole IP for every single ω-position as the crystal is incrementally rotated. This is only feasible if the exposure and measurement of the IP is automated. Devices capable of doing this are now readily available and Fig. 1.8 shows two frames taken from a series of 500 frames recorded using a Mar Research 345 IP Detector. Figure 1.9 shows the experimental set-up used to record this data.

The two frames of data shown in Fig. 1.8 are in fact consecutive exposures from a 3D data collection corresponding to a difference in crystal orientation ω of only $0.36°$. Of particular note is the fact that although the diffuse scattering in the two figures is very similar Fig. 1.8(a) shows an anomalous flaring or blooming around one of the

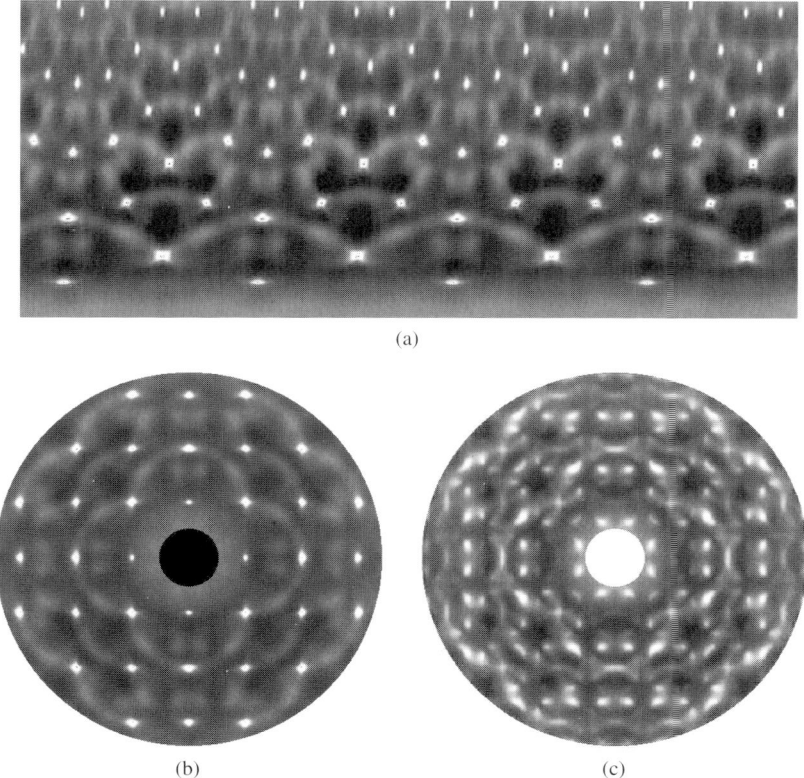

Fig. 1.7 Example sections of diffuse scattering data for Ca-stabilised cubic zirconia recorded using high-energy X-rays at the beam-line 1-ID-C at the Advanced Photon Source using the experimental arrangement shown in Fig. 1.6. (a) Weissenberg data for the 0 05$\mathbf{c}^*$ section. (b) Undistorted reciprocal section obtained from (a). $Q_{\max} = 8.4$, where $Q = 4\pi\sin(\theta)/\lambda$. (c) The corresponding 0.3$\mathbf{c}^*$ section. Note, these sections contain no Bragg peaks.

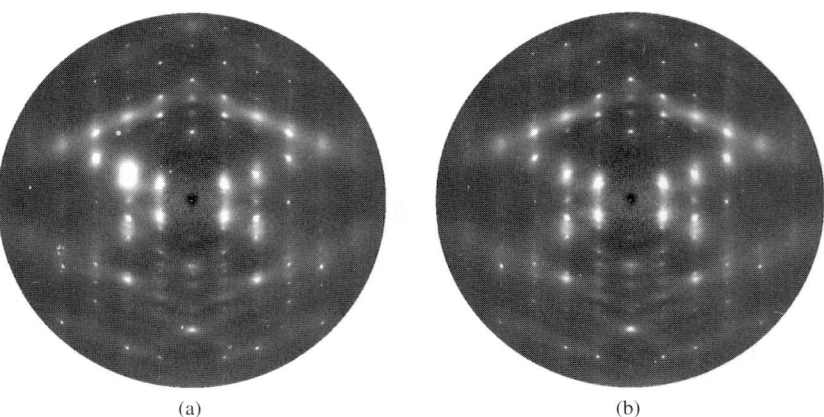

(a)   (b)

Fig. 1.8 Consecutive 2D frames from a 3D data collection using an IP detector. The two exposures correspond to a difference in crystal orientation ω of only 0.36° Note that although the diffuse scattering in the two figures is very similar (a) shows an anomalous blooming around one of the low-order Bragg peak positions. This only occurs when the Bragg peak itself is directly incident on the detector.

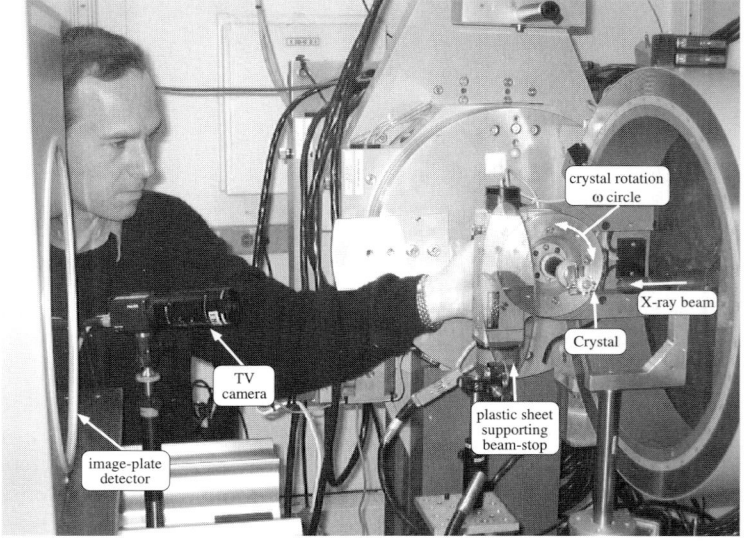

Fig. 1.9 The IP detector set-up.

low-order Bragg peak positions. This only occurs when the Bragg peak itself is directly incident on the detector. Although the ω-axis is approximately aligned with the mono-clinic **b**-axis of the crystal (horizontal in the figure) the alignment is not quite perfect and equivalent reflections in the two halves of the frame do not quite satisfy the Bragg condition simultaneously. The peak displaying the blooming on the left side of (a) has a symmetry equivalent on the right side that displayed a similar blooming in the imme-diately preceding frame.

This blooming, which occurs when strong Bragg peaks are incident on the IP detec-tor, represents a significant problem for the collection of diffuse scattering data. In fact the problem is not confined to only the strong Bragg peaks, although the effect is most prominent for those. If a plane section of reciprocal-space is reconstructed from the 500 individual raw data frames, the presence of a 2D bloom in a single frame produces a lin-ear streak in the reconstruction similar to that which occurred for the linear PSD system (see Fig. 1.3(b)). Figure 1.10(a) shows a zero level Weissenberg plot constructed from the 500 raw frames. Here it is seen that the effect is not just confined to the very strong low-angle peaks. Of the 3D volume of reciprocal-space recorded the regions affected by the blooming make up only a small percentage of the total and realistically speaking lit-tle information would be lost if these regions were simply discarded, but their presence does detract from the appearance of the pattern. In the linear PSD case it was possible to step away from the peaks causing the streaking and to replace much of the lost data by utilising symmetry-related measurements. For the present IP data it is not possible to step away from the offending peaks since a whole frame must be recorded and scanned before their presence is realised. However, such affected regions can be discarded after data collection. For a complete data collection comprising a 180° rotation of ω each re-gion of reciprocal-space or its symmetry-equivalent is recorded four times and so there is sufficient redundancy that it should be possible in principle to replace any such dis-carded data by an alternative measurement, in the way shown for the linear PSD data in Fig. 1.5. Here though, the use of high-energy X-rays has a disadvantage since the Ewald sphere is only slightly curving and the images obtained from the two sides of the sphere may not be sufficiently different to allow the replacement of the affected regions. As an alternative, however, since the streaks are usually only a single pixel wide, the discarded data may be replaced by interpolation using values from neighbouring pix-els. Figure 1.10(b) shows a comparable Weissenberg image after this has been carried out. Although some of the most problematic streaking remains the pattern has generally been considerably cleaned up. Undistorted reciprocal lattice sections obtained from the data shown in Fig. 1.10(b) are shown in Fig. 1.11.

There is, however, a second potential problem associated with the heavily overex-posed peaks. At the centre of each bloomed region, where the actual Bragg peak falls, a significant number of IP pixels are completely saturated. In the normal operation of the detector each pixel is erased when the read-out laser scans the stored intensity value. However, when a pixel is saturated a single passage of the laser is insufficient to com-pletely erase the stored intensity and a 'ghost' image is left that can be seen in the next frame. In fact for badly overexposed images this 'ghost' may remain even after sev-eral subsequent scans. This will have the effect of producing in reconstructed images a

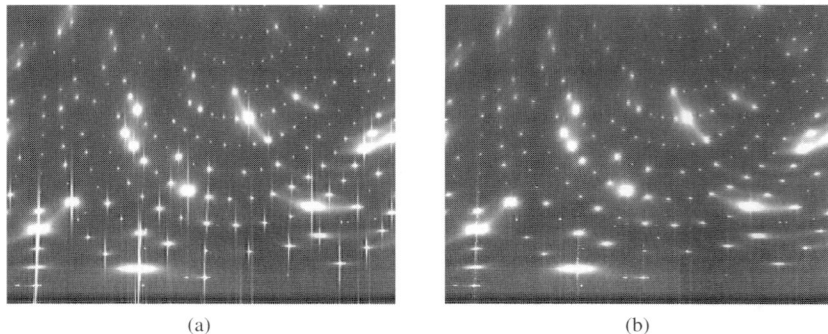

Fig. 1.10 Weissenberg images of the $(h\,0\,l)$ secton of data for $4,4'$-dimethoxybenzil. (a) Data extracted directly from the 500 raw data frames two of which are shown in Fig. 1.8. (b) the same data after processing to remove streaks.

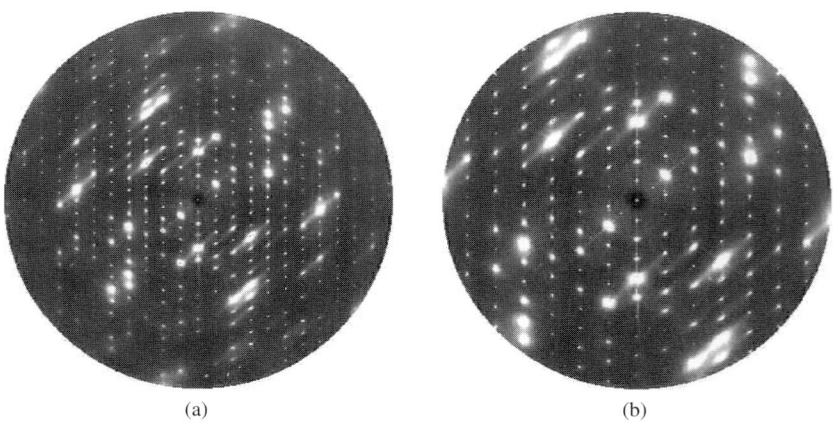

Fig. 1.11 The $(h\,0\,l)$ secton of data for $4,4'$-dimethoxybenzil obtained from the Weissenberg data shown in Fig. 1.10(b). For (a) $Q_{max} = 9.3\,\text{Å}^{-1}$. For (b) $Q_{max} = 5.7\,\text{Å}^{-1}$.

streak emanating from one side of a Bragg peak position and since it occurs at constant 2θ this will have an appearance akin to a powder streak. In the example shown, although this effect was detectable, its magnitude was minor compared to the blooming problem discussed above.

1.3.2 *Using a CCD detector*

An alternative to using an IP detector is to use a CCD detector. The available area of such detectors is generally less than that of the IP detector described in the previous section. This means that to obtain the same coverage of reciprocal space the detector must be brought closer to the sample. In addition the recorded image contains spatial distortions which have to be calibrated and corrected for. However the CCD has one great advantage over the IP detector and that is the recorded data from an individual frame can be read out in a fraction of a second rather than the 1–2 min required for the IP.

Figure 1.12 shows an example of data collected using a CCD detector. The same basic method of collection was used as in the previous IP example but with rather different experimental conditions. In this case much lower energy X-rays (20 keV) were used and the CCD detector (Bruker APEX) was placed at a distance of only 5 cm from the sample. At each crystal position, a curved 2D section of reciprocal-space is projected onto the CCD detector. In the example in the previous section in which much higher energy X-rays and a sample to detector distance of 65 cm were used, the curvature of the Ewald sphere is much less and the layer lines of the aligned sample are very close to straight lines (see Fig. 1.8) on the detector image. Here, with a close detector position and a much greater curvature of the Ewald sphere the layer lines have a hyperbolic shape on each of the individual frames. An example frame is shown in Fig. 1.12(b). The shape of the layer lines on the flat detector plane can be understood in the following terms. The intersection of a given reciprocal lattice plane with the Ewald sphere is a circle. This circle projects outwards from the sample as a cone of scattering and the intersection of this cone with the flat detector plane then produces a line in the form of a general conic section.

With much faster read-out times for the CCD compared to the IP there is very little 'dead time' so it was possible to record frames at much smaller increments in ω. In the example shown 1000 frames at 0.1° intervals were recorded each of 10 s exposure time. Figure 1.12(a) shows a Weissenberg plot obtained from these 1000 frames. To reconstruct such a reciprocal lattice plane, the relevant conic section was calculated and extracted from the CCD frame. This then corresponds to a vertical line on the Weissenberg plot. From this plot the undistorted reciprocal lattice section can be obtained in a straightforward manner and this is shown in Fig. 1.12(c).

Both the Weissenberg plot (Fig. 1.12(a)) and the resulting undistorted reciprocal lattice section (Fig. 1.12(c)) show the same kind of spurious streaking due to 'blooming' that was observed in the IP data of the previous section. In this case the 'blooming' artefact can be associated with the overflow of electric charge from one CCD pixel to another when a strong Bragg reflection saturates one or more pixels of the CCD chip. In this particular case the artefact appears to be rather less of a problem for the CCD

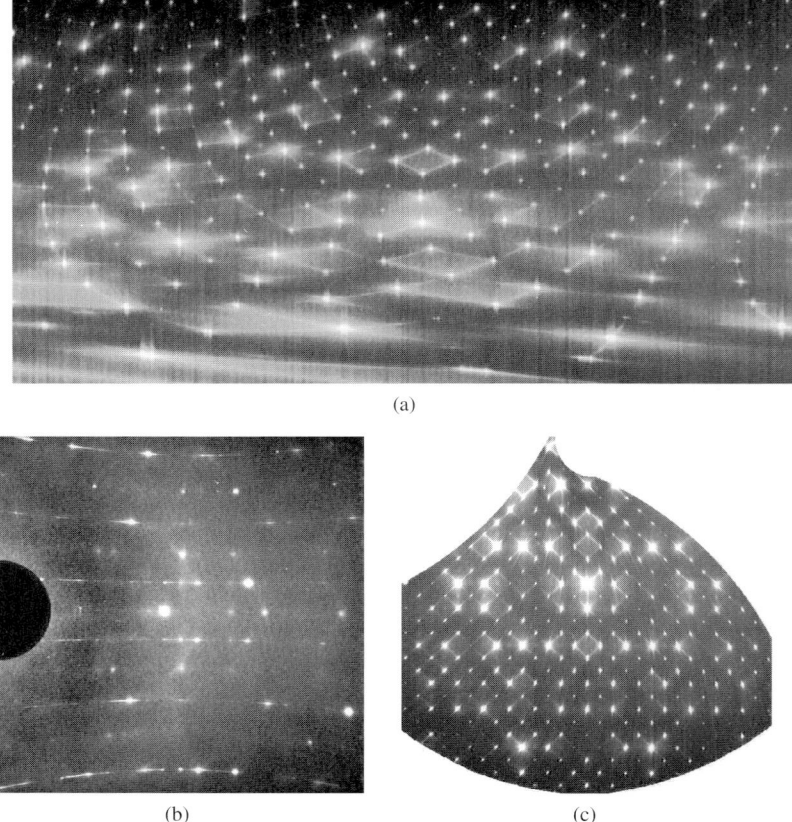

Fig. 1.12 Example of 3D data collection using a CCD detector. (a) A Weissenberg plot of the *hk*5 section for the zeolite mordenite. (b) A single one of the 1000 stationary Laue exposures (recorded at intervals of $0.1°$ in ω) from which (a) was constructed. (c) The derived planar reciprocal lattice section corresponding to the $100°$ total rotation. (Data used in this figure are reproduced with kind permission of Dr Branton Campbell.)

data than for the IP data but there were many differences between the two experiments (X-ray energy, type and size of sample, exposure time etc.) so that it is not possible to say with any certainty that one type of detector is superior to the other. It is clear that the basic problem for all the detectors described is that of trying to measure the very weak diffuse intensities in the presence of the very strong Bragg peaks.

1.4   Diffuse neutron scattering

Although this book is primarily concerned with X-ray diffraction it is nevertheless of in-terest to give here a brief description of the state-of-the-art in the use of neutrons for ob-taining diffuse scattering data. The neutron fluxes available at the best neutron facilities in the World are so low in comparison to the photon fluxes available for X-ray sources that it might be supposed that obtaining diffuse scattering data using neutrons is simply not viable, except perhaps for small selected regions of reciprocal-space. This is in fact not the case and for suitable samples valuable 3D diffuse scattering data can be obtained in a relatively short time. Whereas for X-ray diffuse scattering a single wavelength must be used so that most of the available X-rays are discarded, for neutrons produced at a spallation source it is possible to utilise the whole spectrum of incident neutron wave-lengths and separate their contributions to the scattering via time-of-flight spectroscopy. This in itself provides a large enhancement factor compared to experiments relying on only a single wavelength. When coupled with the possibility of surrounding the sam-ple with detector banks, which cover a large fraction of the complete solid angle, data acquisition rates sufficient to allow the measurement of diffuse scattering data can be achieved for a reasonably sized sample.

Figure 1.13 shows details of the single crystal instrument SXD at the ISIS Facil-ity of the Rutherford Appleton laboratory. The sample chamber is surrounded by 11 detector banks each of which comprises a 64×64 pixel detector array. Each detector array consists of a square array of 4096 $3 \times 3 \, \text{mm}^2$ pixel elements which are encoded by 16,384 optical fibres onto 32 photomultiplier tubes (these may be seen in Fig. 1.13(b) which shows the instrument during construction). As a single pulse of neutrons is inci-dent on the sample a complete time-of-flight (tof) spectrum is recorded for each pixel of each of the 11 detectors—a truly formidable amount of data recorded simultaneously. Figure 1.13(c) shows a typical tof spectrum for a single pixel.

Figure 1.14 shows the scattering geometry for a given stationary setting of a sample on the SXD instrument. The two circles shown are the Ewald spheres for the maximum and minimum wavelengths of neutrons in the incident beam. Consequently all points in reciprocal-space lying between the two circles (the shaded area) satisfy Bragg's law for some wavelength in the incident beam and will simultaneously produce scattering. The points A and B (and all points along the line A–B) satisfy Bragg's law for scattering at an angle $2\theta = 67°$. Similarly A$'$ and B$'$ (and all points along the line A$'$–B$'$) satisfy Bragg's law for a higher angle $2\theta = 113°$. These two angular directions correspond to the lower and upper limits of the range of scattering angle covered by of one of the detector banks on SXD (detector 5 in Fig. 1.13(a)) and this detector can therefore record simultaneously scattering from the whole of the darker shaded region. At the same time

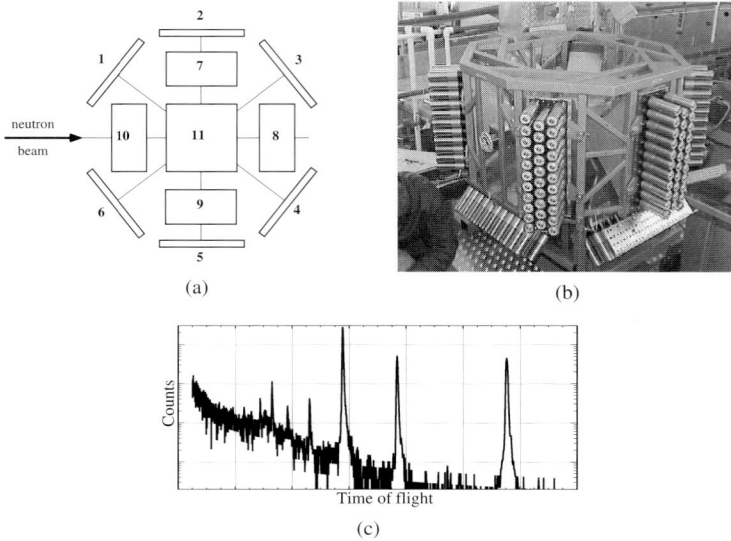

Fig. 1.13   (a) Schematic drawing of the arrangement of the 11 detector banks on the SXD time-of-flight instrument at ISIS. (b) View of the detector banks under construction (reproduced with kind permission of Dr D.A. Keen). (c) A typical time-of-flight spectrum recorded for a single pixel of one of the 64×64 pixel detector banks.

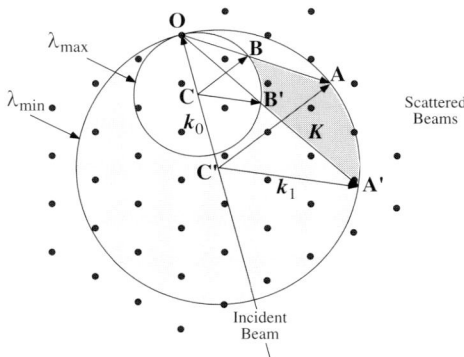

Fig. 1.14   Showing the scattering geometry for a single stationary setting of the crystal on SXD. The two circles are the Ewald spheres for the maximum and minimum wavelengths of neutrons in the incident beam. All parts of reciprocal-space between the two circles (light and dark shading) will satisfy the Bragg condition for some wavelength. All points in the darker shaded region, bounded by the scattering vectors OA and OA$'$, scatter in the range $2\theta = 90° \pm 23°$ and all this scattering is recorded simultaneously within a single detector bank.

the detectors 4 and 6 centred at $2\theta = 45°$ and $2\theta = 135°$ will record other segments of reciprocal space of a comparable area.

Since the equatorial detector banks subtend an angle of $\sim 46°$ of 2θ approximately $23°$ segments of reciprocal-space are mapped out with each single setting of the crystal orientation ω. A relatively small number of crystal settings are therefore required to map out the whole accessible 3D volume.

In the examples shown in Fig. 1.15, in which the crystal (d-benz l, see Welberry *et al.*, 2003*a*) possessed $P\bar{3}m$ Laue symmetry, only 3 settings were required to collect a full unique set of data ($60°$ of ω). Each setting corresponded to about 18 h of counting so that the whole volume of data was collected in less than 3 days. The sample used was a crystal measuring $3 \times 6 \times 12$ mm. Of particular note in these patterns is the fact that strong diffuse scattering can be observed well beyond the Q-range where Bragg peaks occur. This is because the scattering power (cross-section) of atoms for neutrons is not Q-dependent, unlike for X-rays where the atomic scattering factors fall away rapidly at high-Q.

These data plots show that, given a suitable sample, obtaining neutron diffuse scattering data over large regions of reciprocal-space is currently quite feasible at the World's best facilities. Given the special qualities of neutrons such as the ability to scatter at high Q and the fact that neutron cross-sections are not dependent on the atomic number Z, thereby enabling light atoms to be seen in the presence of heavy ones, for example, neutron scattering may be the technique of choice for some studies.

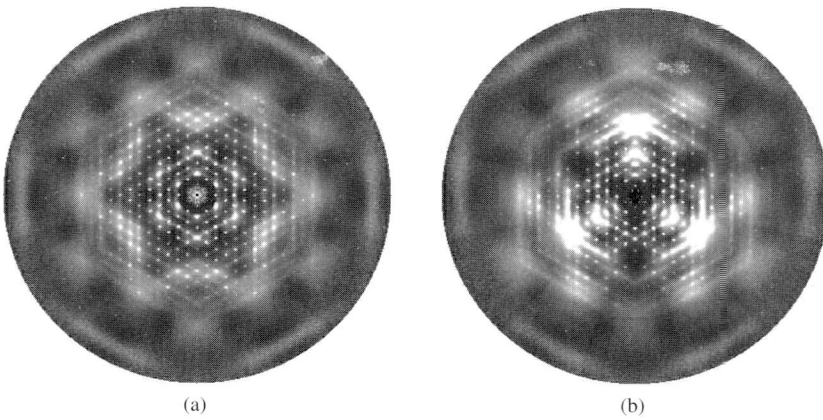

(a)   (b)

Fig. 1.15 Two example sections extracted from the 3D neutron diffuse scattering of d-benzil recorded on SXD. (a) *hk*0. (b) *hk*2.

A final important aspect to neutron scattering has so far not been mentioned. This is the fact that the neutrons have energies which are comparable to the vibrational modes of crystals (phonons) and so the scattering event is not elastic. If the energy of an incident neutron is E_n where

$$E_n = \hbar^2 k_0^2 / 2m_n, \tag{1.1}$$

and E_s and E_p are the energies of the scattered neutron and the phonon respectively, then,

$$E_s = E_n \pm E_p. \tag{1.2}$$

With reference to Fig. 1.14 k_0 is the incident wave-vector and k_1 is the scattered wave-vector. The consequence of eqn (1.2) is that the scattered wave-vector is either shortened or lengthened by an amount Δk_1. The ratio of $(k_1 \pm \Delta k_1)^2$ to k_0^2 must be the same as the ratio of the scattered energy, E_s, to the incident neutron energy, E_n, that is,

$$E_s/E_n = (E_n \pm E_p)/E_n = (k_1 \pm \Delta k_1)^2 / k_0^2. \tag{1.3}$$

If E_p is small in comparison to E_n then the ratio in eqn (1.3) will be close to unity and the resulting scattering will be very little different from the elastic scattering case. As can be seen from Fig. 1.14 each point in reciprocal-space scatters a neutron of a particular energy (though the same point may scatter into a different detector with a different energy neutron). For the d-benzil example shown in Fig. 1.15 most of the diffuse scattering is caused by low-energy phonons of energy ~ 1.1 meV. Since the range of incident neutron energies on the SXD beam-line is $0.3 - 300$ meV, most of the recorded scattering utilises energies much greater than the phonon energy and so to all intents and purposes is comparable to X-ray scattering data. In fact the images in Fig. 1.15 were constructed using only incident energies greater than about 20 meV. In the study of d-benzil (Welberry *et al.*, 2003*a*) inelastic effects were observed for those regions of reciprocal-space which utilised the lowest incident energies.

1.5   Electron diffraction

In the examples of diffuse scattering discussed in Part III there are a few instances in which electron diffraction patterns are shown. This is usually when X-ray scattering patterns are not available because the material under study is difficult to produce in the form of a single crystal of sufficient size for an X-ray study. Electron diffraction has this advantage that patterns may be obtained from very small specimens but in addition it is very sensitive to very weak scattering. This is because the atomic cross-sections for electron-diffraction are very much higher than those in either X-ray or neutron diffraction. However, this same attribute is also responsible for the fact that double (or multiple) diffraction effects are much more prevalent in electron diffraction. That is to say, after the initial scattering event the scattered beam is much more likely to be scattered again before it emerges from the sample. While not diminishing the overall strength of the scattered pattern it does tend to alter the distribution of intensity in different parts of the pattern so that any quantitative analysis is difficult and conclusions drawn from intensity distributions should be treated with caution. In the hands of experienced and expert operators who are able to manipulate samples to minimise the effects of multiple scattering invaluable diffuse scattering patterns can be obtained for materials that could not otherwise be studied.

Part II

Disorder models

2

DISORDER IN ONE DIMENSION

2.1 Diffraction intensity

When an X-ray beam is incident on a diffracting object the amplitude of the scattered wave in a given direction may be expressed in electron units in the form,

$$A(\mathbf{S}) = \sum_m f_m \exp(2\pi i \mathbf{S}.\mathbf{r}_m), \tag{2.1}$$

where f_m is the atomic scattering factor for an atom situated at the end point of a vector $\mathbf{r}_m$. f_m is the Fourier transform of the electron density distribution within an atom and for the purposes of the present discussion may be considered an entirely real quantity. $\mathbf{S}$ is the scattering vector defined by $\mathbf{S} = (\mathbf{s} - \mathbf{s}_0)/\lambda$, where $\mathbf{s}_0$ and $\mathbf{s}$ are unit vectors in the direction of the incident and diffracted beams. If θ is the diffraction angle then $|S| = 2\sin\theta/\lambda$. Using * to signify the complex conjugate, the diffracted intensity is obtained from eqn (2.1) as,

$$I(\mathbf{S}) = A(\mathbf{S}).A^*(\mathbf{S}) \tag{2.2}$$

$$= \sum_m f_m \exp(2\pi i \mathbf{S}.\mathbf{r}_m) \sum_{m'} f_{m'} \exp(-2\pi i \mathbf{S}.\mathbf{r}_{m'}) \tag{2.3}$$

$$= \sum_m \sum_{m'} f_m f_{m'} \exp(2\pi i \mathbf{S}.(\mathbf{r}_m - \mathbf{r}_{m'})) \tag{2.4}$$

$$= \sum_m \sum_j f_m f_{j+m} \exp(2\pi i \mathbf{S}.\mathbf{d}_m). \tag{2.5}$$

Whereas eqn (2.1) shows the scattered amplitude is the Fourier transform of the distribution of electron density, eqn (2.5) expresses the scattered intensity as the Fourier transform of a function involving the vector distances $\mathbf{d}_m = \mathbf{r}_m - \mathbf{r}_{m'}$ between pairs of atoms. This function is variously known as the autocorrelation function, the pair-distribution function, or the Patterson function.

If the object is a crystal with regular repetitive structure these equations may be simplified. Suppose the m atoms are arranged in N identical unit cells each containing n atoms. Each vector $\mathbf{r}_m$ may then be replaced by

$$\mathbf{r}_m = \mathbf{R}_N + \mathbf{r}_n, \tag{2.6}$$

where $\mathbf{R}_N$ defines the position of the Nth unit cell and $\mathbf{r}_n$ are local vectors to each atom in that cell. Equation (2.4) then becomes:

$$I(\mathbf{S}) = \sum_N \sum_{N'} \exp(2\pi i \mathbf{S}.(\mathbf{R}_N - \mathbf{R}_{N'})) \times \sum_n \sum_{n'} f_n f_{n'} \exp(2\pi i \mathbf{S}.(\mathbf{r}_n - \mathbf{r}_{n'})). \quad (2.7)$$

This may be expressed in terms of the structure factor, $F = \sum_n f_n \exp(2\pi i \mathbf{S}.\mathbf{r}_n)$, as

$$I(\mathbf{S}) = \sum_N \sum_{N'} F_N F_{N'} \exp(2\pi i \mathbf{S}.(\mathbf{R}_N - \mathbf{R}_{N'})) \quad (2.8)$$

or

$$I(\mathbf{S}) = \sum_N \sum_j F_N F_{j+N} \exp(2\pi i \mathbf{S}.\mathbf{R}_N). \quad (2.9)$$

For a perfect lattice in which F_N is independent of N and $\mathbf{R}_N$ define points on a regular lattice the summation may be carried out explicitly to yield,

$$I(\mathbf{S}) = N^2 |F|^2 \quad (2.10)$$

for values of $\mathbf{S}$ on the reciprocal lattice and $I(\mathbf{S}) = 0$ elsewhere.

2.2   One-dimensional disorder—layer structures

In discussing examples in which layers of atoms remain as perfect units the summations in eqn (2.1) to (2.5) may be carried out over the atoms within each layer so that in eqn (2.9) F_N would represent a layer form-factor and the summations over N would involve a one-dimensional (1D) array of points defining the origins of each of the layers.

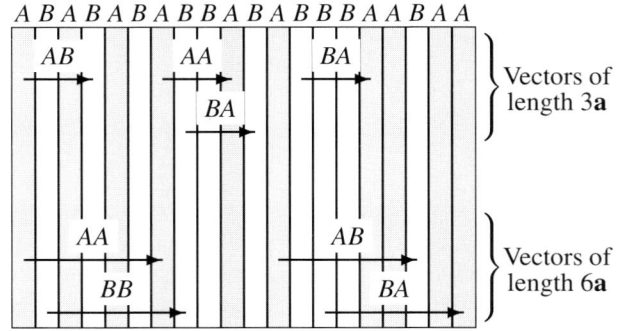

Fig. 2.1   Schematic representation of a layer structure.

Suppose there are two types of layer, A and B, with layer form-factors F_A and F_B and these are stacked with a regular spacing $\mathbf{a}$ so that $\mathbf{R}_N = N.\mathbf{a}$. Also suppose that of the N layers a fraction m_A are of type A and $m_B = 1 - m_A$ are of type B. In order to carry out the summations in eqn (2.8) all pairs of layers having the same separation $\mathbf{R}_N - \mathbf{R}_{N'}$ and hence the same exponent may be grouped together. For any particular value of the layer separation, $n\mathbf{a}$ say, four types of layer-pairs may be encountered, $A \ldots A$, $B \ldots A$, $A \ldots B$, and $B \ldots B$. The respective proportions of each are designated as P_n^{AA}, P_n^{BA}, P_n^{AB}, P_n^{BB}. These proportions are not independent and must satisfy three conditions,

$$P_n^{AA} + P_n^{BA} + P_n^{AB} + P_n^{BB} = 1 \tag{2.11}$$

$$P_n^{AB} + P_n^{AA} = m_A \tag{2.12}$$

$$P_n^{BA} + P_n^{AA} = m_A. \tag{2.13}$$

The second two conditions express the fact that at every lattice site the chance of there being an A-layer is equal to the overall fraction of layers that are type A. Using this notation the intensity may be written as,

$$I(\mathbf{S}) = \sum_{1-N}^{N-1} (N-n)^2 \exp(2\pi i \mathbf{S}.n\mathbf{a}) \tag{2.14}$$
$$\times \left(P_n^{AA} F_A F_A^* + P_n^{BA} F_B F_A^* + P_n^{AB} F_A F_B^* + P_n^{BB} F_B F_B^* \right).$$

Using eqns (2.11)–(2.13) parameters $C_n = \left(P_n^{AA} - m_A^2 \right)/m_B m_B$ may be defined, from which may be obtained,

$$P_n^{AA} = m_A^2 + C_n m_A m_B \tag{2.15}$$

$$P_n^{AB} = P_n^{BA} = m_A m_B - C_n m_A m_B \tag{2.16}$$

$$P_n^{BB} = m_B^2 + C_n m_A m_B. \tag{2.17}$$

The post-exponential factor in eqn (2.14) can be split into two parts, the first being constant and the second involving the parameters C_n. That is, eqn (2.14) may be rewritten,

$$I(\mathbf{S}) = \sum_{1-N}^{N-1} \left(m_A^2 F_A F_A^* + m_A m_B \left(F_A F_B^* + F_B F_A^* \right) + m_B^2 F_B F_B^* \right) \exp(2\pi i \mathbf{S}.n\mathbf{a}) \tag{2.18}$$
$$+ \sum_{1-N}^{N-1} m_A m_B C_n \left(F_A F_A^* - F_A F_B^* - F_B F_A^* + F_B F_B^* \right) \exp(2\pi i \mathbf{S}.n\mathbf{a})$$

$$= \left(m_A F_A + m_B F_B \right) \left(m_A F_A^* + m_B F_B^* \right) \sum_{1-N}^{N-1} \exp(2\pi i \mathbf{S}.n\mathbf{a}) \tag{2.19}$$
$$+ m_A m_B \left(F_A - F_B \right) \left(F_A^* - F_B^* \right) \sum_{1-N}^{N-1} C_n \exp(2\pi i \mathbf{S}.n\mathbf{a}).$$

The first term is recognisable as the intensity expression for a perfect crystal with a layer form-factor of $\bar{F} = m_A F_A + m_B F_B$, that is, the average layer form-factor. The second term, which depends on the difference between the two layer form-factors, gives the diffuse intensity. In order to see the form of this intensity it is necessary to consider the quantities C_n.

2.3   Correlations and short-range order

The C_n which appear in eqn (2.19) are in fact correlation coefficients or short-range order (SRO) parameters. They measure the degree of mutual dependence of the types of layers separated by given numbers of lattice spacings. For $C_n = 0$ there is no dependence, and the chance of finding an A-layer at a distance $n\mathbf{a}$ from a given layer is the same whether that layer is an A-type or B-type. For $0 < C_n < 1$ layers separated by $n\mathbf{a}$ are more likely to be of a like kind, $(A\ldots A$ or $B\ldots B)$ and this is called positive correlation. For $0 > C_n > -1$ layers separated by $n\mathbf{a}$ are more likely to be of the opposite kind, $(A\ldots B$ or $B\ldots A)$ and this is called negative correlation.

It is necessary to consider how the observed values of C_n might have arisen and what possible constraints apply to them. The C_n are derived from averages over the whole of the lattice (in practical terms, over a volume of crystal for which the X-ray beam may be considered coherent). This could of course include the case where a number of quite different effects (for example, growth conditions) occur in different regions of the crystal. However, in the present context only disordered lattices which are homogeneous will be considered, that is, ones in which probabilities and lattice averages are independent of position in the lattice: for example, the quantity m_A which represents the probability of finding an A-type layer is assumed to be a constant over the whole lattice.

This assumption of homogeneity imposes restrictions on the range of permissible values for C_n. Such restrictions start from the short-range properties and extend out to the higher-order (larger n) correlations. The simplest example of this is the restriction imposed on the nearest-neighbour correlation C_1 by the value of m_A. If C_1 is large and negative the lattice must consist of an alternating sequence of layers $ABABAB\ldots$ with few mistakes. This necessarily implies that m_A must have a value close to 0.5. Conversely if $m_A \gg 0.5$ then C_1 must be restricted to positive or small negative values. This can be described in a more quantitative way as follows.

2.4   Restrictions on correlation coefficients

First consider pairs of neighbouring layers for which the proportions of AA, AB, BA and BB pairs are expressed in terms of m_A, m_B, and C_1 from eqn (2.15) to (2.17). Using the fact that P_1^{AA} is a probability which must therefore be in the range $0 \to 1$, from eqn (2.15),

$$0 < m_A^2 + C_1 m_A m_B < 1 \qquad (2.20)$$

or

$$-m_A/m_B < C_1 < (1 + m_A)/m_A \qquad (2.21)$$

and similarly from eqn (2.16),

$$-m_B/m_A < C_1 < (1 + m_B)/m_B. \qquad (2.22)$$

This shows that there is no restriction on C_1 in the positive direction, but C_1 is limited in the negative values that may be obtained when $m_A \neq m_B$.

Next suppose that there is a disordered distribution of layers in which m_A and C_1 are fixed, and it is desired to ascertain the possible range of C_2 and higher correlations. This may be achieved by considering the proportions (or frequencies) of the triplets AAA, AAB, ABA, etc. Suppose,

$$P^{AAA} = f_0 \qquad P^{AAB} = f_1 \qquad P^{ABA} = f_2 \qquad P^{ABB} = f_3$$
$$P^{BAA} = f_4 \qquad P^{BAB} = f_5 \qquad P^{BBA} = f_6 \qquad P^{BBB} = f_7.$$

Then the frequencies f_i may be expressed in terms of m_A, P_1 and P_2 (it is more convenient here to use $P_1 = P_1^{AA}$ rather than the corresponding C_1). Thus,

$$1 = f_0 + f_1 + f_2 + f_3 + f_4 + f_5 + f_6 + f_7 \qquad (2.23)$$
$$m_A = f_0 + f_1 + f_2 + f_3 \qquad (2.24)$$
$$m_A = f_0 + f_1 + f_4 + f_5 \qquad (2.25)$$
$$m_A = f_0 + f_2 + f_4 + f_6 \qquad (2.26)$$
$$P_1 = f_0 + f_1 \qquad (2.27)$$
$$P_1 = f_0 + f_4 \qquad (2.28)$$
$$P_2 = f_0 + f_2. \qquad (2.29)$$

If the probability or frequency of AAA is defined as T it is possible to rearrange these equations to get,

$$f_0 = T \qquad (2.30)$$
$$f_1 = f_4 = P_1 - T \qquad (2.31)$$
$$f_2 = P_2 - T \qquad (2.32)$$
$$f_3 = f_6 = m_A - P_1 - P_2 + T \qquad (2.33)$$
$$f_5 = m_A - 2P_1 + T \qquad (2.34)$$
$$f_7 = 1 - 3m_A + 2P_1 + P_2 - T. \qquad (2.35)$$

For given values of m_A and P_1 these equations restrict the possible range of values that may exist for P_2 and the triplet probability T. To demonstrate this, consider a simple example for which $m_A = 0.5$, $P_1 = 0.4$ (corresponding to $C_1 = +0.6$). Equation (2.31) requires $T \leq 0.4$ while eqn (2.34) requires $T \geq 0.3$, whence the permissible range for P_2 may be seen from eqn (2.32) and (2.33) to be 0.3 to 0.5 (corresponding to a range $0.2 \leq C_2 \leq 1.0$).

For the next highest-order correlation, C_3, it is necessary to consider the frequencies of the 16 different combinations of four nearest neighbouring layers. Similar eqns to (2.30)–(2.35) may be derived in which the f_i are dependent on m_A, three different triplet probabilities, P_1, P_2, P_3, and a quadruplet probability $Q = P^{AAAA}$. Given previous choices of m_A, P_1 and P_2 the value of P_3 is similarly restricted to a narrow range. In general the restriction becomes more stringent as the order of the correlation increases.

3

PARTICULAR DISORDER MODELS

3.1 The simple Markov chain

In the previous chapter it was shown how values of the correlation coefficients in a disordered 1D lattice are subject to some restrictions, irrespective of the manner in which they are introduced into the lattice. In this chapter some particular simple models of the ways in which correlations might be introduced into a crystal lattice are considered, and their dependence on n and the consequent effect on the diffracted intensity investigated.

The simplest of all models is the nearest-neighbour Markov chain. Here it is imagined that layers of two types (A and B) are added to the lattice one at a time and the probability of the new layer being an A or a B is dependent only on the immediately preceding layer. Such a model might be considered reasonably realistic for a crystal into which disorder is introduced at growth and in which only short-range forces are important. For convenience the layers at sites i on the 1D lattice are represented by binary $(0,1)$ random variables x_i; where $x_i = 1$ denotes that site i is occupied by an A-type layer and $x_i = 0$ a B-type layer. The probability of adding an A-type layer at site i may then be expressed as

$$P(x_i|x_{i-1}) = \alpha + \beta x_{i-1}. \tag{3.1}$$

Equation (3.1) is not only a prescription for growing the disordered lattice but it also allows the correlation structure of the disordered lattice to be determined. First, by taking the average of eqn (3.1) over all i,

$$m_A = \alpha + \beta m_A \qquad \text{or} \qquad m_A = \alpha/(1-\beta). \tag{3.2}$$

Since the conditional probability on the left hand side of eqn (3.1) is independent of all $x_j (j < i)$, eqn (3.1) may be multiplied by any such independent variable without affecting its validity. In particular multiplying by x_{i-1},

$$x_{i-1}P(x_i = 1|x_{i-1}) = \alpha x_{i-1} + \beta x_{i-1}. \tag{3.3}$$

Note that $x_i^2 = x_i$ for $(0,1)$ variables. Taking averages over all i for this equation gives,

$$P(x_i = 1, x_{i-1} = 1) = P_1^{AA} = P_1 = \alpha m_A + \beta m_A. \tag{3.4}$$

Substituting for α from eqn (3.2) yields,

$$P_1 = \beta m_A + m_A^2(1-\beta) \tag{3.5}$$

$$\beta = (P_1 - m_A^2)/(m_A - m_A^2) = C_1. \tag{3.6}$$

The quantities α and β in the growth algorithm eqn (3.1) are thus seen to be simply related to the proportion of A-type layers m_A, and the nearest-neighbour correlation C_1, that occur in the resulting disordered lattice.

The general correlation coefficient C_n may be obtained as follows. Equation (3.1) is first multiplied by x_{i-n} and than taking averages as before yields,

$$P_n = \alpha m_A + \beta P_{n-1} \tag{3.7}$$

$$P_{n-1} = \alpha m_A + \beta P_{n-2}. \tag{3.8}$$

Combining eqns (3.7)–(3.8) to eliminate m_A gives,

$$P_n - P_{n-1}(1+\beta) + \beta P_{n-2} = 0. \tag{3.9}$$

Changing from the P_n to the correlation coefficients C_n the form of this equation remains unchanged and becomes,

$$C_n - C_{n-1}(1+\beta) + \beta C_{n-2} = 0. \tag{3.10}$$

This is a simple difference equation which has the solution $C_n = \beta^n$. Hence the result is obtained that for a simple 1D Markov chain the correlation coefficients, C_n, form a geometric series based on the nearest-neighbour correlation $C_1 = \beta$.

To obtain the diffracted diffuse intensity, β^n may be substituted for C_n in eqn (2.19), that is,

$$I(\mathbf{S}) = m_A m_B (F_A - F_B)(F_A^* - F_B^*)\left(1 + 2\sum_{n=1}^{N-1} \beta^n \cos 2\pi \mathbf{S}.n\mathbf{a}\right). \tag{3.11}$$

By allowing the number of layers, N, to tend to infinity and summing, the following form is obtained for the diffuse intensity,

$$I(\mathbf{S})_{\text{diffuse}} = K\frac{1-\beta^2}{1+\beta^2 - 2\beta\cos(2\pi\mathbf{S}.\mathbf{a})}. \tag{3.12}$$

Here K has been used to replace the constant terms involving m_A and F_A etc.

Example plots of the intensity function eqn (3.12) are shown in Fig. 3.1. The function is plotted over two cycles of the variable $2\pi\mathbf{S}.\mathbf{a}$, for different values of the correlation coefficient β. This form of the diffuse intensity peak shape is the single most important formula in the theory of diffuse scattering. It is characteristic of 1D nearest-neighbour systems. From a knowledge of this peak shape the primary correlation coefficient β can be determined by measurement of the peak width at half height (e.g. see Wilson, 1962), and this in turn leads to such concepts as correlation length and domain size. It will be seen in later sections that such concepts as correlation length and domain size are not *simply* transferable to higher-dimensional systems where mutual correlations in different directions are important.

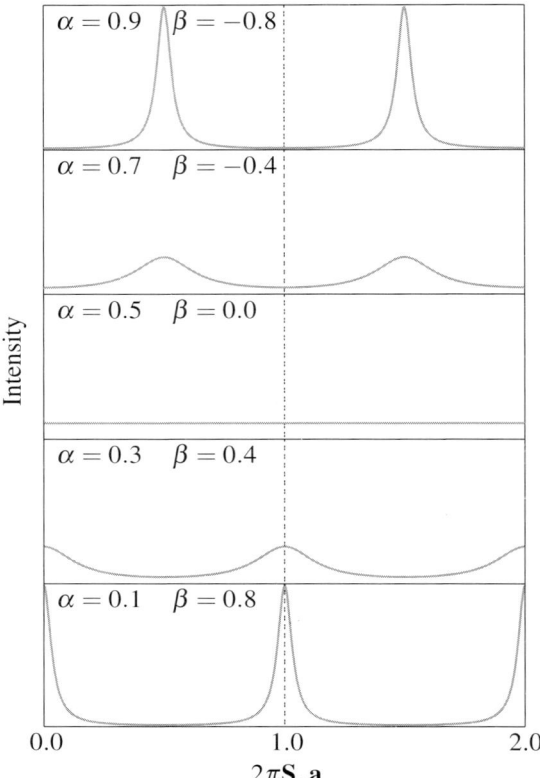

Fig. 3.1   Plots of the diffuse intensity given by eqn (3.12) as a function of the growth parameters α and β.

3.2   Alternative treatment for Markov chains—stochastic matrices

An alternative treatment of the Markov chain involves the use of stochastic matrices. Such treatments are to be found in mathematical textbooks such as Cox and Miller (1965), and are the methods favoured by Kakinoki and Komura (1952, 1954*a,b*, 1965). Here it is shown how the methods may be applied to the same simple model defined by eqn (3.1). It is possible to write the four transition probabilities defined by eqn (3.1) (i.e. $P(1|1)$, $P(1|0)$, $P(0|1)$ and $P(0|0)$) in matrix form as,

$$\mathbf{M} = \begin{pmatrix} P(0|0) & P(1|0) \\ P(0|1) & P(1|1) \end{pmatrix} = \begin{pmatrix} 1-\alpha & \alpha \\ 1-\alpha-\beta & \alpha+\beta \end{pmatrix}. \tag{3.13}$$

The following matrices are also defined

$$\mathbf{f} = \begin{pmatrix} 1-m_A & 0 \\ 0 & m_A \end{pmatrix}, \qquad \mathbf{V} = \begin{pmatrix} F_B F_B^* & F_B F_A^* \\ F_A F_B^* & F_A F_A^* \end{pmatrix}. \tag{3.14}$$

where F_A, F_B are the layer form-factors and * indicates the complex conjugate as in eqn (2.14). Using the previous definitions of the probabilities of pairs of neighbouring layers as P_1^{AA}, P_2^{BB} etc., it can be seen that these are the elements of the matrix $\mathbf{fM}$.

$$\mathbf{fM} = \begin{pmatrix} (1 - m_A)P(0|0) & (1 - m_A)P(1|0) \\ m_A P(0|1) & m_A P(1|1) \end{pmatrix} = \begin{pmatrix} P_1^{BB} & P_1^{BA} \\ P_1^{AB} & P_1^{AA} \end{pmatrix}. \tag{3.15}$$

To obtain the probabilities for the general case of layers separated by n-cell vectors the transition probability matrix is applied n times and this gives,

$$\mathbf{fM}^n = \begin{pmatrix} P_n^{BB} & P_n^{BA} \\ P_n^{AB} & P_n^{AA} \end{pmatrix}. \tag{3.16}$$

It is then seen that the quantity occurring in brackets in the general expression for the intensity, eqn (2.14), may be obtained as the *trace* or *spur* of the matrix $\mathbf{VfM}^n$, that is,

$$\mathrm{Tr}(\mathbf{VfM}^n) = \left(P_n^{AA} F_A F_A^* + P_n^{BA} F_B F_A^* + P_n^{AB} F_A F_B^* + P_n^{BB} F_B F_B^* \right). \tag{3.17}$$

To evaluate $\mathbf{M}^n$ the matrix $\mathbf{Q}$ is found such that

$$\mathbf{M} = \mathbf{Q} \begin{pmatrix} \lambda_1 & 0 \\ 0 & \lambda_2 \end{pmatrix} \mathbf{Q}^{-1}. \tag{3.18}$$

The eigenvalues of $\mathbf{M}$ are found to be $\lambda_1 = 1$ and $\lambda_2 = \beta$ and the matrices $\mathbf{Q}$ and $\mathbf{Q}^{-1}$ are,

$$\mathbf{Q} = \begin{pmatrix} 1 & \alpha \\ 1 & \alpha + \beta - 1 \end{pmatrix}, \qquad \mathbf{Q}^{-1} = \frac{1}{1 - \beta} \begin{pmatrix} 1 - \alpha - \beta & \alpha \\ 1 & -1 \end{pmatrix}. \tag{3.19}$$

The value of $\mathbf{M}^n$ is then,

$$\mathbf{M}^n = \mathbf{Q} \begin{pmatrix} \lambda_1^n & 0 \\ 0 & \lambda_2^n \end{pmatrix} \mathbf{Q}^{-1} = \mathbf{Q} \begin{pmatrix} 1 & 0 \\ 0 & \beta^n \end{pmatrix} \mathbf{Q}^{-1}. \tag{3.20}$$

It is also necessary to determine the equilibrium or steady-state probability of getting an A or a B at a single site. This is expressed as a vector, $\mathbf{m} = (1 - m_A,\ m_A)$, and must satisfy the equation,

$$\mathbf{mM} = \mathbf{m}. \tag{3.21}$$

Hence it is found

$$\alpha(1 - m_A) + (\alpha + \beta)m_A = m_A. \tag{3.22}$$

or

$$m_A = \frac{\alpha}{1 - \beta}. \tag{3.23}$$

Using this expression it is found,

$$\mathbf{Q} = \begin{pmatrix} 1 & m_A(1-\beta) \\ 1 & (1-m_A)(1-\beta) \end{pmatrix} \tag{3.24}$$

and

$$\mathbf{Q}^{-1} = \frac{1}{1-\beta} \begin{pmatrix} (1-m_A)(1-\beta) & m_A(1-\beta) \\ 1 & -1 \end{pmatrix}. \tag{3.25}$$

Evaluating $\mathbf{M}^n$ with the use of these matrices an expression which separates into two terms is found, the first independent of n and the second varying with n.

$$\mathbf{M}^n = \frac{1}{1-\beta} \begin{pmatrix} 1-\alpha-\beta & \alpha \\ 1-\alpha-\beta & \alpha \end{pmatrix} + \frac{\beta^n}{1-\beta} \begin{pmatrix} \alpha & -\alpha \\ \alpha+\beta-1 & 1-\alpha-\beta \end{pmatrix}. \tag{3.26}$$

Substituting for α from eqn (3.23) and multiplying by $\mathbf{f}$ leads to,

$$\mathbf{fM}^n = \begin{pmatrix} (1-m_A^2) & m_A(1-m_A) \\ m_A(1-m_A) & m_A^2 \end{pmatrix} + \beta^n \begin{pmatrix} m_A & -m_A \\ -(1-m_A) & (1-m_A) \end{pmatrix}. \tag{3.27}$$

This equation corresponds to eqns (2.15)–(2.17) with the general correlation coefficient $C_n = \beta^n$ and consequently this treatment leads to the same result for the diffuse intensity, that is, eqn (3.12). It is interesting to note that the eigenvalues of the stochastic matrix $\mathbf{M}$ only involve the correlation coefficient $C_1 = \beta$ and do not depend on the single-point properties which are described by $\mathbf{m} = (1-m_A, \ m_A)$ These latter only occur in the intensity equation as a result of the application of the $\mathbf{f}$ and $\mathbf{Q}$ matrices.

3.3   The 1D Ising model

An important model with a seemingly quite different physical basis is the linear-chain Ising model (Ising, 1925). Here a 1D chain is considered in which objects occurring at the lattice sites (the same A- and B-type layers as before may be considered) inter-act only with nearest-neighbours on either side. A given configuration of the lattice (a disordered sequence of A's and B's) is assumed to occur with a probability given by a Boltzmann partition,

$$P_{\text{config}} = \frac{\exp(-E_c/kT)}{\sum_c \exp(-E_c/kT)}, \tag{3.28}$$

where the summation is over all configurations c, and the interaction energy E_c is given by,

$$E_c = \sum_i J\sigma_i\sigma_{i-1}, \tag{3.29}$$

where σ_i are random variables at the sites i and represent the two states A and B. Follow-ing the usual convention in the Ising field, σ_i is a binary variable with states $(-1, +1)$, and consequently is related to the x_i variables used previously by $x_i = (\sigma_i + 1)/2$.

The Ising model (and this includes higher-dimension Ising models—see later), might be considered applicable in quite diverse circumstances. These range from use in magnetic materials where σ_i represents a magnetic spin, in molecular crystals such as p-chlorotoluene (e.g. Reynolds, 1975) where σ_i might represent one of two possible molecular orientations, to solid solutions of many kinds where different atomic or molecular species may occupy a given lattice site. In the first two examples the disorder is envisaged as a dynamical process involving a continual shuffling of the configuration of the random variables. In the last example different configurations are only achieved by diffusion processes, so only after a very long time can the particular static configuration that exists, be taken as representative of an equilibrium distribution.

In fact the pair interaction 1D Ising model, eqn (3.28), is exactly equivalent to the simple Markov chain of eqn (3.1) with $m_A = 0.5$ and $\alpha = (1 - \beta)/2$. This can be demonstrated as follows. First eqn (3.28) is rewritten in terms of the σ_i variables,

$$P(\sigma_i = 1|\sigma_{i-1}) = \tfrac{1}{2}(1 + \sigma_{i-1}). \tag{3.30}$$

For a given configuration produced using eqn (3.30) the total probability is obtained as a product of the individual probabilities invoked at the growth of each lattice site, that is,

$$P_{\text{total}} = \prod_i P(\sigma_i = 1|\sigma_{i-1}). \tag{3.31}$$

If the same configuration had arisen as a result of the Ising model the probability would have been expressed as,

$$P_{\text{total}} = \exp(-E_c/kT)/Z. \tag{3.32}$$

where Z is the normalising constant (partition function). Equating these expressions and taking logarithms, whence each side of the equation is in an analogous form of a sum over the lattice sites, the following is obtained,

$$\sum_i \ln\left(P(\sigma_i = 1|\sigma_{i-1})\right) = -\frac{1}{kT}\sum_i J\sigma_i\sigma_{i-1} - \ln(Z). \tag{3.33}$$

If the zero of the energy scale is chosen to correspond to the energy of the configuration in which the $\sigma_i = -1$ the $\ln(Z)$ term can be removed,

$$\sum_i \ln\left(P(\sigma_i = 1|\sigma_{i-1})\right) - \ln\left(\tfrac{1}{2}(1+\beta)\right)^N = -\frac{1}{kT}\sum_i J\sigma_i\sigma_{i-1}, \tag{3.34}$$

where N is the total number of sites.

To determine the relationship between the interaction J in the Ising formulation and the variable β in the Markov process, a particular configuration (see Fig. 3.2) is considered in which all but one of the σ_i variables are -1. The left-hand side of eqn (3.34) becomes,

$$\ln\left(\tfrac{1}{2}(1+\beta)\right)^{N-2} + \ln\left(\tfrac{1}{2}(1+\beta)\right)^2 - \ln\left(\tfrac{1}{2}(1+\beta)\right)^N. \tag{3.35}$$

In changing the single variable from the zero energy configuration, two interactions have been changed each of which yields $2J/kT$. It is then found,

$$2\ln\left((1-\beta)/(1+\beta)\right) = 4J/kT \tag{3.36}$$

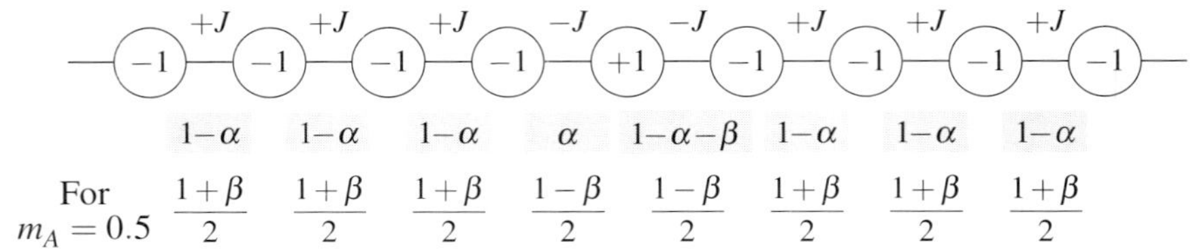

Fig. 3.2 Configuration in which all but one of the σ_i variables is -1. The top row of symbols indicates the energy contribution for each interaction. The bottom two rows are the individual Markov transition probabilities—the lower of the two shows the probabilities for $m_A = 0.5$, that is, α has been eliminated.

whence we obtain,

$$\beta = \tanh(-J/kT). \tag{3.37}$$

Consideration of other more complex configurations yields the same single relationship and it may be concluded that the Ising and Markov (or growth) descriptions are entirely equivalent.

3.4   Models involving second-nearest-neighbour and more distant interactions

A number of disorder problems encountered by early workers could not be explained purely in terms of nearest-neighbour interactions and hence theories were developed to allow the range of interaction to extend to second and more distant neighbours. Again, a comparison of the Markov or growth-disorder formulation with the Ising formulation is instructive.

In the growth or Markov formulation, account is taken of second-nearest-neighbours by allowing the growth probabilities to depend on two predecessors. In the most general way for a binary lattice it is possible to write in terms of the $(0,1)$ variables x_i used previously,

$$P(x_i = 1|x_{i-1}, x_{i-2}) = \alpha + \beta x_{i-1} + \gamma x_{i-2} + \delta x_{i-1} x_{i-2}. \tag{3.38}$$

3.4.1   *Linear form*

In order to obtain the correlation structure and hence the diffuse intensity from this model a simpler case is first considered in which the non-linear term δ is omitted, that is,

$$P(x_i = 1|x_{i-1}, x_{i-2}) = \alpha + \beta x_{i-1} + \gamma x_{i-2}. \tag{3.39}$$

The same procedure is followed as before in premultiplying by particular site variables and taking averages over the lattice. This yields,

$$
\begin{aligned}
m_A &= \alpha + \beta m_A + \gamma m_A \\
P_1 &= \alpha m_A + \beta m_A + \gamma P_1 \\
P_2 &= \alpha m_A + \beta P_1 + \gamma m_A \\
\vdots \quad &\quad \vdots \qquad \vdots \qquad \vdots \\
P_{n-1} &= \alpha m_A + \beta P_{n-2} + \gamma P_{n-3} \\
P_n &= \alpha m_A + \beta P_{n-1} + \gamma P_{n-2}.
\end{aligned}
\tag{3.40}
$$

By using the last two equations to eliminate αm_A the following difference equation is obtained,

$$
P_n - (1 + \beta)P_{n-1} + (\beta - \gamma)P_{n-2} + \gamma P_{n-3} = 0 \tag{3.41}
$$

or after substituting for P_n in terms of C_n as before,

$$
C_n - (1 + \beta)C_{n-1} + (\beta - \gamma)C_{n-2} + \gamma C_{n-3} = 0. \tag{3.42}
$$

The solution of this difference equation is again of the form $C_n = \lambda^n$, where now λ is a root of,

$$
\lambda^3 - (1 + \beta)\lambda^2 + (\beta - \gamma)\lambda + \gamma = 0. \tag{3.43}
$$

In this case the three roots are

$$
1 \qquad \text{and} \qquad \frac{1}{2}\left(\beta \pm \sqrt{\beta^2 + 4\gamma}\right) \tag{3.44}
$$

and the general solution for C_n is found to be,

$$
C_n = c_1 \left(\frac{1}{2}\left(\beta + \sqrt{\beta^2 + 4\gamma}\right) \right)^n + c_2 \left(\frac{1}{2}\left(\beta - \sqrt{\beta^2 + 4\gamma}\right) \right)^n. \tag{3.45}
$$

The (complex) constants c_1 and c_2 are found by using the values for the near-neighbour correlations C_1 and C_2.

Figure 3.3 shows a series of diffraction patterns illustrating this intensity expression, which have been calculated from computer simulations. In the examples, each horizontal row of points is an independent realisation of the model eqn (3.39). The diffuse intensity occurring as continuous vertical bands of scattering then has a profile given by eqn (3.45). The two terms of C_n in eqn (3.45) each contribute to the diffuse intensity expression a term of the form eqn (3.12). For $\beta^2 + \gamma \geq 0$ the two terms are both real and give rise to diffuse peaks at 0 or 0.5 in $2\pi\mathbf{S}.\mathbf{a}$, as for example, in Fig. 3.3(i) and Fig. 3.3(j). For $\beta^2 + \gamma < 0$ the two terms are complex but conjugate, so maintaining a real positive value for the intensity. Such a pair of complex terms gives rise to diffuse peaks with maxima at non-integral values of $2\pi\mathbf{S}.\mathbf{a}$. Note that this occurs when γ is relatively large and negative, as for example, in Fig. 3.3(a) and Fig. 3.3(b). In Figs 3.4 and 3.5 some calculated peak profiles are also shown.

　　　　　　　PARTICULAR DISORDER MODELS

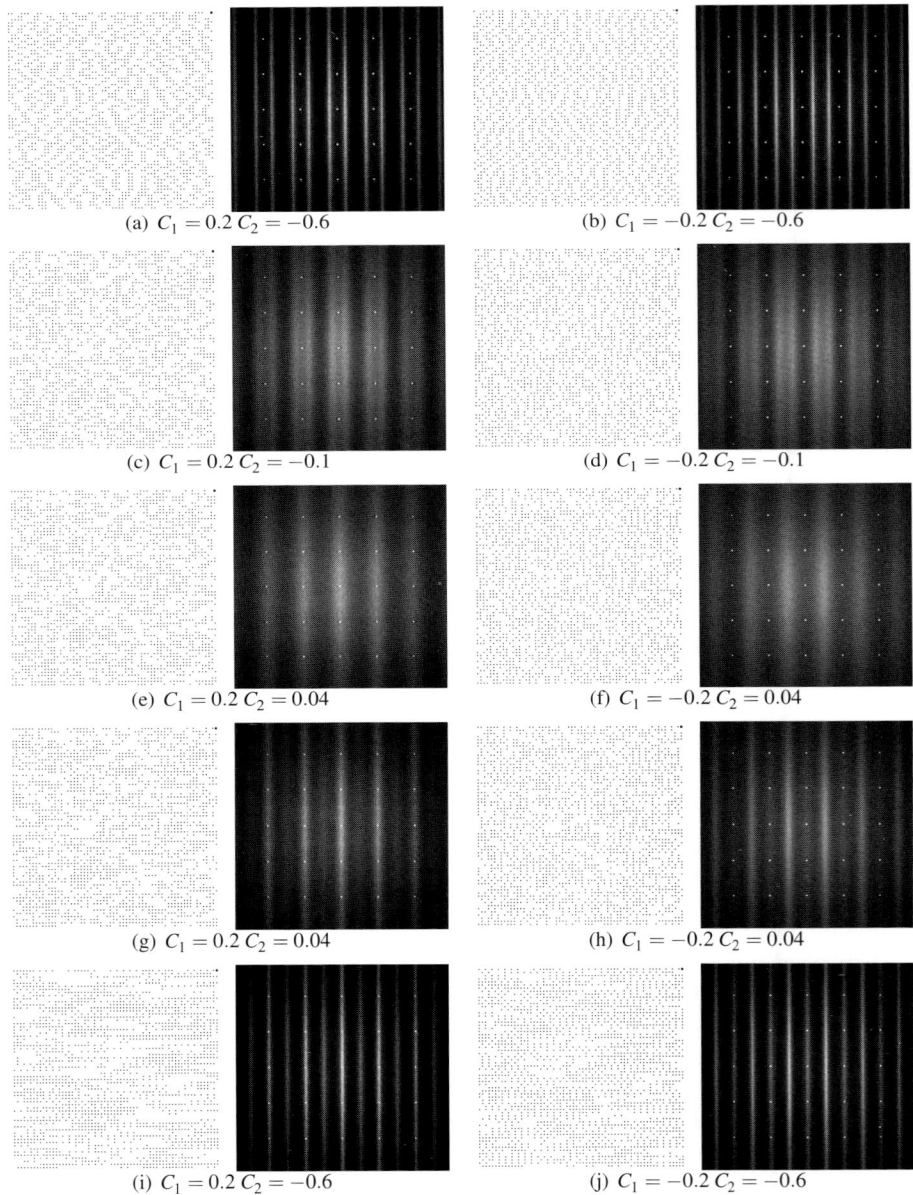

Fig. 3.3　Examples illustrating the second-neighbour linear Markov model.

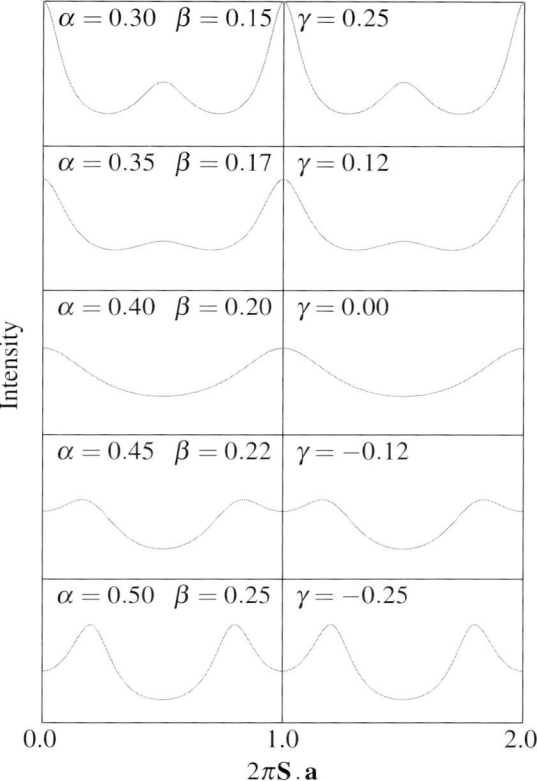

Fig. 3.4 Calculated diffuse peak profiles.

3.4.2 *Non-linear form*

Returning to the more general form, eqn (3.38), for second-neighbour interactions it is found on taking averages,

$$
\begin{aligned}
m_A &= \alpha + \beta m_A + \gamma m_A + \delta P_1 \\
P_1 &= \alpha m_A + \beta m_A + \gamma P_1 + \delta P_1 \\
P_2 &= \alpha m_A + \beta P_1 + \gamma m_A + \delta P_1 \\
T_2 &= \alpha P_1 + \beta P_2 + \gamma P_1 + \delta P_1 \\
&\ \ \vdots \qquad \vdots \qquad \vdots \qquad \vdots \qquad \vdots \\
P_{n-1} &= \alpha m_A + \beta P_{n-2} + \gamma P_{n-3} + \delta T_{n-2} \\
P_n &= \alpha m_A + \beta P_{n-1} + \gamma P_{n-2} + \delta T_{n-1},
\end{aligned}
\tag{3.46}
$$

where $T_{n-1} = \langle x_i x_{i-1} x_{i-n+1} \rangle$ or $T_{n-1} = \langle x_{i-1} x_{i-2} x_{i-n} \rangle$.

The first four of these equations may either be used to determine the low-order lattice averages m_A, P_1, P_2, T_2 in terms of the parameters α, β, γ, δ or vice versa. To obtain the solution for the general lattice average P_n eqn (3.38) may be multiplied by $x_{i-n+1} x_{i-1}$ and then averages taken to obtain,

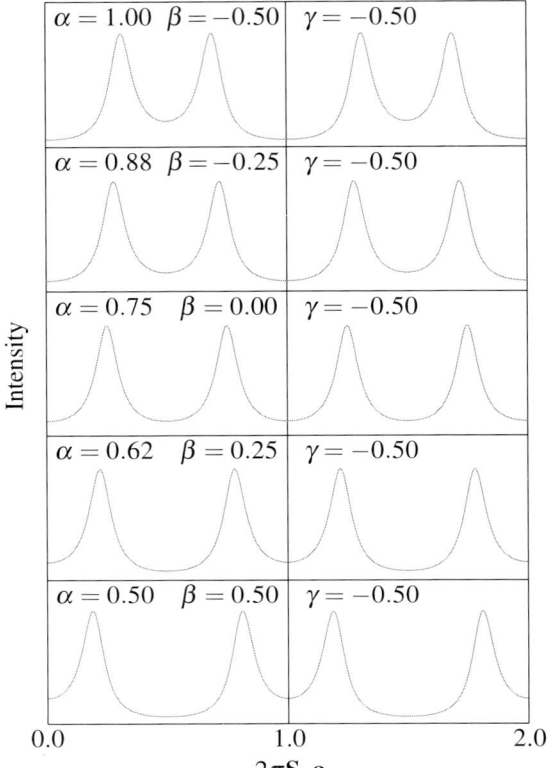

Fig. 3.5 Calculated diffuse peak profiles.

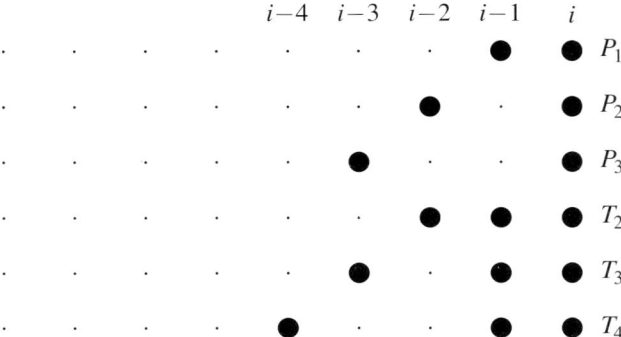

Fig. 3.6 Illustrating the lattice averages P_1, P_2, P_3, T_1, T_2, T_3 and T_4.

$$T_{n-1} = (\alpha + \beta)P_{n-2} + (\gamma + \delta)T_{n-2}. \tag{3.47}$$

Substituting for T_{n-1} and T_{n-2} from eqn (3.46) it is possible to obtain after some manipulation,

$$P_n = (\beta + \gamma + \delta)P_{n-1} + (\gamma + \alpha\delta - \beta\gamma)P_{n-2} \\ - \gamma(\gamma + \delta)P_{n-3} + m_a\alpha(1 - \gamma - \delta). \tag{3.48}$$

Comparing this equation with eqn (3.41) it may be seen that it could equally well have arisen from a linear model in which the growth probability depended on three predecessors, that is,

$$P(x_i = 1 | x_{i-1}, x_{i-2}, x_{i-3}) = a + bx_{i-1} + cx_{i-2} + dx_{i-3}, \tag{3.49}$$

where $a = \alpha(1 - \gamma - \delta)$, $b = \beta + \gamma + \delta$, $c = \gamma + \alpha\delta - \beta\gamma$ and $d = -\gamma(\gamma + \delta)$. In an analogous way as was found in the treatment of eqn (3.39) it is found that the solution of eqn (3.49) is again of the form $C_n = \lambda^n$ where λ satisfies a quartic. Again one root is unity (a property of stochastic matrices) and the three other roots each contribute a term of the form eqn (3.12) to the diffuse intensity.

Although eqn (3.49) gives rise to the same diffracted intensity as eqn (3.38), that is not to say that the actual disordered distributions are the same. The intensity is obtained from the difference equation, eqn (3.48), involving only the 1- and 2-point properties m_A and P_n, and multi-point averages such as T_n do not appear. It can be seen that there is a substantial difference between the two models by considering the value of $T_2 = \langle x_i x_{i-1} x_{i-2} \rangle$. Premultiplying eqn (3.49) by $x_{i-1} x_{i-2}$ and averaging gives

$$T_2 = \frac{a + b + c}{1 - d} P_1, \tag{3.50}$$

that is,

$$T_2 = \frac{\alpha + \beta + \gamma + \delta + \gamma(1 - \alpha - \beta)}{1 + \gamma(\gamma + \delta)} P_1 \tag{3.51}$$

while treating eqn (3.38) in a similar manner yields,

$$T_2 = (\alpha + \beta + \gamma + \delta)P_1. \tag{3.52}$$

To emphasise this point it is worth considering a simple example. Consider a particular case where the values of α, β, γ, δ in eqn (3.38) have the values,

$$\alpha = 0.18571, \qquad \beta = 0.78095, \qquad \gamma = 0.38095, \qquad \delta = -0.34763 .$$

The resulting lattice averages for this example may be shown to be

$$m_A = 0.5, \qquad P_1 = 0.35, \qquad P_2 = 0.29, \qquad P_3 = 0.29647, \qquad T_2 = 0.205 .$$

The corresponding values of a, b, c, d in eqn (3.48) are then,

$$a = 0.25646, \qquad b = 0.4, \qquad c = -0.05804, \qquad d = 0.14512$$

for which the resulting lattice averages can be shown to be

$$m_A = 0.5, \qquad P_1 = 0.35, \qquad P_2 = 0.29, \qquad P_3 = 0.29647, \qquad T_2 = 0.245 .$$

It is thus found that for a second-neighbour model the intensity distribution is characterised by the low-order lattice averages m_A, P_1, P_2 and T_2. While the triplet probability T_2 cannot itself directly affect the intensity, it has the indirect effect of influencing the higher-order 2-point properties P_3, P_4, etc. For a given choice of T_2 the resulting value of P_3 is sufficient to define the whole 2-point correlation field (and hence the intensity) via eqn (3.48). Such an intensity distribution could also have been produced by a linear model involving three nearest-neighbours, as given by eqn (3.49). This interesting result shows that interpretation of diffuse intensity data may be ambiguous. If the fit to observed data in a real experiment utilises the model eqn (3.49) instead of the model eqn (3.38) a quite different interpretation as to the underlying cause of the disorder is implied, even though the fit is identical.

The question arises as to whether the intensity from a non-linear model of any order (i.e. involving direct interaction with n nearest-neighbours) can alternatively be represented by that of a linear model of correspondingly higher order. To investigate this possibility it is worth considering the next-highest order non-linear model, in which the growth probabilities depend on three preceding neighbours. In its most general form this is,

$$P(x_i = 1 | x_{i-1}, x_{i-2}, x_{i-3}) = a + b x_{i-1} + c x_{i-2} + d x_{i-3} + e x_{i-1} x_{i-2}$$
$$+ f x_{i-1} x_{i-3} + g x_{i-2} x_{i-3} + h x_{i-1} x_{i-2} x_{i-3}. \quad (3.53)$$

Analogous to eqns (3.46) and (3.47) four equations involving general lattice averages are now obtained,

$$P_n = a m_A + b P_{n-1} + d P_{n-3} + e T_{n-1} + f S_{n-1} + f T_{n-2} + h Q_{n-1}, \qquad (3.54)$$
$$T_n = (a+b) P_{n-1} + (c+e) T_{n-1} + (d+f) S_{n-1} + (g+h) Q_{n-1}, \qquad (3.55)$$
$$S_n = (a+c) P_{n-2} + (b+e) T_{n-1} + (d+g) T_{n-2} + (f+h) Q_{n-1}, \qquad (3.56)$$
$$Q_n = (a+b+c+e) T_{n-1} + (d+f+g+h) Q_{n-1}, \qquad (3.57)$$

where, $P_n = \langle x_i x_{i-n} \rangle$, $T_n = \langle x_i x_{i-1} x_{i-n} \rangle$, $S_n = \langle x_i x_{i-2} x_{i-n} \rangle$ and $Q_n = \langle x_i x_{i-1} x_{i-2} x_{i-n} \rangle$.

Although the multi-point properties T_n, S_n, Q_n etc. can be eliminated from these equations to yield a single recurrence equation involving only the 2-point properties P_n, this equation no longer involves a finite number of terms as was found for the simpler case eqn (3.48). This means that arbitrary specification of the low-order properties T_2, T_3, S_3 and T_3 together with a given choice of m_A, P_1, P_2 and P_3 yields an intensity distribution which cannot be fitted exactly with a linear model of finite order. However, since in general the coefficients of P_{n-i} will decay with i, from a practical viewpoint a linear model involving seven neighbours may well be chosen to produce an intensity distribution that is virtually indistinguishable from that given by the non-linear model eqn (3.53).

3.5 Second-neighbour Ising model

It was shown earlier that the simple nearest-neighbour pair interaction 1D Ising model and a simple Markov chain in which the lattice average m_A was 0.5 are equivalent. In this section an Ising model in which the interaction energy is dependent on longer-range effects is considered in a similar fashion. For interactions involving up to second-nearest-neighbours the most general expression for the energy, E_c in eqn (3.28) becomes,

$$E_c = \sum_i H\sigma_i + J_1\sigma_i\sigma_{i-1} + J_2\sigma_i\sigma_{i-2} + K\sigma_i\sigma_{i-1}\sigma_{i-2}. \tag{3.58}$$

Here, in addition to the original nearest-neighbour pair interaction J_1, a term H involving individual sites, a second-neighbour pair interaction J_2, and a three-body interaction K are now included. Note that this model is defined by four parameters H, J_1, J_2 and K, which corresponds to the same number of parameters as the Markov formulation, eqn (3.38).

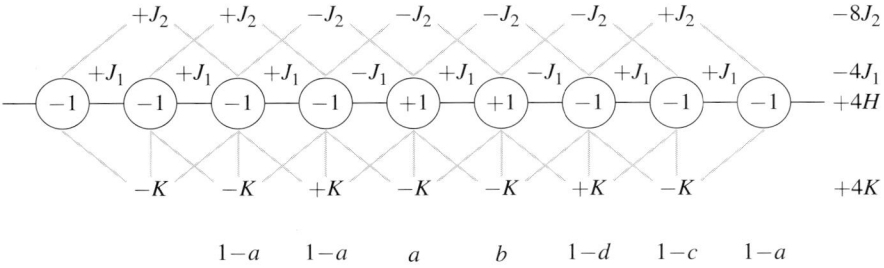

Fig. 3.7 Configuration in which all but two σ_i variables are -1. The top two rows of symbols show the first and second-nearest-neighbour two-body energies of interaction (J_1 and J_2), and the row directly below shows the three-body energy, K. To the right of the figure are the differences in Ising energy of this configuration from that in which all variables are -1. The bottom row of symbols indicates the growth transition probabilities for the corresponding Markov formulation.

To determine the relationship between H, J_1, J_2 and K and the Markov growth parameters α, β, γ and δ the same procedure as before is followed in considering particular lattice configurations. For convenience the configuration in which all $\sigma_i = -1$ is assumed to correspond to the zero of the energy scale and four particular configurations of the lattice in which a group of three variables takes a different local arrangement (see eqn (3.60)) are considered. For convenience also a set of parameters for the Markov description defined by,

$$\begin{aligned}
a &= \alpha = P(+1|-1,-1)\\
b &= \alpha + \beta = P(+1|-1,+1)\\
c &= \alpha + \gamma = P(+1|+1,-1)\\
d &= \alpha + \beta + \gamma + \delta = P(+1|+1,+1)
\end{aligned} \tag{3.59}$$

is used. One example configuration is shown in Fig. 3.7. Compared to the configuration in which all σ_i were -1 it is seen that the effect of flipping two of the variables to $+1$ has resulted in four of the J_2 Ising interactions changing from $+$ to $-$, two of the J_1 interactions changing from $+$ to $-$ and two of the K interactions changing from $-$ to $+$ as well as the two H interactions changing from $-$ to $+$. This makes the total change in energy equal to $4H - 4J_1 - 8J_2 + 4K$. In terms of the Markov transition probabilities four of the original $(1-a)$ factors have been replaced by $ab(1-c)(1-d)$. Equating the logarithms of the probability of the whole lattice configuration as before for this example and three others, four different equations are obtained.

Configuration

$-1\ +1\ -1$ $\quad \ln\left(\dfrac{a(1-b)(1-c)}{(1-a)^3}\right) = (2H - 4J_1 - 4J_2 + 6K)/kT$

$-1\ +1\ +1$ $\quad \ln\left(\dfrac{ab(1-c)(1-d)}{(1-a)^4}\right) = (4H - 4J_1 - 8J_2 + 4K)/kT$

$+1\ -1\ +1$ $\quad \ln\left(\dfrac{ab(1-b)^2(1-c)}{(1-a)^5}\right) = (4H - 8J_1 - 4J_2 + 8K)/kT$

$+1\ +1\ +1$ $\quad \ln\left(\dfrac{abd(1-c)(1-d)}{(1-a)^5}\right) = (6H - 4J_1 - 8J_2 + 6K)/kT.$

$$(3.60)$$

Consideration of further lattice configurations yields no new equations—all reduce to one of these four. From these equations the following relationships can be obtained,

$$H = \frac{1}{8}kT \ln\left(\frac{bcd^3(1-d)}{(1-a)^3 a(1-b)(1-c)}\right)$$

$$J_1 = \frac{1}{8}kT \ln\left(\frac{(1-a)^2 d^2}{(1-b)^2 c^2}\right)$$

$$J_2 = \frac{1}{8}kT \ln\left(\frac{ab(1-b)^2(1-c)}{(1-a)^5}\right)$$

$$K = \frac{1}{8}kT \ln\left(\frac{a(1-b)(1-c)d}{(1-a)bc(1-d)}\right).$$

$$(3.61)$$

Although in considering the form of the growth model, eqn (3.61), and the Ising energy, eqn (3.58), it appears that the parameters α and H are intimately associated with the 1-point lattice average m_A, and likewise the parameters β, γ and J_1, J_2 with the 2-point correlations C_1, C_2, and δ and K with the 3-point probability T_2, it is evident from eqn (3.61) that the relationships are not simple. In fact the Ising parameters H, J_1, J_2 and K are most directly linked to the lattice averages m_A, C_1, C_2, and T_2 that result, since, for example, altering K will affect T_2 without affecting the 2-point correlations. Increasing J_1 will directly translate into an increased value for C_1 but will also result in a change to C_2, since for $J_2 = K = 0$ the model reverts to the simple nearest-neighbour Ising model for which geometric correlations exist. It is also interesting to note that applying the constraint that the model be independent of spin-inversion (i.e. interchanging $\sigma = 1$ with $\sigma = -1$) translates to the conditions $a = (1 - d)$; $b = (1 - c)$ for the growth model and results in a zero value for both H and K.

4

DISPLACEMENTS IN ONE DIMENSION

4.1 General

The preceding two chapters were concerned with developing theory (in 1D only) for occupancy or substitutional disorder—that is, where a particular lattice site is occupied by one or other of two different atomic or molecular species. Before proceeding to consider corresponding models in higher dimensions (2D and 3D) it is useful to consider analogous 1D models which can be used to describe atomic displacements. In this case instead of the (0,1) binary random variables x_i or the (-1,+1) binary variables σ_i used in the previous chapter, the random variables here are continuous variables representing the displacement of an atom away from its mean position.

4.2 Perturbed regular lattice

Consider a 1D lattice of spacing a_0 and consider random perturbations x_i about each site where the x_i are longitudinal displacements. The spacing d_i between the $(i-1)$th and ith points is then $x_i - x_{i-1} + a_0$. Next consider a simple model where all x_i are identically normally distributed and the joint distribution of two neighbouring variables is also normal, that is,

$$P(x_i) = K \exp\left(-\frac{x_i^2}{2\sigma_L^2}\right),$$
(4.1)

$$P(x_{i-1}, x_i) = K \exp\left(-\frac{1}{2\sigma_L^2}\frac{\left(x_{i-1}^2 + x_i^2 - 2rx_{i-1}x_i\right)}{\left(1 - r^2\right)}\right),$$
(4.2)

where σ_L is the standard deviation of displacements from the underlying regular lattice points, and r is a correlation coefficient,

$$r = \frac{\langle x_{i-1}x_i \rangle}{\sigma_L^2}.$$
(4.3)

Given eqns (4.1) and (4.2), the conditional probability of x_i given x_{i-1} can be determined as:

$$P(x_i|x_{i-1}) = \frac{P(x_{i-1}, x_i)}{P(x_i)} = K \exp\left(-\frac{1}{2\sigma_L^2}\frac{\left(x_i - rx_{i-1}\right)^2}{\left(1 - r^2\right)}\right).$$
(4.4)

Equation (4.4) is analogous to the simple binary growth model or Markov chain defined in eqn (3.1). Here the position of the ith site is entirely defined by the position of the

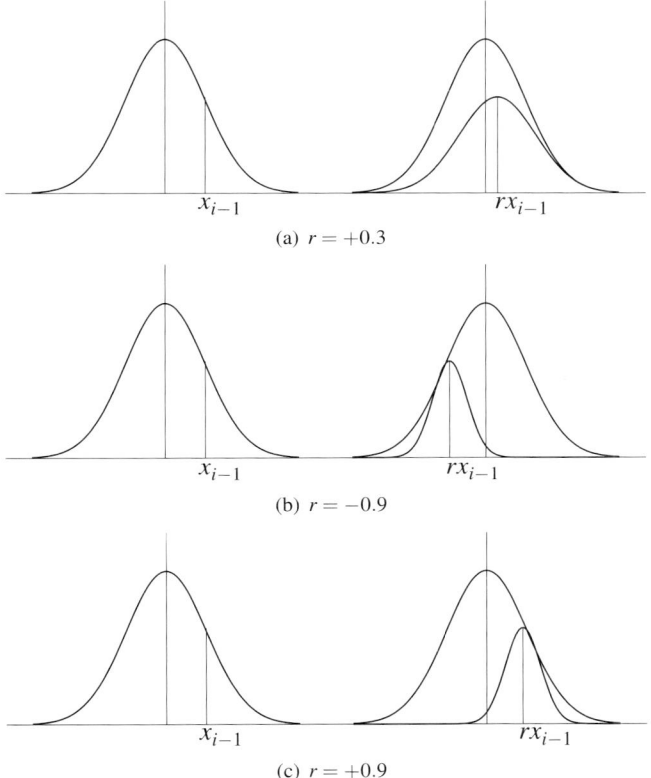

(a) $r = +0.3$

(b) $r = -0.9$

(c) $r = +0.9$

Fig. 4.1 Probability distributions.

preceding site. In the same way as for the binary variables a realisation of the model can be produced by using eqn (4.1) to generate the first point and then eqn (4.4) for successive points. The lattice produced will be immediately stationary with properties given by eqns (4.1) and (4.2).

Figure 4.1 illustrates how the conditional growth probability, eqn (4.4), and the single site Gaussian distribution, eqn (4.1), are related. Once the particular value of x_{i-1} is chosen from the left-hand Gaussian distribution the choice for the right-hand variable is made from the smaller Gaussian centred rx_{i-1} away from the new origin. If r is small the smaller Gaussian is broad but is narrow if r is large. Note also that if r is negative the centre of the smaller Gaussian is on the opposite side of the origin. In all cases, irrespective of the magnitude of r, the same site distribution, eqn (4.1), is generated when averaged over different choices of x_{i-1}.

It is also interesting to consider the distribution of the lengths of nearest-neighbour vectors, $d_i = x_i - x_{i-1} + a_0$. This is readily shown to be (see Welberry *et al.*, 1980).

$$P(d_i) = \exp\left(-\frac{1}{2\sigma_L^2}\frac{(d-a_0)^2}{2(1-r)}\right). \tag{4.5}$$

A property of the model defined by eqns (4.1) and (4.4), which is in fact a simple Markov chain, is that the correlation coefficients between successively distant variables go as r^n; so eqn (4.5) can be generalised as

$$P(d_i) = P(x_i - x_{i-1} + na_0)$$
$$= \exp\left(-\frac{1}{2\sigma_L^2}\frac{(d-na_0)^2}{2(1-r^n)}\right). \tag{4.6}$$

From this it can be seen that for a perturbed regular lattice the variance increases with n as the effect of the correlation diminishes but reaches a bounded value of twice the single-site variance, σ_L^2.

4.3 Diffraction from a perturbed regular lattice

In this section the diffraction pattern of a perturbed regular lattice defined by eqns (4.1) and (4.2) is derived. Suppose for simplicity that the lattice constant a_0 is unity so that the nth point is at the position $z_n = n + x_n$ and the mth point at $z_m = m + x_m$ where x_n, x_m are random values of the displacement. The scattered intensity then is

$$\left|\sum_n \exp(ik(n+x_n))\right|^2 = \sum_n \exp(ik(n+x_n))\sum_m \exp(ik(m+x_m)). \tag{4.7}$$

The ensemble average of this is proportional to

$$\sum_l \exp(ikl)\langle\exp(ik(x_m-x_n))\rangle, \tag{4.8}$$

where $l = m - n$.

The correlation coefficient between x_n and x_m is $S = r^{|l|}$ and x_n and x_m are identically distributed as in eqn (4.1). Thus

$$\langle\exp(ik(x_m-x_n))\rangle = K\iint \exp(ik(x_m-x_n))$$
$$\times \exp\left(\frac{x_m^2+x_n^2}{2\sigma^2(1-S^2)}\frac{Sx_mx_n}{\sigma^2(1-S^2)}\right)dx_m\,dx_n \tag{4.9}$$
$$= \exp(|-(1-S)k\sigma|)$$

and the diffracted intensity becomes,

$$I(k) = \sum_l \exp(ikl)\exp\left(-(1-r^{|l|})(k\sigma)^2\right)$$
$$= \exp(-k^2\sigma^2)\sum_l \exp(\sigma^2k^2r^{|l|})\exp(ikl). \tag{4.10}$$

If $|r| < 1$ this can be split into two parts,

$$I(k)_{\text{Bragg}} = \exp(-k^2\sigma^2) \sum_l \exp(ikl) \qquad (4.11)$$

and

$$
\begin{aligned}
I(k)_{\text{Diffuse}} &= \exp(-k^2\sigma^2) \sum_{l=-\infty}^{\infty} \left(\exp(\sigma^2 k^2 r^{|l|}) - 1\right) \exp(ikl) \\
&= \exp(-k^2\sigma^2) \sum_{P=1}^{\infty} \frac{(\sigma^2 k^2)^P}{P!} \sum_{l=-\infty}^{\infty} r^{P|l|} \exp(ikl) \qquad (4.12) \\
&= \exp(-k^2\sigma^2) \sum_{P=1}^{\infty} \frac{(\sigma^2 k^2)^P}{P!} \frac{\left(1 - r^{2P}\right)}{\left(1 + r^{2P} - 2r^P \cos(k)\right)}.
\end{aligned}
$$

The Bragg intensity consists of peaks of magnitude $\exp(-k^2\sigma^2)$ at positions for which k is a multiple of 2π. For σ equal to unity (i.e. the same as the cell spacing) the intensity of the first-order Bragg peak relative to the origin peak is $7 \times 10^{-18} : 1$ while for $\sigma = 0.5$ the ratio is $5 \times 10^{-5} : 1$. It is thus virtually undetectable for these values of σ.

For very small values of σ for which $\sigma^4 k^4$ is negligible, terms in the diffuse intensity with $P > 1$ may be neglected and the familiar formula for short-range-order diffuse scattering (see eqn (3.12)) is obtained. For values of σ much greater than this many such diffuse curves corresponding to higher values of P must be included in the summation. Each of these will represent successively broader more diffuse peaks as r^{2P} approaches zero. The factor $(\sigma^2 k^2)^P / P!$ eventually goes to zero as P increases for any given σ but for values of $\sigma \approx 1$ many terms must be included. Note that since $(\sigma^2 k^2)$ is zero at $k = 0$ the intensity goes to zero at the origin, a property characteristic of displacement disorder.

4.4   Real example of a 1D perturbed regular lattice

Figure 4.3(a) shows a real example of a diffuse diffraction pattern where the intensity of the vertical diffuse bands varies in a way similar to that described by eqn (4.12). This is the $(0kl)$ section of the diffraction pattern of the dibromodecane-urea (DBD-urea) inclusion compound. In urea inclusion compounds the urea molecules form a hydrogen-bonded network containing hexagonal channels which run along the c-direction. The channels can accommodate various kinds of long-chain molecules such as DBD and these pack end-to-end in the individual channels to form pseudo 1D crystals. The diffuse bands arise from these chains of molecules. Each band is virtually uniform in intensity because there are no correlations linking one channel with the next. It is seen that the first diffuse band is very narrow but higher order ones get progressively broader. Note also that the spacing of the bands is incommensurate with the Bragg peak spacing. [It should be noted that the second band is absent due to the particular form of the molecular structure factor (omitted from eqn (4.12)).]

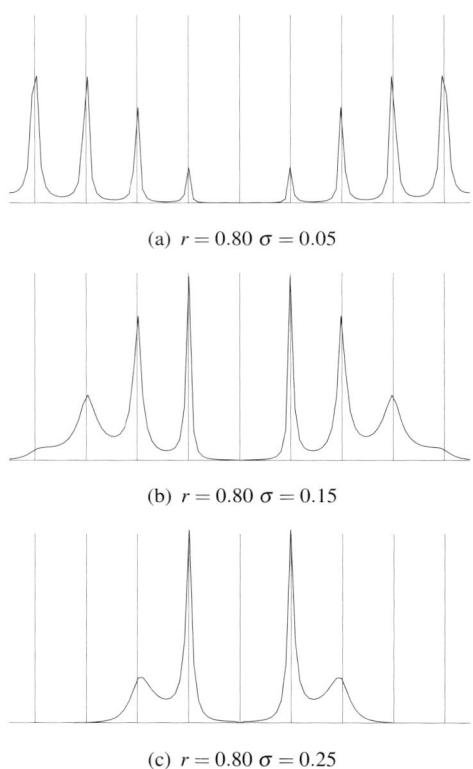

(a) $r = 0.80$ $\sigma = 0.05$

(b) $r = 0.80$ $\sigma = 0.15$

(c) $r = 0.80$ $\sigma = 0.25$

Fig. 4.2 Example plots of the intensity distribution, eqn (4.12), for different values of the nearest-neighbour correlation, r, and the single-site e.s.d., σ.

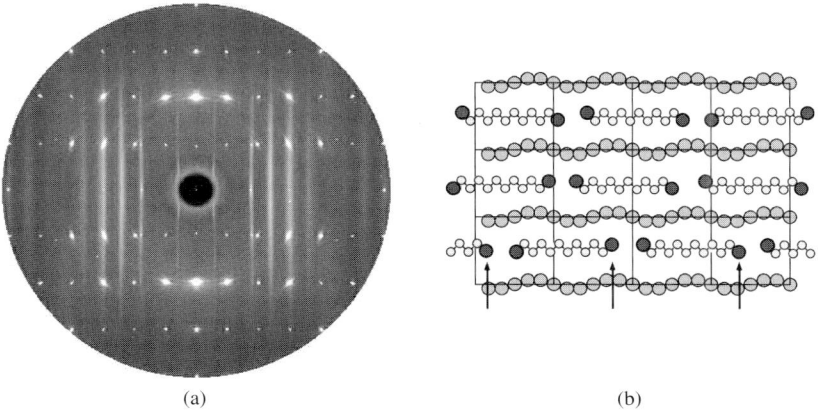

(a)   (b)

Fig. 4.3 (a) The 0*kl* diffraction pattern of dibromodecane/urea inclusion compound showing a sequence of progressively broader diffuse planes of scattering due to the dibromodecane
molecules which form pseudo 1D crystals within the urea channels. (b) Schematic diagram of the
molecules in the channels.

5

DISORDER IN HIGHER DIMENSIONS

5.1 General considerations

The diffracted intensity for systems which are disordered in more than one dimension may be obtained just as simply as for one-dimension. All that is required is that the summations in equations such as eqn (2.5) be replaced by summations over points in two- and three-dimensions. For example, the expression for the intensity from a binary substitutionally disordered crystal, eqn (2.19), may equally apply to a 3D solid if the index n is taken over the three spatial dimensions. If the distribution of the disordered species defining the set of correlations, C_n, is known, the higher dimensionality does not present any impediment to a determination of the intensity distribution. Conversely an analysis of the diffracted intensity distribution leads in an equally straightforward way to a 3D set of C_n values. For displacement disorder also, using for example the 1D Gaussian distributed random variables discussed in Chapter 4, there is little difficulty in calculating the diffraction pattern if the appropriate lattice averages in 2D or 3D are known, although in this context the higher dimension adds the complication that both the direction of the displacements and the direction of the correlations must be considered.

However, in going to higher dimensions considerable difficulties do arise which are not present in 1D systems. As a measure of this difficulty the development of the nearest-neighbour pair interaction Ising model (arguably the simplest of all disorder models) is a good indicator. While the properties of the 1D model were established by Ising (1925), results for the 2D model did not appear until the work of Onsager (1944), and the 3D model still remains quite intractable. Even for the 2D model, calculation of the general correlation coefficient C_n involves evaluation of a large number of integrals (see Kaufmann and Onsager, 1949; Hurst and Green, 1960; Montroll *et al.*, 1963). Not only is the mathematics involved in dealing with disorder in higher dimensions more complex but there are other difficulties too such as the increase in complexity of symmetry specification. While there are only two 1D symmetry groups, there are 17 2D plane groups and 230 3D space groups (e.g. see Phillips, 1963).

However, in going to high dimensions, the most significant problem relevant to the field of crystallography and disordered materials is concerned with the fact that the diffraction pattern contains only information about the 1-point properties (atom positions, mean-square displacements, site occupancies) and 2-point properties (pair correlations) of a lattice. This stems directly from the form of the basic diffraction equation, eqn (2.9), which is the Fourier transform of the *pair* correlation function. In 1D the properties of a nearest-neighbour model are completely specified by the pair-correlations and ambiguities of interpretation only arise where longer-range interactions are concerned (see Section 3.4). In 2D, specification of the 1-point and 2-point properties is no-longer sufficient to define unambiguously the local spatial arrangement of atoms and molecules, and in 3D the problems are compounded.

An observed diffuse scattering pattern does not contain all the information one would like to obtain about a disordered system. Such a pattern can give a lot of information over and above that obtainable solely from the Bragg experiment but equally there is a lot of information that is simply not accessible. The aim in this chapter, by exploring some simple higher dimensional models, is to show the reader how different features of a diffraction pattern might arise and the information that can consequently be extracted but also to give some feeling regarding what information cannot be obtained from a diffraction pattern.

Disorder in 2D and 3D may be formulated in terms of sets of random variables as for 1D and these may similarly be discrete (e.g. binary) or continuous (e.g. Gaussian) according to the problem. A study of disorder in higher dimensions, therefore, is concerned with the way in which such variables may mutually interact. The study of spatially interacting random variables in higher dimensions than 1D is a very difficult and complex field, and, it remains imperfectly understood and few explicit results are still available. Systems of spatially interacting random variables have application in a wide and diverse range of fields, and much of the present understanding is derived from fields other than that of crystallography and disordered crystals with which this book is concerned. For example, the early work of Whittle (1954) dealt with the distribution of grain yield in a wheat crop.

One key result that has been established with varying degrees of generality by a number of workers is the equivalence between the Gibbsian ensemble models such as the Ising model which are characterised by joint probability distributions, and Markov random fields which are characterised by conditional probability distributions (Dobrushin, 1968; Spitzer, 1971; Grimmett, 1973; Preston, 1973; Sherman, 1973; Moussouris, 1974; Speed, 1978). This equivalence has already been demonstrated and used in 1D for simple cases, and in discussing higher-dimension models the relationship will be further exploited.

5.2   A simple 2D model of disorder

The advantage of the Markov growth process in 1D was that it was extremely easy to use the formulation such as eqn (3.1) both to generate a realisation of a particular model and to calculate the resulting lattice averages that contribute to the diffracted intensity distribution. At the same time those results may equally well have arisen from an Ising model formulation, for which realisations are not directly obtainable—indirect methods, such as Monte Carlo (MC) simulation (Metropolis *et al.* (1953)), which rely on the fact that such distributions arise as equilibrium solutions to certain spatial–temporal processes must be used (see Ogita *et al.*, 1969). Consequently it might seem an attractive proposition for disorder in 2D and 3D to be generated by some simple extension of eqn (3.1). Indeed for disorder that is introduced into a crystal at growth, this growth description would appear more physically appropriate than the corresponding Ising-like description where each atomic or molecular species is in equilibrium with its surroundings on all sides.

The simplest of all such possible 2D growth-disorder models has been discussed by Welberry and Galbraith (1973, 1975), Welberry (1977a), Welberry and Miller (1977) and Galbraith and Walley (1976, 1980). It may be written in the form,

$$P(x_{i,j}|\text{all predecessors}) = \alpha + \beta x_{i-1,j} + \gamma x_{i,j-1} + \delta x_{i-1,j} x_{i,j-1}. \tag{5.1}$$

Here $x_{i,j}$ is a $(0,1)$ random variable defined at the i,jth point of a 2D lattice. Such a formulation as eqn (5.1) may be considered to correspond to growth on an infinite 1D surface of a 2D crystal (see Fig. 5.1). In the figure the As and Bs might be envisaged to represent two different atomic or molecular species or one molecule in either of two different orientations. An A or B may add to the lattice at any one of the prospective sites indicated by the black dot according to probabilities defined by the two nearest-neighbours already forming part of the crystal lattice.

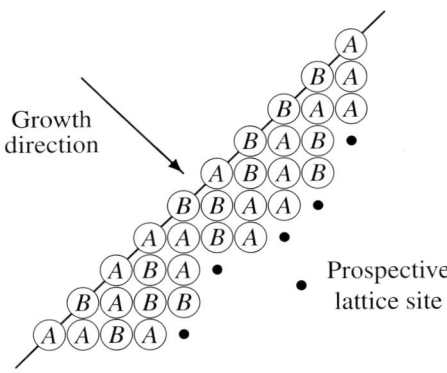

Fig. 5.1   Growth on an infinite 1D surface of a 2D crystal.

Although eqn (5.1) is simple in form, a general solution even for the concentrations, m_A and m_B, of the two disordering species has not been found. The difficulty arises because the general procedures which were used to obtained the lattice averages in 1D, typified by eqns (3.40) and (3.46), and which enabled a closed set of equations to be made, do not work in 2D. For each new lattice average which may be generated by premultiplying eqn (5.1) by an appropriate random variable already incorporated in the crystal, new undefined lattice averages appear on the right-hand side of the equations. For example, the equations relating some of the simple lattice averages may be written thus,

$$m_A = \alpha + \beta m_A + \gamma m_A + \delta P_{1,\bar{1}}$$
$$P_{1,0} = \alpha m_A + \beta m_A + \gamma P_{1,\bar{1}} + \delta P_{1,\bar{1}}$$
$$P_{0,1} = \alpha m_A + \beta P_{1,\bar{1}} + \gamma m_A + \delta P_{1,\bar{1}}$$
$$P_{1,1} = \alpha m_A + \beta P_{0,1} + \gamma P_{1,0} + \delta T_1$$

(5.2)

where, $m_A = \langle x_{i,j} \rangle$, $P_{1,1} = \langle x_{i,j} x_{i-1,j-1} \rangle$, $P_{1,0} = \langle x_{i,j} x_{i-1,j} \rangle$, $P_{1,\bar{1}} = \langle x_{i-1,j} x_{i,j-1} \rangle$, $P_{0,1} = \langle x_{i,j} x_{i,j-1} \rangle$ and $T_1 = \langle x_{i-1,j-1} x_{i,j-1} x_{i-1,j} \rangle$.

It is useful to express eqn (5.2) in a graphical form where the various lattices averages defined above are written in the form of a square of either occupied (black dot) or vacant sites. In each case the black dot represents the position of the particular variable $x_{i,j}$, $x_{i-1,j}$, $x_{i,j-1}$ or $x_{i-1,j-1}$.

(5.3)

The bottom right-most corner of the square represents the growth point (i, j), while the other 3 corners correspond to the sites $(i, j-1)$, $(i-1, j)$ and $(i-1, j-1)$. Thus the symbol $\square$ corresponds to m_A, the overall probability that $x_{i,j} = 1$ irrespective of what the values of the three other sites are. Because of the translational symmetry the symbols $\square$, $\square$ and $\square$ also equal m_A. Similarly the symbols $\square$ and $\square$ represent the average $P_{0,1}$.

Despite the fact that a general solution to the model eqn (5.1) has not been found solutions have been obtained in a number of special cases. These are discussed in the following sections.

5.2.1 Simple linear growth model, $\delta = 0$

The first of these was described by Welberry and Galbraith (1973) and occurs when the non-linear term $\delta = 0$. The solution of the model eqn (5.1) in this case was obtained following theory described by Whittle (1954) and Bartlett (1967, 1968). The results for the various lattice averages were found to be,

$$m_A = (1 - \beta - \gamma)^{-1}$$

(5.4)

and

$$C_{\bar{r},\bar{s}} = C_{r,s} = \begin{cases} X^r Y^s & \text{for } r,s \geq 0 \\[2em] \displaystyle\sum_{j=1}^{r} \binom{r+s-j-1}{s-1} \beta^{r-j} \gamma^s X^j & \\[1.5em] \quad + \displaystyle\sum_{k=1}^{s} \binom{r+s-k-1}{r-1} \beta^r \gamma^{s-k} Y^k & \text{otherwise} \end{cases} \qquad (5.5)$$

where

$$X = C_{1,0} = (P_{1,0} - m_A^2)/(m_A - m_A^2) = (1 + \beta^2 - \gamma^2 - \Delta)/2\beta$$
$$Y = C_{0,1} = (P_{0,1} - m_A^2)/(m_A - m_A^2) = (1 - \beta^2 + \gamma^2 - \Delta)/2\gamma$$

and

$$\Delta = ((1 + \beta + \gamma)(1 - \beta + \gamma)(1 + \beta - \gamma)(1 - \beta - \gamma))^{\frac{1}{2}}.$$

Fig. 5.2 shows some realisations of this model together with their calculated diffraction patterns. In these and in subsequent similar figures the growth direction is from the top left to the bottom right. The important results coming from this work were that:-

1. Choice of the three parameters α, β, γ enabled lattice realisations to be generated with various values of the three primary lattice averages m_A, $C_{1,0}$, $C_{0,1}$ over the whole of their respective permissible ranges.
2. Values of the diagonal correlations $C_{1,1}$, $C_{1,\bar{1}}$ were dependent on these primary values.
3. The correlation structure in the quadrant corresponding to the growth direction was substantially different from the structure in the quadrant normal to the growth direction. (see especially Fig. 5.2(a), (c) and (e)).
4. The correlation structure in the quadrant normal to the growth direction followed the same geometric progression typical of 1D nearest-neighbour systems. The correlations in the growth, $[1\,1]$, direction decayed less rapidly. The result in the diffraction pattern is that the diffuse scattering is more sharply peaked when scanned in the $[1\,1]$ than in the $[1\,\bar{1}]$ direction. This slower decay of correlations is more characteristic of truly 2D disorder. A similar asymmetry in the correlation field has been noted by Hammersley (1967) in discussing *harnesses*. The essential difference between the two quadrants which may be called the 'growth' and 'non-growth' quadrants, is that if two molecules are separated by a vector in the growth quadrant the choice of the molecule added at the earlier time directly affects the later molecule via direct sequences of growth steps. For molecules separated by a vector in the non-growth quadrant no such direct transmission of interaction is possible, and any correlation between them derives purely from their having mutual predecessors.

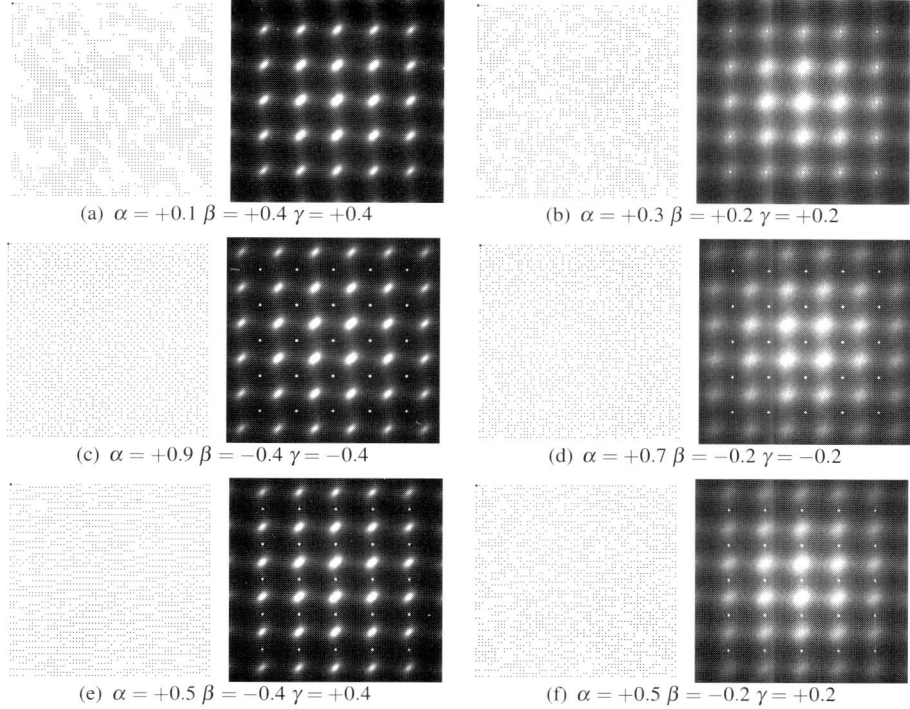

(a) $\alpha = +0.1\ \beta = +0.4\ \gamma = +0.4$ (b) $\alpha = +0.3\ \beta = +0.2\ \gamma = +0.2$

(c) $\alpha = +0.9\ \beta = -0.4\ \gamma = -0.4$ (d) $\alpha = +0.7\ \beta = -0.2\ \gamma = -0.2$

(e) $\alpha = +0.5\ \beta = -0.4\ \gamma = +0.4$ (f) $\alpha = +0.5\ \beta = -0.2\ \gamma = +0.2$

Fig. 5.2 Examples of 2D linear growth model.

5.2.2 *Simple growth model with constraint,* $\gamma(1-\beta) = -\alpha\delta$

A second special case solution of eqn (5.1) was reported by Welberry and Galbraith
(1975). Some example realisations and their calculated diffraction patterns for this case
are shown in Fig. 5.4. This solution referred to as SC1 was discovered initially, when
certain stochastic matrices were found to commute if the growth parameters satisfied
the constraint

$$\gamma(1-\beta) = -\alpha\delta. \tag{5.6}$$

Later (Welberry, 1977a) it was shown that this SC1 solution corresponded to a situation where the growth-model, eqn (5.1), was constrained to have statistical mirror
symmetry when the variables were placed on a triangular lattice. Deferring for the moment a discussion of how this symmetry arises, it will be seen with reference to Fig. 5.3
and eqn (5.2) that the symmetry requires that $P_{1,\bar{1}} = P_{0,1}$ and hence the expressions in
eqn (5.2) may be used to calculate lattice averages since they now form a closed set. In
fact it was found that to obtain the lattice averages up to the fourth order 2-point correlation along the axis normal to the mirror plane, that is, $C_{4,0}$, 32 such equations were
required and for each additional more distant correlation the number of such equations

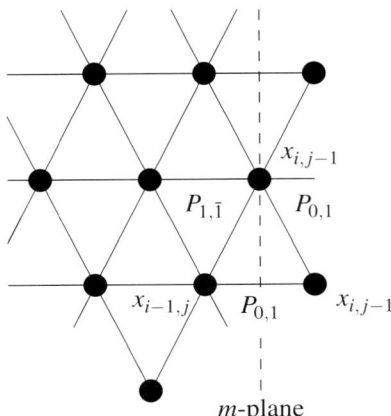

m-plane

Fig. 5.3 Schematic diagram illustrating imposition of a mirror plane on the lattice averages when the model eqn (5.1) is placed on a triangular lattice.

approximately doubles. It was found that all off-axis correlations $C_{n,m}$, $m \neq n$ were zero for this model.

This SC1 solution illustrates another important aspect of 2D disorder, even though the non-zero correlations are confined to a 1D row. The three growth parameters remaining after the constraint from eqn (5.6) has been applied allow lattice distributions to be produced for which the low-order lattice averages m_A, $P_{1,0}$ and $P_{2,0}$ may be chosen independently. In this respect the SC1 model has a great similarity to the second-neighbour 1D model given by eqn (3.39). This comes about because the second dimension allows an alternative pathway for correlations to be transmitted to the second-nearest-neighbour (other than simply via the nearest-neighbour). Fig. 5.4(b) and (c) were chosen to illustrate this indirect effect. These should be compared to the patterns in Fig. 3.3 which were obtained from the 1D model, eqn (3.39). It is evident from the realisations in these examples that the triplet probability $\langle x_{i,j} x_{i,j-1} x_{i-1,j} \rangle$ plays an important part in transmitting the additional correlation effects to the second-neighbour.

5.3 Ising models and growth models in 2D

The imposition of symmetry on a growth-model such as eqn (5.1) requires the equivalence of such a conditional probability model to a joint probability (Ising) model to be invoked. In order to derive the relationship between the growth parameters and the equivalent Ising model parameters a more general model than eqn (5.1), in which the growth probabilities depend on the three previous points forming the corners of the generic unit cell, will be considered.

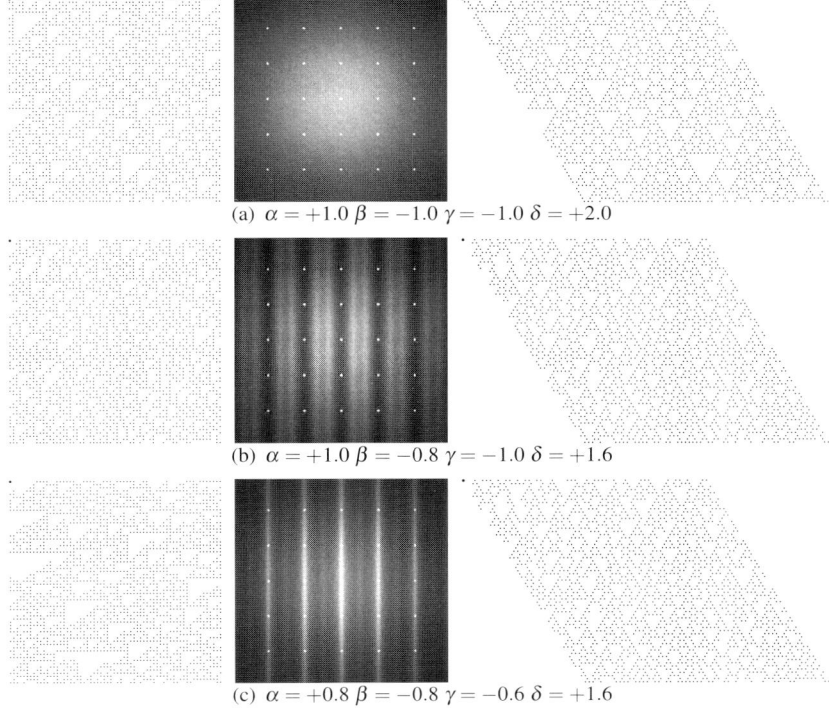

(a) $\alpha = +1.0 \; \beta = -1.0 \; \gamma = -1.0 \; \delta = +2.0$

(b) $\alpha = +1.0 \; \beta = -0.8 \; \gamma = -1.0 \; \delta = +1.6$

(c) $\alpha = +0.8 \; \beta = -0.8 \; \gamma = -0.6 \; \delta = +1.6$

Fig. 5.4   Examples of the SC1 constrained model.

$$P(x_{i,j}|\text{all predecessors}) = \alpha + \beta x_{i-1,j} + \gamma x_{i,j-1} + \delta x_{i-1,j-1}$$
$$+ \varepsilon x_{i-1,j} x_{i,j-1} + \zeta x_{i-1,j-1} x_{i-1,j} + \eta x_{i-1,j-1} x_{i,j-1} \quad (5.7)$$
$$+ \xi x_{i-1,j} x_{i,j-1} x_{i-1,j-1}.$$

This model includes the model eqn (5.1) as a special case when $\varepsilon = \zeta = \eta = \xi = 0$. To make the comparison with the Ising formulation it is more convenient to use the alternative transition probabilities

$$
\begin{aligned}
a &= P(1|000) = \alpha \\
b &= P(1|001) = \alpha + \beta \\
c &= P(1|010) = \alpha + \delta \\
d &= P(1|100) = \alpha + \gamma \\
e &= P(1|101) = \alpha + \beta + \gamma + \varepsilon \\
f &= P(1|110) = \alpha + \beta + \delta + \zeta \\
g &= P(1|011) = \alpha + \gamma + \delta + \eta \\
h &= P(1|111) = \alpha + \beta + \gamma + \delta + \varepsilon + \zeta + \eta + \xi.
\end{aligned}
\qquad (5.8)
$$

The equivalent Ising model has a value of E_c dependent upon interactions between the corresponding $\sigma_{i,j}$ $(1,-1)$ spin variables making up the generic unit cell. The general expression for this energy is given by,

$$
\begin{aligned}
E_c = {} & H\sigma_{i,j} + J_1\sigma_{i,j}\sigma_{i-1,j} + J_2\sigma_{i,j}\sigma_{i,j-1} + J_3\sigma_{i,j}\sigma_{i-1,j-1} + J_4\sigma_{i-1,j}\sigma_{i,j-1} \\
& + K_1\sigma_{i,j}\sigma_{i-1,j}\sigma_{i,j-1} + K_2\sigma_{i,j}\sigma_{i-1,j}\sigma_{i-1,j-1} + K_3\sigma_{i,j}\sigma_{i,j-1}\sigma_{i-1,j-1} \qquad (5.9) \\
& + K_4\sigma_{i,j-1}\sigma_{i-1,j}\sigma_{i-1,j-1} + L\sigma_{i,j}\sigma_{i-1,j}\sigma_{i,j-1}\sigma_{i-1,j-1}.
\end{aligned}
$$

It is immediately noticeable that the Ising formulation defined thus has 10 parameters— two more than the growth or Markov model, eqn (5.3). This means that for the growth model, eqn (5.3), to be equivalent to the Ising model, eqn (5.8), two constraints must be applied to the Ising parameters. This difference in generality between the Ising and Markov formulations arises because the Markov formulation in this case is 'one-sided', that is, the conditional probability depends only on neighbours on one side of the target point. 'Two-sided' conditional probability models which are entirely equivalent to the corresponding Ising formulation have been discussed for example by Bartlett (1971) but these cannot be used directly to produce lattice realisations.

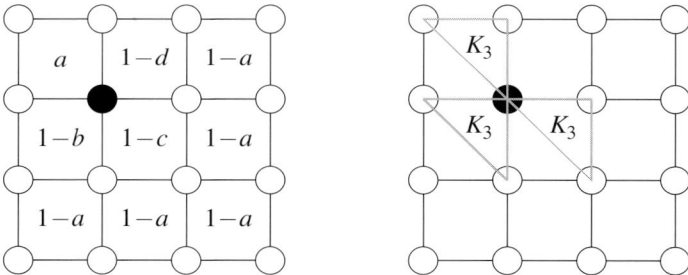

Fig. 5.5   Configuration with all but one of the σ_i variables equal to -1. The left figure correspond to the growth description and the right to the Ising descripton.

In order to derive the relationships between the parameters of the two formulations the same strategy employed in the 1D case is used. The total crystal configuration probability is considered for different local configurations of one unit cell surrounded by an otherwise perfect lattice for which $\sigma_{i,j} = -1$. Fig. 5.5 shows one such configuration where one of the variables in the central unit cell has been flipped from -1 (white) to $+1$ (black). In the left-most figure each cell contains the growth probability that would have been invoked to arrive at this total configuration. The ratio of the growth probabilities for this configuration compared to the configuration representing the chosen zero of energy in which all spins are -1, is thus

$$
\text{probability ratio} = \frac{a(1-b)(1-c)(1-d)}{(1-a)^4}. \qquad (5.10)
$$

The right-most figure illustrates how just one term (the K_3 term) of the Ising formulation is altered when the single site is flipped. In this case there are three instances for which the sign of the K_3 is changed (which therefore alters the Ising energy by $6K_3$). Taking into account the effect of the flip on all the terms the total difference in energy in the Ising formulation is,

$$\Delta E = 2H - 4(J_1 + J_2 + J_3 + J_4) + 6(K_1 + K_2 + K_3 + K_4) - 8L. \quad (5.11)$$

Equating this relative energy to the logarithm of the probabilities as in eqn (3.34), and making use of nine other such relationships obtained from consideration of other different cell configurations, simultaneous equations which may be solved to give the 10 Ising interaction parameters in terms of the eight growth probabilites are obtained. This analysis has been carried out by Enting (1977a). In terms of the variables used in the present context he found,

$$H = \frac{1}{8}kT \ln\left(\frac{efgh^2(1-h)}{a(1-a)^2(1-b)(1-c)(1-d)}\right)$$

$$J_1 = \frac{1}{8}kT \ln\left(\frac{(1-a)(1-f)bh}{(1-g)(1-e)cd}\right)$$

$$J_2 = \frac{1}{8}kT \ln\left(\frac{(1-a)(1-g)dh}{(1-e)(1-f)bc}\right)$$

$$J_3 = \frac{1}{8}kT \ln\left(\frac{(1-a)(1-b)(1-d)(1-e)cgfh}{(1-c)(1-f)(1-g)(1-h)abde}\right)$$

$$J_4 = \frac{1}{8}kT \ln\left(\frac{(1-a)(1-c)(1-e)(1-h)aceh}{(1-b)(1-d)(1-f)(1-g)bdfg}\right)$$

$$K_1 = \frac{1}{16}kT \ln\left(\frac{(1-b)(1-d)(1-f)(1-g)aceh}{(1-a)(1-c)(1-e)(1-h)bdfg}\right)$$

$$K_2 = \frac{1}{16}kT \ln\left(\frac{(1-b)(1-c)(1-e)(1-f)adhg}{(1-a)(1-d)(1-h)(1-g)bcef}\right)$$

$$K_3 = \frac{1}{16}kT \ln\left(\frac{(1-c)(1-d)(1-e)(1-g)abfh}{(1-a)(1-b)(1-f)(1-h)cdeg}\right)$$

$$K_4 = \frac{1}{16}kT \ln\left(\frac{(1-b)(1-c)(1-d)(1-h)bcdh}{(1-a)(1-e)(1-f)(1-g)aefg}\right)$$

$$L = \frac{1}{16}kT \ln\left(\frac{(1-a)(1-e)(1-f)(1-g)bcdh}{(1-b)(1-c)(1-d)(1-h)aefg}\right).$$

$$(5.12)$$

Several important points arise out of a consideration of these results.

1. The most general model that can be grown in a sequential fashion according to some growth algorithm such as eqn (5.3) or eqn (5.1) is less general than the corresponding Ising joint-probability distribution. In the present case the growth imposes two restrictions on the 10 Ising parameters. The simpler model, eqn (5.1), may be obtained from this model by setting $a = c$; $b = g$; $d = f$; $e = h$, which means that the conditional probabilities in eqn (5.8) are not dependent on $x_{i-1,j-1}$. The consequence of this simplification is that the Ising interactions J_3, K_2, K_3, K_4 and L are zero. Thus the simpler model, eqn (5.1), defined by four growth probabilities is equivalent to an Ising model with five parameters which are accordingly subject to one constraint.

2. Since, given the two growth constraints, the two formulations are entirely equivalent the symmetry of the growth model distribution must be the same as that of the Ising model and in particular must have the symmetry of the interaction energy E_c. Symmetry may be imposed on the growth model by setting the appropriate Ising interactions to be equal. Thus for example a vertical mirror plane may be obtained by setting $J_3 = J_4$; $K_1 = K_2$; $K_3 = K_4$ and additionally to obtain a horizontal mirror $K_1 = K_3$. When such symmetry is imposed, as was shown earlier, expressions for the various lattice averages may be obtained in closed form. The explicit solutions obtained by Welberry (1977b,a) could all be explained in terms of such imposition of symmetry. This was discussed by Enting (1978a,b). For some cases the symmetry applied when the random variables were placed on a triangular lattice, while in others the lattice was rectangular.

3. The most general such symmetrised growth model that was discussed by Welberry (1977b) was the single mirror-plane solution mentioned previously, for which three constraints ($J_3 = J_4$; $K_1 = K_2$; $K_3 = K_4$) were imposed in addition to the two growth constraints. This left five free parameters to be chosen independently. It was found that the disordered lattice realisations of this model could be categorised in terms of the five low-order lattice averages $\langle x_{i,j} \rangle$, $\langle x_{i,j} x_{i,j-1} \rangle$, $\langle x_{i,j} x_{i-1,j} \rangle$, $\langle x_{i,j} x_{i-1,j} x_{i,j-1} \rangle$ and $\langle x_{i,j} x_{i,j-1} x_{i-1,j} x_{i-1,j-1} \rangle$; that is a 1-point, two 2-point, a 3-point and a 4-point probability. Addition of the second mirror-plane resulted in the loss of the freedom to choose the 3-point property.

Although the relationship between the five free growth parameters and the general coefficient $C_{n,m}$ had not been established at this stage, it was clear that some of the multi-point properties could be changed without affecting the 2-point correlations at all. In particular, when the model eqn (5.9) is constrained to have rectangular (mm) symmetry the 4-point probability $\langle x_{i,j} x_{i,j-1} x_{i-1,j} x_{i-1,j-1} \rangle$ could be varied independently of all 2-point probabilities, over a range that was large for low values of the primary correlations $C_{1,0}$ and $C_{0,1}$ and progressively smaller for higher correlation values. Figure 5.6 shows some computer simulations and their corresponding calculated diffraction patterns, which illustrate this effect.

A similar sort of effect is exhibited by the simpler model eqn (5.1) when the Ising model is constrained to have triangular symmetry. That is, starting with the five remaining interactions in eqn (5.12), (H, J_1, J_2, J_4, K_1) the two further constraints $J_1 = J_2 = J_4$

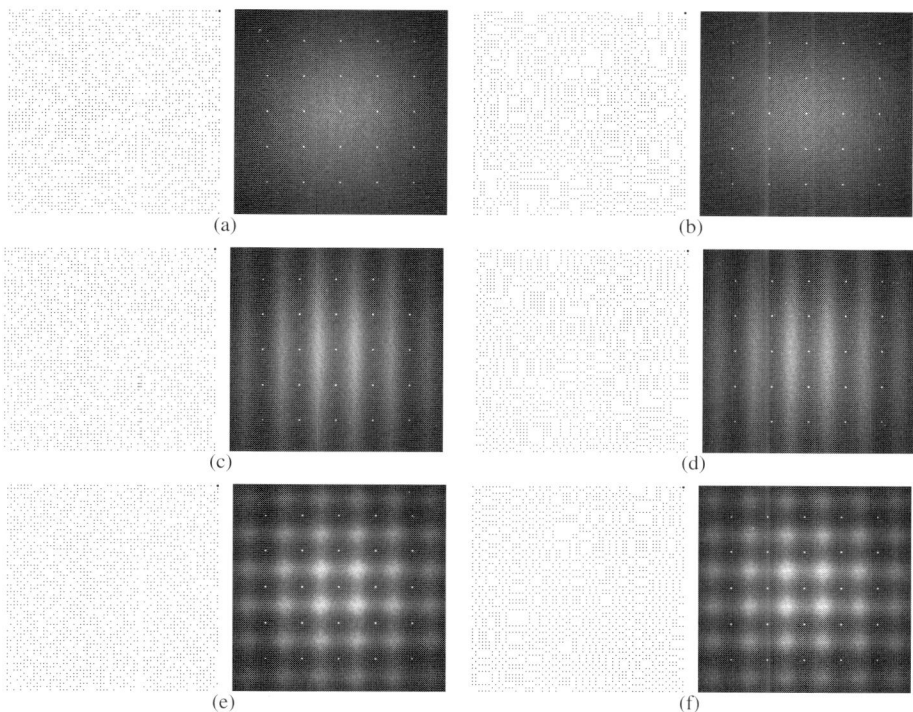

Fig. 5.6 Examples of rectangular-symmetric (mm) model. The figures on the left all have a low 4-point correlation while those on the right have a high 4-point correlation. Despite this the 2-point correlations (and hence the diffraction patterns) of the two sets of examples are indistinguishable. The growth probabilities for these figures are:-

(a) $a=e=f=g=0.1$ $b=c=d=h=0.9$,
(b) $a=g=0.9$ $b=c=0.4$ $d=h=0.1$ $e=f=0.6$,
(c) $a=g=0.9$ $b=c=0.4$ $d=h=0.1$ $e=f=0.6$,
(d) $a=g=0.1$ $b=c=0.933$ $d=h=0.9$ $e=f=0.067$,
(e) $a=0.9$ $b=d=e=0.4$ $c=f=g=0.6$ $h=0.1$,
(f) $a=0.1$ $b=d=0.933$ $c=0.956$ $e=0.044$ $f=g=0.067$ $h=0.9$.

together with one growth constraint results only two free parameters remaining. In realisations of this model it is found that all 2-point correlations are zero and the two
free parameters define the values of m_A and the triple probability $T = \langle x_{i,j} x_{i,j-1} x_{i-1,j} \rangle$.
Figure 5.4(a) shows an example of this. The dominance of the triplet probability is very
evident and yet the diffraction pattern is indistinguishable from the 'Laue monotonic
scattering' pattern of a random distribution. Compare this pattern also with Figs 5.6(a)
and (b).

5.4   An alternative approach to growth-disorder models

In a series of papers following the previously described rather heuristic results, Pickard
(1977, 1978, 1980) put the solution of symmetrised growth-disorder models on a more
rigorous footing. In these papers he also showed the relationship of these models to
the vector Markov chain approach of Verhagen (1977). A brief description of these
methods is given here, albeit with no attempt made to give a rigorous account, and the
most important results that ensue are simply quoted.

   The distribution of variables on the lattice is described in terms of the joint probability $P(A,B,C,D)$ on the generic unit cell.

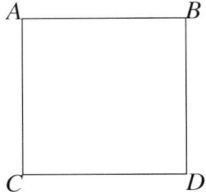

Here $A = x_{i-1,j-1}$, $B = x_{i,j-1}$, $C = x_{i-1,j}$ and $D = x_{i,j}$. For a stationary distribution
the joint probability, $P(A,B,C,D)$, must have translational symmetry. This requires the
following relationships between the marginals of P:

$$P(A) = P(B) = P(C) = P(D)$$
$$P(A,C) = P(B,D) \tag{5.13}$$
$$P(A,B) = P(C,D).$$

For a distribution in which $x_{i,j}$ is a binary variable, these five conditions plus a sixth
which ensures the total probability is unity, reduces the number of independent parameters specifying $P(A,B,C,D)$ from 16 to 10, corresponding to the 10 parameters of
the Ising formulation, eqn (5.12). This 10 parameter model is the most general consistent with a homogeneous distribution involving only interactions within the unit cell.
The aim here is to find subsets of this distribution which are consistent with unilateral
growth typified, for example, by eqn (5.3).

   Any joint probability can be decomposed into conditional probabilities as follows,
for example,

$$P(A,B,C,D) = P(A).P(B|A).P(C|A,B).P(D|A,B,C) \tag{5.14}$$

From eqn (5.14) it is readily seen why in this most general form the model cannot be used as a growth model, since a point being added to the lattice would have to simultaneously satisfy, for example, $P(C|A,B)$ and $P(D|A,B,C)$ in different cells.

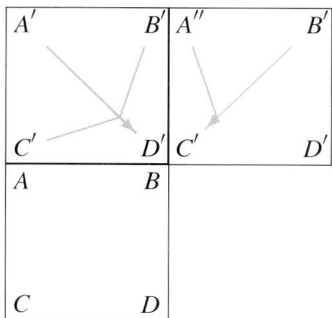

However, if the following conditional independence relation is imposed,

$$P(C|A,B) = P(C|A) \tag{5.15}$$

there is no such conflict and eqn (5.14) becomes,

$$P(A,B,C,D) = P(A).P(B|A).P(C|A).P(D|A,B,C) \tag{5.16}$$

In this form it would appear suitable as a prescription for growth as follows. The first point (uppermost left corner) on a finite rectangular lattice may be chosen according to the single point distribution $P(A)$. The upper boundary row is generated using the conditional probability $P(B|A)$ and the left boundary using the probability $P(C|A)$. The rest of the array is then constructed using $P(D|A,B,C)$.

Equation (5.15) yields the same 'growth conditions' on the Ising parameters for the model to become equivalent to the growth model, and eqn (5.16) is thus an alternative to the formulation in eqn (5.3). Pickard, however, showed that this is insufficient to ensure that eqn (5.16) generates a stationary distribution. A minimum of one further condition was found necessary to ensure stationarity, either

$$P(C|B,D) = P(C|D) \tag{5.17}$$

or

$$P(D|A,C) = P(D|C) \qquad \text{and} \qquad P(D|A,B) = P(D|B). \tag{5.18}$$

When either of these conditions is satisfied the resulting distributions were proven to be stationary Markov random fields. Equation (5.17) yielded a seven parameter solution and eqn (5.18) a five parameter solution. In both cases the form of the correlation field in the non-growth quadrant was geometric, that is,

$$C_{r,\bar{s}} = C_{\bar{r},s} = C_{1,0}^r C_{1,0}^s \qquad \text{for } r,s \geq 0. \tag{5.19}$$

For eqn (5.18) the growth quadrant was also geometric with

$$C_{\bar{r},\bar{s}} = C_{r,s} = C_{1,0}^{r}C_{1,0}^{s} \qquad \text{for } r,s \geq 0 \qquad (5.20)$$

but for eqn (5.17) the form of the correlations in the growth quadrant was more complex. In fact $C_{1,1}$ was available as one of the variable parameters of the model. The rectangular-symmetry (mm) model of Welberry (1977b) was shown to correspond to the intersection of the two solutions eqns (5.17) and (5.18), that is, when all three conditional independence relations were satisfied. This therefore provided the proof that the realisations shown in Fig. 5.6 which have the same nearest-neighbour 2-point correlations but different 4-point properties, in fact have the same total 2-point correlation fields as given by eqns (5.19) and (5.20). It should be noted, however, that although the expectation values for the 2-point correlations are the same, they will in general have different variances. In finite samples of different realisations the expected lattice averages would be achieved with different degrees of accuracy. This would probably affect the small-angle scattering pattern, but this has not been investigated.

The single m-plane solution reported by Welberry (1977b) and the SC1 solution (Welberry, 1977a), however, do not satisfy the stationarity conditions, eqns (5.17) and (5.18). This evidently must be so, since the correlation structure of $C_{n,0}$ for the SC1 solution was found to be much more complex than that of the simple Markov chain which is used for the boundaries in eqn (5.14). Pickard showed, however, that although on a finite lattice the construction did not produce immediate stationarity as for the mm-symmetry lattices, for suitably large lattices the distribution converged to stationarity, and the correlations along the axis normal to the mirror plane were then described by a homogeneous vector Markov chain. This Markov chain is a simple nearest-neighbour chain of the four-state variable $Z_i = (x_{i,j}x_{i,j-1})$, and as such yields results for the lattice averages in a straightforward manner, although these results were not given by Pickard.

5.5   A useful parameterisation of growth-disorder models

Although for given values of the growth transition probabilities in eqn (5.8), the values of the various resulting lattice averages could be evaluated readily, the reverse procedure of obtaining growth transition probabilities for given values of the lattice averages was not easily handled using the parameterisations of Welberry (1977b,a) or Enting (1977b,a,c). An account is therefore given here of the parameterisation of Pickard (1978) which enables easy computation of the necessary transition probabilities required to generate a realisation with particular properties.

Considering the $(0,1)$ $x_{i,j}$ variables, the frequencies of the 16 possible configurations on the unit cell may be assigned as follows:

$$f_0 = \begin{smallmatrix} 0 & 0 \\ 0 & 0 \end{smallmatrix} \qquad f_1 = \begin{smallmatrix} 0 & 0 \\ 1 & 0 \end{smallmatrix} \qquad f_2 = \begin{smallmatrix} 1 & 0 \\ 0 & 0 \end{smallmatrix} \qquad f_3 = \begin{smallmatrix} 1 & 0 \\ 1 & 0 \end{smallmatrix}$$

$$f_4 = \begin{smallmatrix} 0 & 1 \\ 0 & 0 \end{smallmatrix} \qquad f_5 = \begin{smallmatrix} 0 & 1 \\ 1 & 0 \end{smallmatrix} \qquad f_6 = \begin{smallmatrix} 1 & 1 \\ 0 & 0 \end{smallmatrix} \qquad f_7 = \begin{smallmatrix} 1 & 1 \\ 1 & 0 \end{smallmatrix}$$

$$\text{(5.21)}$$

$$f_8 = \begin{smallmatrix} 0 & 0 \\ 0 & 1 \end{smallmatrix} \qquad f_9 = \begin{smallmatrix} 0 & 0 \\ 1 & 1 \end{smallmatrix} \qquad f_{10} = \begin{smallmatrix} 1 & 0 \\ 0 & 1 \end{smallmatrix} \qquad f_{11} = \begin{smallmatrix} 1 & 0 \\ 1 & 1 \end{smallmatrix}$$

$$f_{12} = \begin{smallmatrix} 0 & 1 \\ 0 & 1 \end{smallmatrix} \qquad f_{13} = \begin{smallmatrix} 0 & 1 \\ 1 & 1 \end{smallmatrix} \qquad f_{14} = \begin{smallmatrix} 1 & 1 \\ 0 & 1 \end{smallmatrix} \qquad f_{15} = \begin{smallmatrix} 1 & 1 \\ 1 & 1 \end{smallmatrix}.$$

These frequencies may be expressed in terms of the following definitions of the primary lattice averages in an analogous way to that presented for 1D in eqn (2.23) to eqn (2.35). The following definitions are used:

$$m_A = \langle x_{i,j} \rangle \qquad\qquad \bullet \qquad\qquad T_{123} = \langle x_{i-1,j} x_{i-1,j-1} x_{i,j-1} \rangle$$

$$P_{1,0} = \langle x_{i,j} x_{i-1,j} \rangle \qquad\qquad \bullet\!-\!\bullet \qquad\qquad T_{234} = \langle x_{i-1,j-1} x_{i,j-1} x_{i,j} \rangle$$

$$P_{0,1} = \langle x_{i,j} x_{i,j-1} \rangle \qquad\qquad T_{134} = \langle x_{i-1,j} x_{i,j-1} x_{i,j} \rangle \qquad\qquad \text{(5.22)}$$

$$P_{1,1} = \langle x_{i,j} x_{i-1,j-1} \rangle \qquad\qquad T_{124} = \langle x_{i-1,j} x_{i-1,j-1} x_{i,j} \rangle$$

$$P_{1,\bar{1}} = \langle x_{i,j-1} x_{i-1,j} \rangle \qquad\qquad Q = \langle x_{i,j} x_{i-1,j} x_{i,j-1} x_{i-1,j-1} \rangle$$

The values of the f_i are then given by,

$$f_{15} = Q$$
$$f_{14} = T_{234} - Q$$
$$f_{13} = T_{134} - Q$$
$$f_{11} = T_{124} - Q$$
$$f_7 = T_{123} - Q$$
$$f_{10} = P_{1,1} - T_{124} - T_{234} + Q$$
$$f_5 = P_{1,\bar{1}} - T_{123} - T_{134} + Q$$
$$f_9 = P_{1,0} - T_{124} - T_{134} + Q$$
$$f_6 = P_{1,0} - T_{123} - T_{234} + Q$$
$$f_3 = P_{0,1} - T_{123} - T_{124} + Q$$
$$f_{12} = P_{0,1} - T_{134} - T_{234} + Q$$
$$f_1 = m_A - P_{1,\bar{1}} - P_{1,0} - P_{0,1} + T_{134} + T_{123} + T_{124} - Q$$
$$f_2 = m_A - P_{0,1} - P_{1,0} - P_{1,1} + T_{123} + T_{124} + T_{234} - Q$$
$$f_4 = m_A - P_{1,\bar{1}} - P_{1,0} - P_{0,1} + T_{123} + T_{134} + T_{234} - Q$$
$$f_8 = m_A - P_{0,1} - P_{1,0} - P_{1,1} + T_{134} + T_{124} + T_{234} - Q$$
$$f_0 = 1 - 4m_A + 2P_{0,1} + 2P_{1,0} + P_{1,1} + P_{1,\bar{1}}$$
$$- T_{123} - T_{124} - T_{134} - T_{234} + Q.$$

(5.23)

Specification of the lattice averages, eqn (5.22), allows calculation of the frequencies, eqn (5.21), via these eqns (5.23). From the values of the frequencies the growth model parameters defined by eqn (5.8) may readily be calculated, for example, $a = P(1|000) = f_8/(f_0 + f_8)$, $b = P(1|001) = f_{12}/(f_4 + f_{12})$ etc. It must be emphasised that the derivation of such growth probability parameters is only valid when the conditional probability relationship, eqn (5.15), is satisfied. Moreover, to obtain growth models which give stationary distributions either eqn (5.17) or eqn (5.18) must also be satisfied. Pickard (1978) reported that the growth condition, eqn (5.15), was equivalent to two constraints imposed in the lattice averages:

$$\frac{T_{123}}{P_{1,0}} = \frac{P_{0,1}}{m_A} \qquad \text{and} \qquad \frac{P_{1,\bar{1}} - T_{123}}{m_A - P_{1,0}} = \frac{m_A - P_{0,1}}{1 - m_A}.$$

(5.24)

This means that for a given choice of $P_{1,0}$, $P_{0,1}$ and m_A the values of $P_{1,\bar{1}}$ and T_{123} are defined and hence not free to be chosen independently. Pickard (1978) also gave relationships between the lattice averages which are equivalent to the conditions eqn (5.17) and eqn (5.18). These are respectively,

$$T_{123} = T_{134}$$

(5.25)

$$T_{123} = T_{124} = T_{234} \qquad \text{and} \qquad P_{1,\bar{1}} = P_{1,1}.$$

(5.26)

Note also that combining eqns (5.25) and (5.26) produces the *mm*-symmetry in the lattice averages used by Welberry (1977b) and that the single *m*-symmetry requires the rather different conditions:

$$T_{123} = T_{234} \qquad T_{124} = T_{134} \qquad P_{1,\bar{1}} = P_{1,1}. \qquad (5.27)$$

In order to give an appreciation of the restriction on the general probability distribution

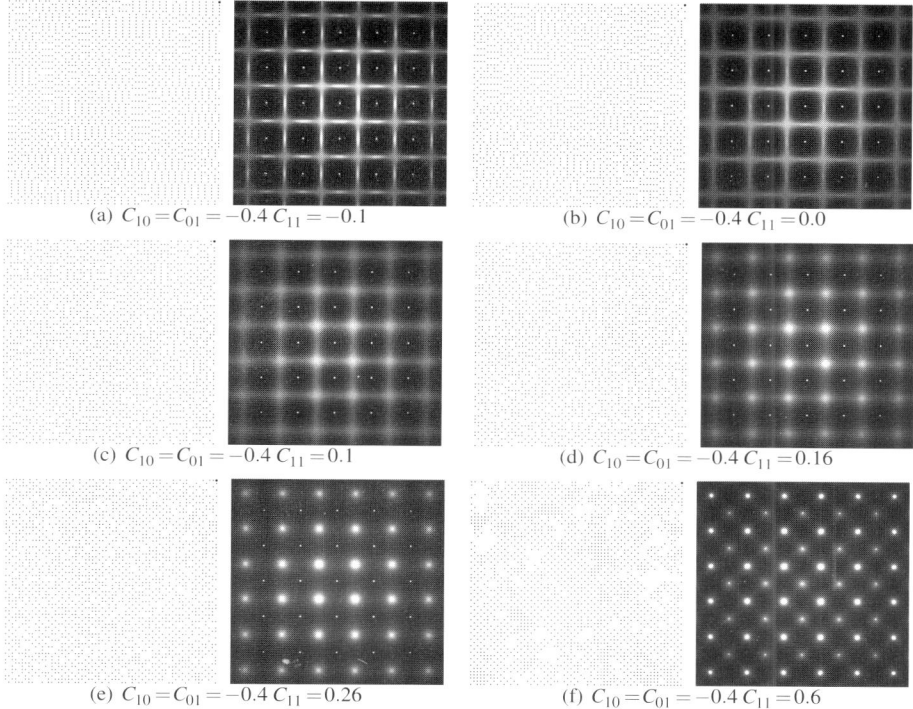

(a) $C_{10} = C_{01} = -0.4 \; C_{11} = -0.1$

(b) $C_{10} = C_{01} = -0.4 \; C_{11} = 0.0$

(c) $C_{10} = C_{01} = -0.4 \; C_{11} = 0.1$

(d) $C_{10} = C_{01} = -0.4 \; C_{11} = 0.16$

(e) $C_{10} = C_{01} = -0.4 \; C_{11} = 0.26$

(f) $C_{10} = C_{01} = -0.4 \; C_{11} = 0.6$

Fig. 5.7   Examples of Monte Carlo generated square-symmetric (mm) model.

that the growth constraints impose Fig. 5.7 shows some example lattice realisations and their corresponding diffraction patterns which have been produced from the general Ising model using Monte Carlo simulation. The examples shown are ones in which vertical and horizontal mirror planes have been imposed. All the examples are ones in which the correlations $C_{10} = C_{01} = -0.4$. The difference between the different frames is in the value of the diagonal correlations $C_{11} = C_{1\bar{1}}$. Note that for Fig. 5.7(d) $C_{11} = 0.16$ which corresponds to the case where the growth condition is satisfied. The growth model case lies in the middle of the range of possible distributions and it can be seen that this range is extensive. At one extreme (Fig. 5.7(a)) the distribution appears to have segregated into two different kinds of domains with a doubled repeat in either the horizontal or the vertical direction. At the other extreme (Fig. 5.7(f)) there appears to be phase segregation into three different types of domain.

5.6   General discussion of binary models

In two dimensions or more the treatment of even simple models of disorder becomes
very complex. Consequently a class of 2D models which are less general than is possible but which are considerably more tractable, have been discussed in some detail
in order to demonstrate some of the effects which can occur in higher dimensions.
Several properties exhibited by these models are of note, because of their difference
from 1D. First, there is the ability for correlation effects to be transmitted to second-
nearest-neighbours even when only nearest-neighbour interactions are involved. Second, the geometric decay of correlation effects which is characteristic of 1D disorder
and which gives rise to the familiar form, eqn (3.12), of a diffuse peak, may be substantially altered in 2D, correlations in general decaying less rapidly with a corresponding
increased sharpness of diffuse peaks. Third, the models exhibit the property that substantial differences in the multi-point lattice averages can exist without affecting the
2-point correlations and the consequent diffraction pattern. Realisations of these models (e.g. Figs 5.4 and 5.6) demonstrate that the textural qualities or visual appearance
of the disordered distributions is in fact more dependent on the multi-point properties
than the pair-correlations which are the only information obtainable from a diffraction
experiment. This fact brings into question some of the elementary concepts of disorder such as 'domain-size' which has its origins in the 1D treatment of disorder with
its relationship to 'correlation length'. Lattice realisations such as Fig. 5.4(a) appear to
contain quite large ordered 'domains' and yet the 2-point correlations for this example
are all zero, and the only diffuse scattering is the Laue monotonic scattering term. Similarly, some of the lattice realisations of Fig. 5.6 have quite large ordered regions but
this is seen to be more a function of the 4-point lattice average than the 2-point correlations. Despite this, discussion of multi-point properties has not received much attention
in the literature although Clapp (1969, 1971) described a probability variation method
which he used to generate n-site probabilities from experimentally determined pair-
probabilities on the assumption that only pairwise interactions contribute to the energy.
Cowley (1968) has shown that multi-point properties will contribute to the intensity
distribution if 'Size-Effect' displacements are present.

   An aspect of growth-disorder models that is worthy of further consideration is the
fact that many real examples of disorder must approximate quite closely to such a unilateral growth mechanism for introducing the disorder rather than the more general Ising
type of model; for example, in some molecular crystals a molecule may occupy a particular site in either of two alternative orientations with very little difference in energy, but
once incorporated into the lattice there are large potential barriers preventing subsequent
rearrangement. It is therefore possible that the asymmetry noted in the very simplest
model, eqn (5.1), could well be a common feature of real disordered crystals, an effect
which might well go unnoticed in a diffraction experiment because of the presence of
different but symmetry-related growth-sectors in the crystal sample. There is some evidence in support of this possibility. For example Schlössen and Lang (1965) reported
differences in chemical composition between different growth-sectors in amethyst crystals containing 1% Fe. On the other hand if disorder is introduced into a crystal at growth
and no asymmetry ensues then it is necessary that relationships between the intermolec-

ular forces must exist which provide just those symmetry conditions which it was found necessary to apply to obtain solutions for the various lattice averages. This may involve some sort of equilibrium process occurring in the surface layer as a crystal forms. Thus while the joint-probability Ising formulation of disorder describes the situation where an atom or molecule is in equilibrium with all of the surrounding lattice, it might be considered that growth-models represent the situation where equilibrium is only achieved on the surface. Interest has also been aroused by another class of models which could be classified as describing entirely non-equilibrium situations, and which may also have application in the area of crystal growth. These models are called 'cellular automata' and have been reviewed, for example, by Wolfram (1983). Cellular automata have some similarities to growth-disorder models since they involve a similar growth process in which each new cell is occupied by one of a number of different species depending upon the occupants of previous cells. The main difference, however, is that this process is not random but each transition involves a fixed set of rules by which each new cell is generated from its predecessors. The only randomness that exists in the lattice is derived solely from the arbitrary boundary, and is perpetuated in an often complex and unpredictable way by the particular set of rules. In fact the pattern of the SC1 lattice shown in Fig. 5.4(a) is an example of a cellular automaton since it is a limiting case of a growth-disorder model for which all of the growth probabilities are either 0 or 1.

Although growth-disorder models in 2D have been used to demonstrate some of the key differences between disorder in higher dimensions, the growth process imposes sufficiently severe restriction on the generality of the distributions that other key properties may be eliminated. One of these is the phase transition which is present in even the simplest 2D Ising model and which is lost by applying the growth-restrictions in 2D. The examples of the general model shown in Fig. 5.7 clearly show a behaviour much more complex than can be achieved with the simple growth-disorder models. Welberry and Miller (1978), however, have shown that in 3D even growth-disorder models can have a phase transition.

5.7 Some symmetry considerations

It was mentioned briefly in the previous section that because real crystals grow in a unilateral fashion, albeit in several directions simultaneously, there was no necessity in a situation where disorder occurs at growth for the resulting distributions to conform to a single particular space group symmetry, since the direction parallel to each growth direction is inherently different from other directions. Nevertheless structures of disordered materials are published and assigned a space group symmetry which is usually determined by observation of the systematic absences of Bragg reflections and confirmed by the structure refinement which again uses the Bragg reflections. Since the Bragg reflections are only dependent on the 1-point properties m_A and m_B of the distribution, it is worth considering to what extent the concept of 'Space group' needs to be amended in considering disorder.

This discussion is focussed by consideration of a simple example involving a hypothetical structure of the molecule, p-Cl-toluene. Consider the simple structure shown in Fig. 5.8. Each molecular site in this average structure shown is related to other sites by

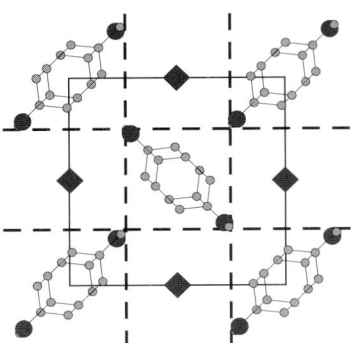

Fig. 5.8 Example used to illustrate the possibility that the symmetry of the Bragg intensities may not be the same as the symmetry of diffuse scattering. The figure shows the average structure where each molecular site contains two superimposed p-Cl-toluene molecules related by a centre of symmetry.

a four-fold rotation axis as well as two intersecting sets of glide-planes. It is imagined that at each site in the actual crystal the molecule is allowed to take one or other of two different orientations (related by a $180°$ rotation). It is further assumed that the two orientations are equally likely so that $m_A = 0.5$. For the average lattice, therefore, the molecular site has a centre of symmetry, and whatever the state of order in the structure the Bragg diffraction will exhibit the full $4mm$ symmetry.

If the 2-point correlations are considered, however, a number of different possibilities arise. Each molecular site has four contacts to nearest-neighbours. In the context of an equilibrium Ising-like situation with the given site equally affected by all the surrounding lattice sites, it might be expected that these contacts are equivalent and thus have the same degree of correlation along them. In a situation where growth has given preference to different directions, it can no longer be assumed that the correlations will be the same along the four vectors. In Fig. 5.9 the effect of allowing different combinations of the interactions k_1, k_2, k_3, k_4 along the four different contact vectors is shown. In the two diffraction patterns shown the diffuse scattering has quite different symmetry, mm, and 4 according to the values chosen, but the Bragg reflections always show the full $4mm$ symmetry.

Although this sort of behaviour is theoretically quite possible, in practice it is usually safe to assume that the diffuse scattering distribution has the same symmetry as the average structure.

5.8   Direct synthesis of disordered distributions

An alternative way of treating diffuse scattering is to consider that the intensity that occurs at any point in reciprocal-space arises from a periodic modulation of the real-space structure. This idea may be used to *synthesise* a real-space distribution of atomic scatterers which will have a given diffraction pattern. In simple terms the real-space lattice is constructed by applying modulations with wave-vectors corresponding to each

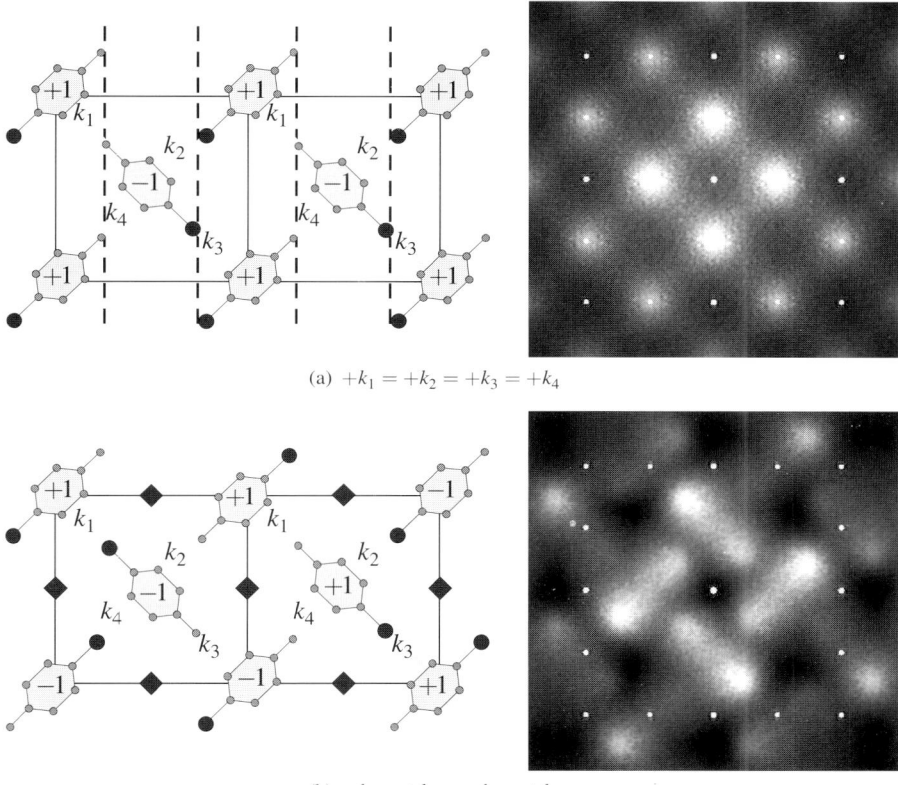

(a) $+k_1 = +k_2 = +k_3 = +k_4$

(b) $-k_1 = +k_2 = -k_3 = +k_4$

Fig. 5.9 Example used to illustrate the possibility that the symmetry of the Bragg intensities may not be the same as the symmetry of diffuse scattering. Although the average structure of the two distributions has 4*mm* for (a) the values of the interactions k_1, k_2, k_3 and k_4 are chosen to ensure the disordered distribution has *mm* symmetry, while for (b) the distribution has 4-fold symmetry.

elemental volume in the first Brillouin zone of the diffraction pattern. Each modulation is given an amplitude which reflects the intensity at that point, and a phase that is chosen at random. For the case of concentration waves the modulations can be written in the form of a variation from cell to cell of the atomic scattering factors. If the point in reciprocal space is infinitesimally small then the modulation wave must extend over the whole of real-space. In practice it is more convenient to consider that a modulation extends over a limited region in real-space and that this contributes to a small (but finite) region of the diffuse distribution in reciprocal-space. Then the atomic scattering factors in real-space may be expressed as a sum of all such modulations:

$$f_\mu(\mathbf{T}) = \langle f_\mu \rangle \left(1 + \sum_{\mathbf{q}} A_{\mathbf{q}} \cos\left(2\pi\mathbf{q} . \mathbf{T} + \phi_{\mu,\mathbf{q}} \right) \exp\left(-\frac{(\mathbf{t} - \mathbf{t}_c)^2}{2\sigma_c^2} \right) \right). \qquad (5.28)$$

Here $\mathbf{T}$ is a real-space lattice vector. $f_\mu(\mathbf{T})$ is the scattering factor of the atom in the site μ of the unit cell with origin $\mathbf{T}$. The amplitude, $A_{\mathbf{q}}$, of each modulation wave is proportional to the amplitude of the scattering at the point $\mathbf{G} + \mathbf{q}$ in reciprocal-space, where $\mathbf{G}$ is a reciprocal space lattice vector. $\phi_{\mu,\mathbf{q}}$ is the phase of the modulation of wave-vector $\mathbf{q}$ at the site μ. The summation is over all wave-vectors in the first Brillouin zone. The final term is a Gaussian whose standard deviation σ_c defines the extent of the region in real-space which is modulated (in practice the modulation is truncated at 2.5σ). $\mathbf{t}$ is a general vector in real-space and $\mathbf{t}_c$ defines the randomly chosen centre of the region of modulation. The effect of such a modulation is to contribute to the diffraction pattern a diffuse peak at the reciprocal point $\mathbf{q}$ with a width inversely proportional to σ_c.

If random phases, $\phi_{\mu,\mathbf{q}}$, are used in the synthesis (and this is reasonable, since for an incommensurate wave the choice of origin is arbitrary) the value of the atomic scattering factor, $f_\mu(\mathbf{T})$, at a given site that will be obtained from eqn (5.28) will be a continuous variable and not just a binary one (representing either atom A or atom B). This is overcome in practice by converting the continuous variables representing the scattering factors into binary ones by comparing the value of $f_\mu(\mathbf{T})$ with a threshold, f_T say. For all those sites for which $f_\mu(\mathbf{T}) > f_T$ a scattering factor f_A is assigned and for those at which $f_\mu(\mathbf{T}) < f_T$ a scattering factor f_B is assigned.

The meaning of the different variables in eqn (5.28) is illustrated schematically in Fig. 5.10. In this figure the required diffraction pattern is seen to consist of a diffuse locus in the form of a circle. The wave-vector $\mathbf{q}$ defines a single point on the locus and this corresponds to a modulation in real-space centred at the point $\mathbf{t}_c$. In Fig. 5.10(b)–(d) the various stages of the synthesis of an example real-space distribution are shown. These figures show a region of real-space corresponding to 512×512 lattice sites. A given site contains either a black or a white atom. In this case the form of the diffuse diffraction pattern is a diffuse ring centred at the origin with a radius of 0.1 reciprocal lattice units. The value of sc was chosen to be 16 lattice repeats so that the area of the region covered by the $2.5\sigma_c$ limit for each modulation was ~ 5026 unit cells, or $\sim 2\%$ of the total area. Figure 5.10(b) shows the resulting distribution after 30 modulations have been added. Note that for this example the spacing of the modulation planes is the same in each case but the phases are all different. Fig. 5.10(c) shows the distribution after 300 modulations, and Fig. 5.10(d) after 30,000 modulations. It is clear that Fig. 5.10(c)

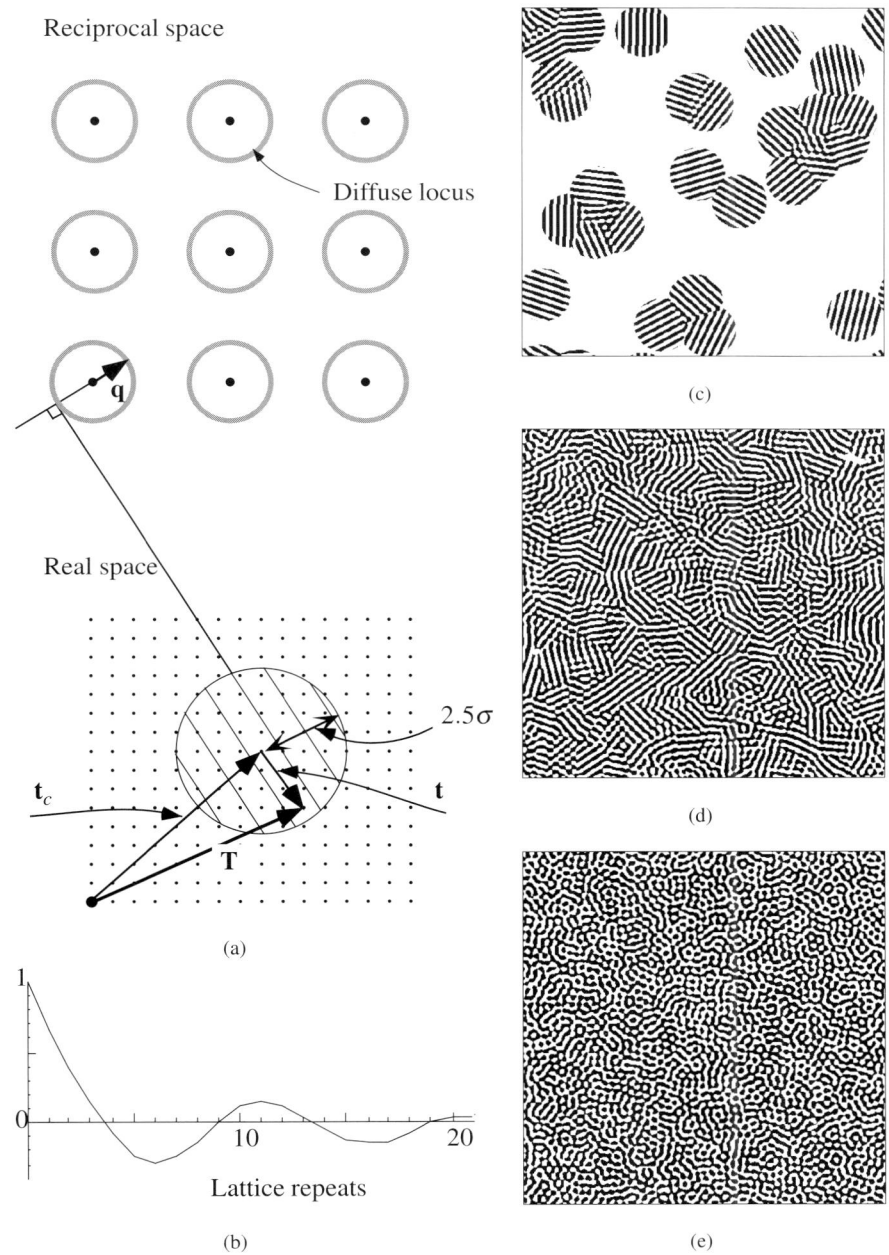

Fig. 5.10 Schematic illustrating the modulation-wave synthesis technique

still shows texture relating to the modulation region size, but Fig. 5.10(d) shows a completely homogeneous texture. In this latter case each real space site is subject to ~ 600 modulations on average. In Fig. 5.10(e) a plot of the correlation coefficients calculated from the distribution shown in Fig. 5.10(d) is shown. The correlations are plotted against the number of lattice repeats, since the correlation field is circularly symmetric, that is, $C_{m,n}^{ij}$ depends only on $|\mathbf{r}_{m,n}|$.

6

DISPLACEMENTS IN TWO OR THREE DIMENSIONS

6.1 Gaussian growth-disorder models in 2D

Many of the results that were obtained for binary models considered in earlier sections can be adapted for use with Gaussian variables. In particular it is possible to consider the joint distribution of variables on the four vertices of the generic unit cell $ABCD$ discussed in Section 5.4 equally well if the variables in question are continuous (Gaussian) rather than binary variables. The most general form having rectangular symmetry for a Gaussian distribution of four variables is:

$$P(x_A, x_B, x_C, x_D) = K \exp\left(-\frac{1}{L}\left(x_A^2 + x_B^2 + x_C^2 + x_D^2 - 2r'(x_A x_B + x_C x_D)\right.\right.$$

$$\left.\left. - 2s'(x_A x_C + x_B x_D) - 2r'(x_A x_D + x_B x_C)\right)\right) \quad (6.1)$$

where r', s', t' and L are simple functions of the horizontal, vertical and diagonal correlation coefficients, r, s and t.

It is found that that in order to factorise this in the form of a growth algorithm like eqn (5.16), it is necessary to impose the constraint $t = rs$, that is, the diagonal correlation is the product of the two axial correlations. This is exactly the same condition on the correlation coefficients as occurs for the binary variable models. Given this restriction the various factors corresponding to eqn (5.16) become

$$P(x_A) = K \exp\left(-\frac{x_A^2}{2\sigma^2}\right)$$

$$P(x_B|x_A) = K \exp\left(-\frac{(x_B - rx_A)^2}{2\sigma^2(1 - r^2)}\right)$$

$$P(x_C|x_A) = K \exp\left(-\frac{(x_C - sx_A)^2}{2\sigma^2(1 - s^2)}\right)$$

$$P(x_D|x_A, x_B, x_C) = K \exp\left(-\frac{(x_D - sx_B - rx_C + rsx_Bx_C)^2}{2\sigma^2(1 - r^2)(1 - s^2)}\right) \quad (6.2)$$

$$P(x_A, x_B, x_C, x_D) = K \exp\left(-\left(2\sigma^2(1 - r^2)(1 - s^2)\right)^{-1}\left(x_A^2 + x_B^2 + x_C^2 + x_D^2\right.\right.$$

$$\left.\left. - 2r'(x_A x_B + x_C x_D) - 2s'(x_A x_C + x_B x_D) - 2r'(x_A x_D + x_B x_C)\right)\right).$$

Here σ is the standard deviation of the single site variable and K is a normalising constant which has a different value in each equation. Just as for the equivalent binary model this Gaussian model has Markov chains embedded along every pathway in the lattice whose steps along each of the axes are always in the same direction. Because of this the correlation field has the simple form

$$\rho_{m,n} = r^{|m|} s^{|n|}.$$ (6.3)

6.2   Gaussian growth-disorder models—examples

6.2.1   *Simple lattice with small σ*

First some simple examples are considered in which the value of σ is a relatively small fraction of the lattice repeat. A value of $\sigma = 0.05$ has been used for convenience. It should be noted that root mean-square atomic displacements in real crystals may be ~ 0.05 of the cell spacing. The random variables may be used to represent displacements in a given direction, either the x-direction, y-direction or some intermediate direction. In Fig. 6.1 examples are shown in which more than one set of random variables are used.

In Fig. 6.1(a) the correlation structure of the random variables has $r = -0.5$ and $s = -0.5$, and this exists independently for both the x-displacements and the y-displacements. In Fig. 6.1(b) the same correlation structure is used, but only x-displacements are involved. In Fig. 6.1(c) the same correlation structure is applied to displacements in a direction $30°$ to the horizontal. In the final example, Fig. 6.1(b), the displacements in the x-direction and y-direction have different correlation structures. For the x-displacements (horizontal) the correlations are $r = -0.5$ and $s = -0.5$ while for the y-displacements (vertical) they are $r = -0.5$ and $s = +0.5$.

In Fig. 6.1(a)–(c) it is seen that the two negative correlation values produce a diffuse peak at the centres of the reciprocal unit cells. Only the intensity of these is different in the three figures. In all three cases the intensity goes to zero at the origin, which is indicative of displacement disorder. For Fig. 6.1(b) it is seen that the horizontal displacement direction results in the intensity being zero along a vertical line through the origin, that is, normal to the displacement direction. For Fig. 6.1(c) the displacement direction is inclined $30°$ to the horizontal and here the line of zero intensity is inclined $30°$ to the vertical direction, that is, normal to the displacement direction.

Figure 6.1(b) may be seen to be comprised of two sets of peaks originating from the two different correlation fields. For the peaks at the centres of the reciprocal unit cells the intensity is strong on the left-side and the right-side of the pattern and zero along a vertical line through the origin. For this correlation field the displacement is therefore horizontal. For the peaks centred along the edges of the reciprocal unit cells the intensity is strong at the top and the bottom of the figure and is zero along a horizontal line through the origin. For this correlation field the displacement must therefore be vertical.

6.2.2   *Lattices with large σ—paracrystals*

The expression for the variation of the nearest-neighbour distances, eqn (4.5), indicates that as the correlation coefficient becomes larger the variation of d_i becomes smaller for a given value of σ_L. It is clear from this that it is possible to have a large value of σ_L

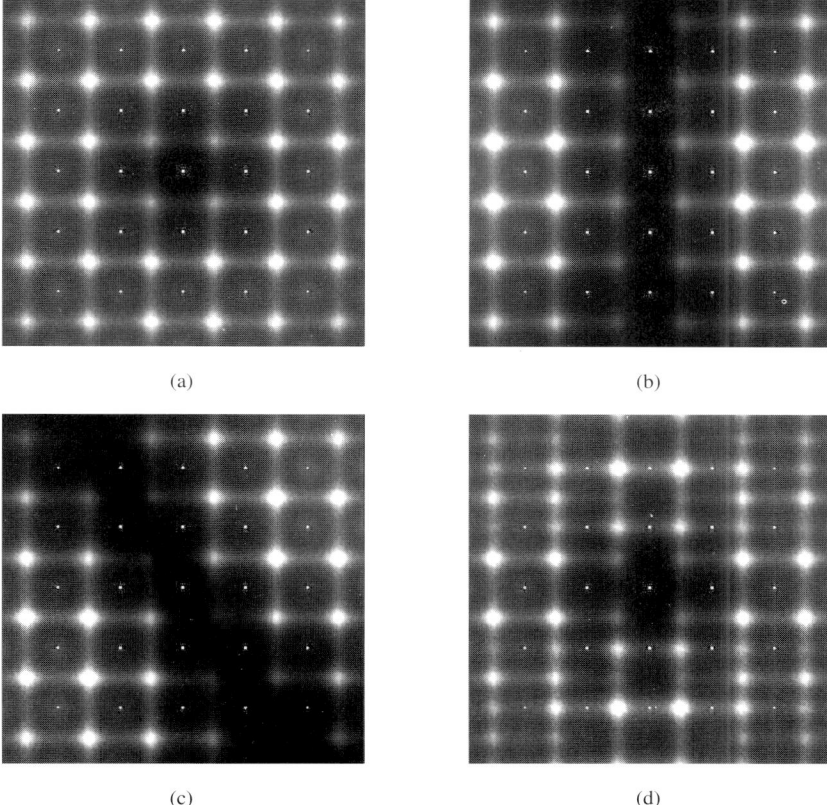

(a) (b)

(c) (d)

Fig. 6.1 Examples of diffraction patterns from gaussian growth disorder models

but still have a small variation of the nearest-neighbour distance by having a sufficiently large value of the correlation coefficient. Thus, for example, if σ_L is equal to unity (i.e. the cell repeat) eqn (4.11) indicates that the lattice will be so distorted that there will be no Bragg peaks present, but by having a value of r and s of ~ 0.95 or 0.99 the variation of nearest-neighbour distance is relatively small and what is produced is a sort of *paracrystal* (see Hosemann and Bagghi, 1962).

A variety of such paracrystals can be generated according to the strength of the correlations between nearest-neighbour displacements in the transverse or longitudinal direction. Figure 6.2 illustrates the formation of a paracrystalline lattice in terms of these displacements. In the examples discussed in this chapter ρ_L and ρ_T represent longitudinal and transverse correlations, respectively. A strong longitudinal correlation indicates a small variation of cell length whereas a strong transverse correlation indicates a small variation in the direction of the cell edge. Fig. 6.3 shows an example paracrystalline lattice together with its calculated diffraction pattern for the case where longitudinal and

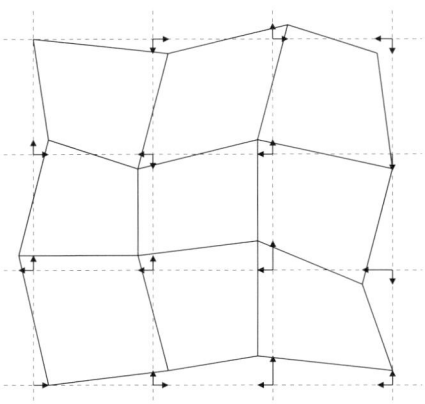

Fig. 6.2 Schematic showing a distorted paracrystalline lattice: $\rho_L = \left\langle x_{i,j} x_{i-1,j} \right\rangle = \left\langle y_{i,j} y_{i,j-1} \right\rangle$ and $\rho_T = \left\langle x_{i,j} x_{i,j-1} \right\rangle = \left\langle y_{i,j} y_{i-1,j} \right\rangle$

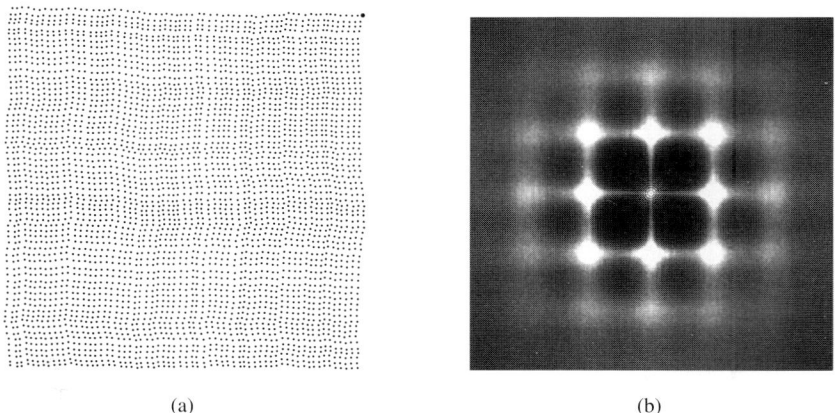

(a)   (b)

Fig. 6.3 Example paracrystal generated using $r = s = 0.99$ and $\sigma = 1.0$. (a) Section of real-space crystal (b) Corresponding diffraction pattern.

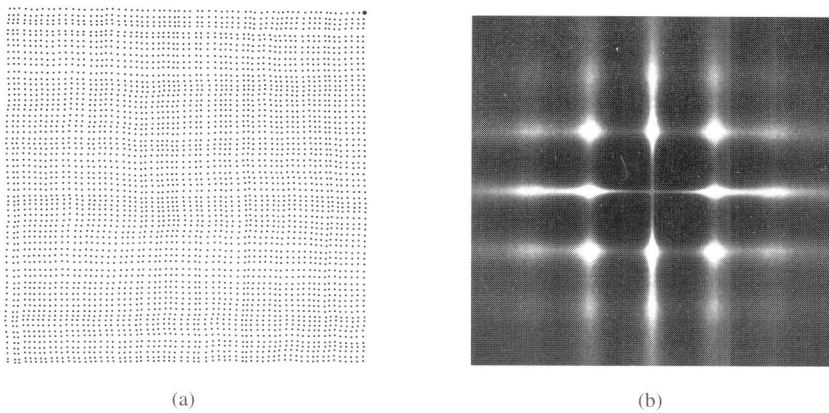

(a)   (b)

Fig. 6.4   Example paracrystal generated using $\rho_L = 0.95$, $\rho_T = 0.99$ and $\sigma = 0.5$. (a) Section of real space crystal (b) Corresponding diffraction pattern.

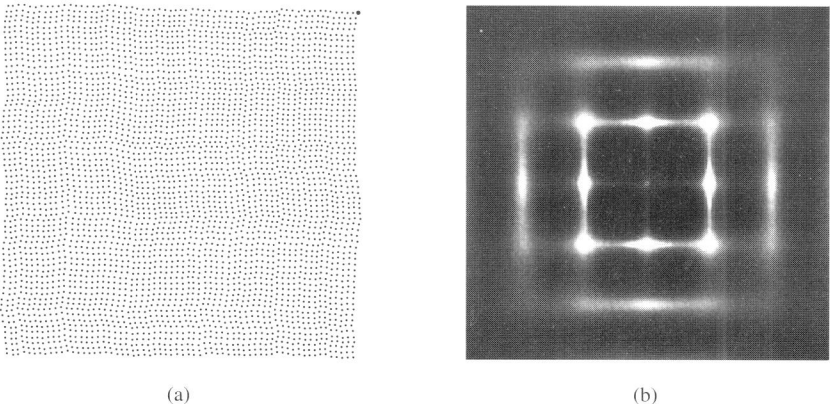

(a)   (b)

Fig. 6.5   Example paracrystal generated using $\rho_L = 0.99$, $\rho_T = 0.95$ and $\sigma = 0.5$. (a) Section of real space crystal (b) Corresponding diffraction pattern.

transverse correlations are equal. In this case $\sigma = 1.0$ and the large variation in position of the individual points away from either vertical or horizontal straight lines may readily seen, particularly near the edges of the plot. Figures 6.4 and 6.5 show examples in which the transverse and longitudinal correlations dominate respectively. The diffraction patterns for these examples were calculated from a much larger sample (512×512 points) than the region shown.

6.3   Generalised Gaussian models

Just as growth disorder models proved to be special cases of more general Ising models in the case of binary random variables, similar results pertain for Gaussian models. The Gaussian growth-disorder models described above can be considered as special cases of a more general joint-probability model with an energy E_c defined by

$$E_c = \sum_i \sum_j x_{i,j}\left(Ax_{i,j} + 2Bx_{i+1,j} + 2Cx_{i,j+1} + 2D(x_{i-1,j+1} + x_{i+1,j+1})\right). \qquad (6.4)$$

Here the random variables $x_{i,j}$ are Gaussian distributed with zero mean. Note that in comparison to the general Ising model, eqn (5.9), there are no terms involving the products of more than two random variables. This is because Gaussian variables do not possess higher moments than two and averages over such products will always equal zero. The four variables A, B, C amd D can be shown to be directly related (though not simply) to the lattice averages σ^2, r, s and t, where r, s, and t are the correlations in the x-direction, the y-direction and along the diagonal of the square respectively. The condition for this distribution to be identical to the Growth-model distribution is again that $r \times s = t$. Thus the extra freedom that this joint probability model provides over the growth-model is the ability to choose t independently.

Some example paracrystal-like distributions generated via Monte Carlo simulation of this more general model are shown in Fig. 6.6. The top-most example corresponds to the growth-model case where $r \times s = t$ and the other examples show the effect of increasing t for the same values of r and s. For these examples the value of σ is 0.35 but this is still sufficient to guarantee that the Bragg peaks due to the underlying regular lattice have virtually zero intensity. Note how in (d) the diffuse peaks are isotropic and note also in (f) that the diagonal correlation is now dominating and the diffraction pattern has the appearance of a paracrystal rotated through $45°$.

6.4   Gaussian growth-disorder models in 3D and higher

The prescription for Growth-disorder models, eqn (5.16), is capable of simple extension to higher dimensions. For example, if the eight points A, B, C, D, E, F, G, H define a cubic unit cell the equivalent prescription for a growth model (having mmm) symmetry) is,

$$\begin{aligned} P(A,B,C,D,E,F,G,H) = {} & P(A).P(B|A).P(C|A).P(E|A).P(D|A,B,C) \\ & \times P(F|A,B,E).P(G|A,C,E).P(H|A,B,C,D,E,F,G). \end{aligned} \qquad (6.5)$$

Note here that each successive factor in this expression uses points already specified. Hence the growth scheme is to first fix A, then the three edges AB, AC, AE, followed by the four faces $DABC$, $FABE$, $GACE$ and finally the completed cube, $HABCDEFG$.

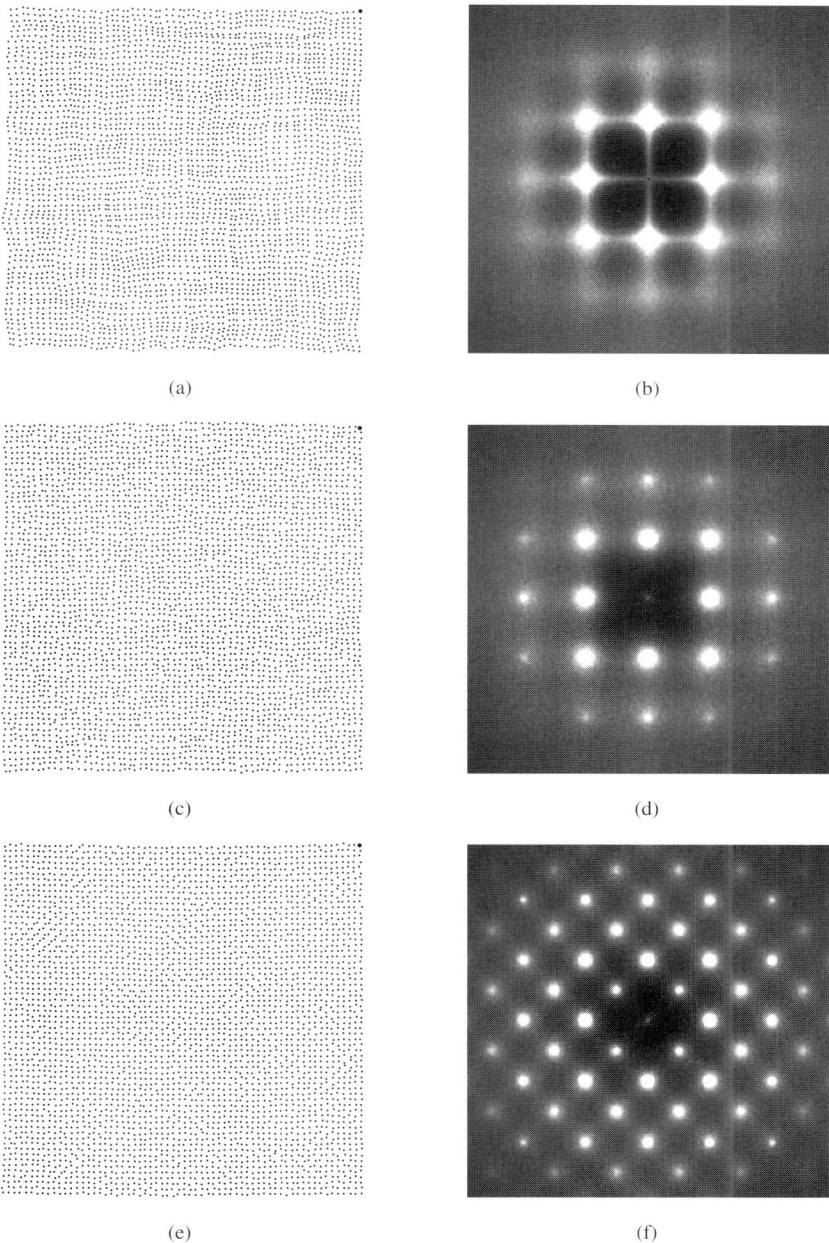

Fig. 6.6 Example paracrystals, with a section of the real-space representation on the left, and right is the corresponding diffraction pattern. (a) and (b) $r, s = 0.9$, $t = 0.81$ and $\sigma = 0.35$; (c) and (d) $r, s = 0.9$, $t = 0.87$ and $\sigma = 0.35$; (e) and (f) $r, s = 0.9$, $t = 0.95$ and $\sigma = 0.35$.

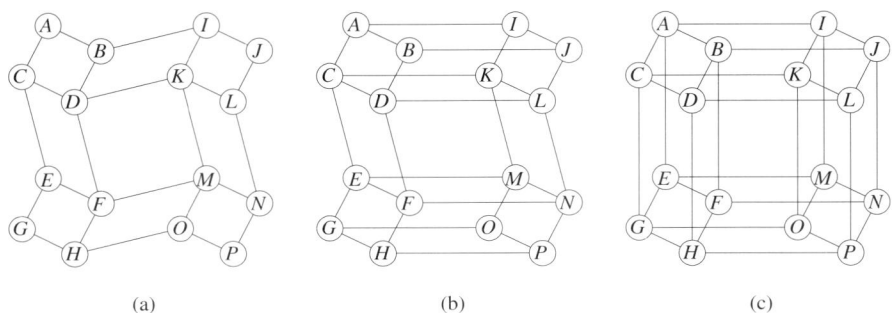

Fig. 6.7 Different topologies of growth model (see text for details).

The other useful property of the formulation eqn (5.16), is that provided the marginal distributions in neighbouring cells match, different growth prescriptions can be combined to produce more complex topologies. For example,

1. In Figure 6.7(a) the points A, B, C, D can be generated using eqn (5.16) and then the lattice can be extended to the points I, K and E, F using a different set of conditions, provided these maintain the distributions on the bounding edges B, D and C, D.

2. Figure 6.7(b) illustrates a quite different topology in which the distribution on the 16 points is factorised in such a way that two different 3D growth models are used.

3. Figure 6.7(c) illustrates the use of a 4D growth formulation.

In this latter the 16 points A, B, C, D, E, F, G, H, I, J, K, L, M, N, O, P defining a hypercubic cell are considered and the factorisation becomes,

$$
\begin{aligned}
P(A,B,C,D,E,F,G,H,I,J,K,L,M,N,O,P) = {} & P(A).P(B|A).P(C|A) \\
\times {} & P(E|A).P(I|A).P(D|A,B,C).P(F|A,B,E).P(G|A,C,E) \\
\times {} & P(J|A,B,I).P(K|A,C,I).P(M|A,E,I).P(H|A,B,C,D,E,F,G) \\
\times {} & P(L|A,B,C,D,I,J,K).P(N|A,B,E,F,I,J,M).P(O|A,C,E,G,I,K,M) \\
\times {} & P(P|A,B,C,D,E,F,G,H,I,J,K,L,M,N,O). \quad (6.6)
\end{aligned}
$$

All of the above factorisations apply equally to binary, Gaussian or any other variable. However, for binary models in dimensions greater than 2D the task of imposing the constraints necessary for this factorisation to be valid are rather daunting and it has been found in 3D, for example, that the range of near-neighbour correlations that is available is rather limited. For Gaussian variables there is no problem either in formulating the factorisation or in the range of correlations that can be produced. The various factors for a 3D Gaussian Growth-disorder model, $P(A)$, $P(B|A)$, $P(C|A)$, $P(E|A)$, $P(D|A,B,C)$, $P(F|A,B,E)$, $P(G|A,C,E)$, $P(H|A,B,C,D,E,F,G)$, can be written down by inspection.

The expression for the distribution on the 16 points of the hypercubic cell depicted above consists of conditional probability factors which refer to four 3D volumes, six 2D faces and four 1D edges, along which the primary correlations r, s, t and u run. The 3D volume conditional probability factors are typified by that involving A, B, C, D, E, F, G, H,

$$P(x_H|x_A,x_B,x_C,x_D,x_E,x_F,x_G) =$$
$$K\exp\left(\frac{-(x_H - rx_G - sx_F - tx_D + rtx_C + stx_B + rsx_E - rstx_A)^2}{2\sigma^2(1-r^2)(1-s^2)(1-t^2)}\right) \quad (6.7)$$

while that involving the 4D volume is,

$$P(x_H|x_A,\ldots,x_O) = K\exp\Big(-(x_P - rx_O - sx_N - tx_L - ux_H + sux_F$$
$$+ stx_J + tux_D + rux_G + rsx_M + rtx_K - rstx_I$$
$$- rsux_E - rutx_C - sutx_B + rstux_A)$$
$$\times\left(2\sigma^2(1-r^2)(1-s^2)(1-t^2)(1-u^2)\right)^{-1}\Big). \quad (6.8)$$

Examples of distributions generated using the three different factorisations given above are shown in Fig. 6.8 and their corresponding diffraction patterns in Fig. 6.9. For these examples the Gaussian variables have been converted after generation to binary oc-cupancy variables. In these examples it is useful to consider the different variables as representing four atomic sites of a molecule within a unit cell. For the 4D growth-model the correlations r and s may then be considered to be intra-molecular correlations and the correlations t and u to be inter-molecular correlations. For the left-most column of Figs 6.8 and 6.9 the internal correlations r and s are -0.9, for the central column r and s are 0.0 and for the right-most column they are $+0.9$. Note in the latter case how many unit cells contain the complete group of four dots forming a square, while in the left col-umn the molecule is most frequently a pair of dots on one or other of the two diagonals of the square. For the top row of figures the inter-molecular correlations t and u are zero but for the other three rows t and u are both set to be -0.5. For the 4D growth model these correlations are clearly cell-to-cell correlations but for the 2D and 3D models they combine with the internal correlations differently. For instance in (d) the negative value for r and s combines with the negative value of t and u to give a positive cell-to-cell correlation. For the 3D model this happens along one direction but not the other.

6.5   Conversion of Gaussian to binary variables

In the preceding sections it has been seen that the Gaussian variables offer many benefits over the binary variables described earlier. In particular it is straightforward to extend models to higher dimensions and also the range of nearest-neighbour correlations that is available is greater (except in 2D). Since for some purposes, (e.g. for representing site occupancy) it is necessary to have distributions of binary variables, it is therefore worth considering the conversion of a distribution of Gaussian variables to a binary distribution. The simplest way in which this might be done is to take each Gaussian variable $x_{i,j}$ and convert to a binary variable $y_{i,j}$ say where,

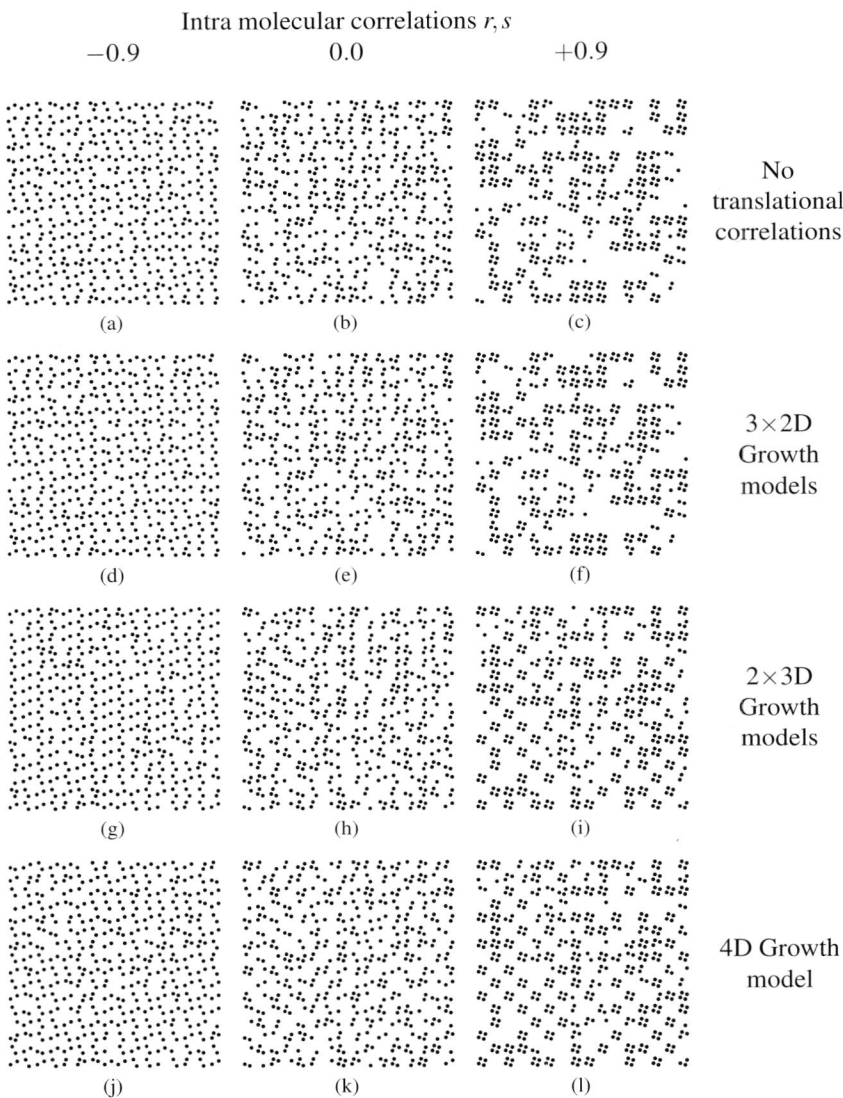

Fig. 6.8   Various distributions generated with 2D, 3D and 4D Gaussian growth models.

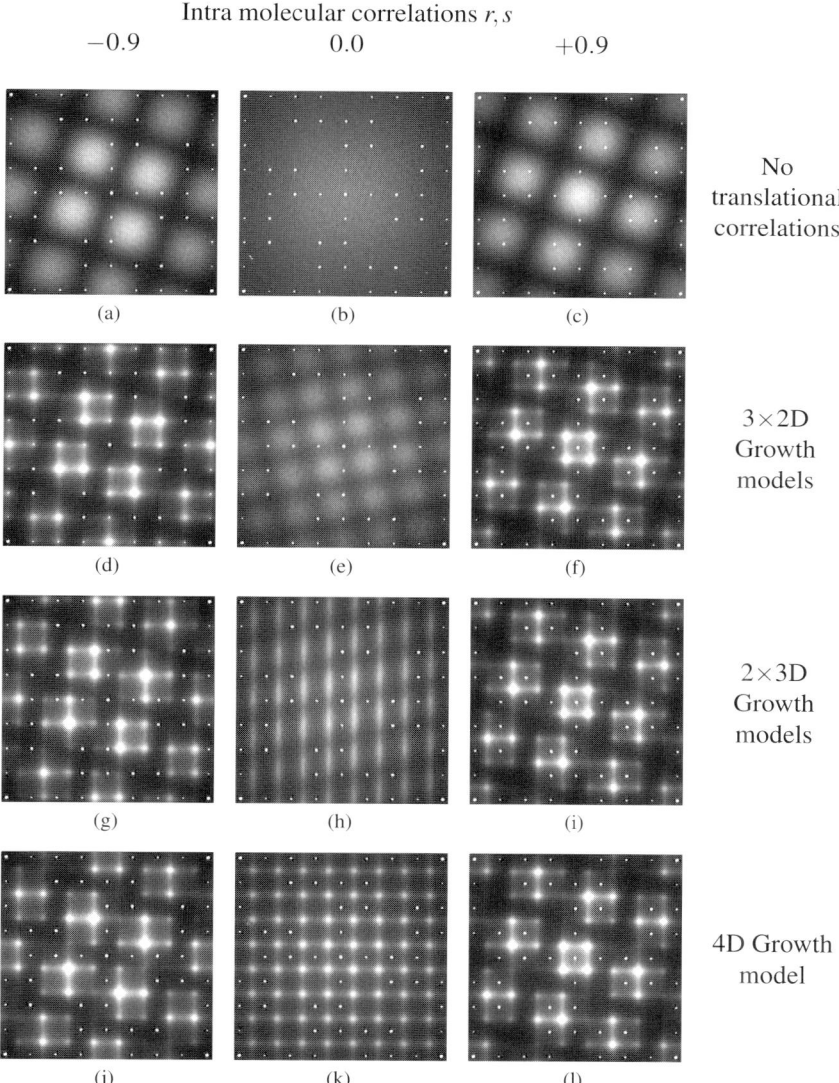

Fig. 6.9 Diffraction patterns of various distributions generated with 2D, 3D and 4D Gaussian growth models.

$$y_{i,j} = \begin{cases} -1 & \text{for } x_{i,j} < c \\ +1 & \text{for } x_{i,j} > c \end{cases} \tag{6.9}$$

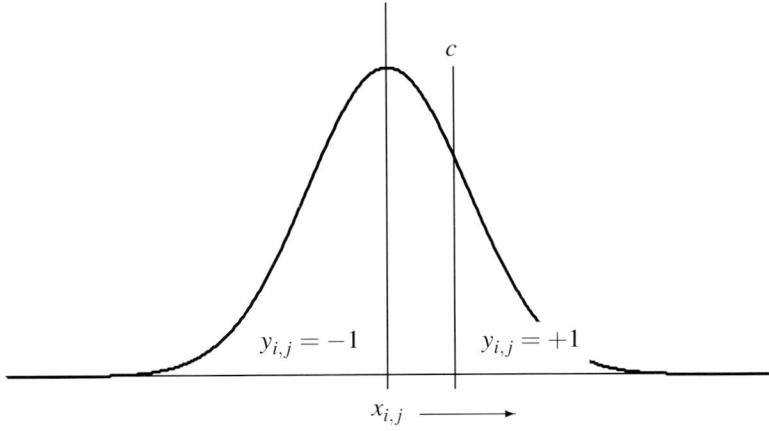

Fig. 6.10   Converting a Gaussian random variable to a $(+1, -1)$ binary variable.

For the simple case when $c = 0$ the concentration of the resulting binary variables, m_A is clearly 0.5. The Gaussian probability density for two real variables, which are correlated with a correlation coefficient r_g, and which have zero mean and unit variance is

$$P(x,y) = \frac{1}{2\pi\sqrt{1-r^2}} \exp\left(-\frac{x^2+y^2}{2(1-r_g^2)} + \frac{r_g xy}{(1-r_g^2)}\right). \tag{6.10}$$

The correlation coefficient between the resulting binary variables is

$$r_b = \frac{1}{2\pi\sqrt{1-r^2}} \int\!\!\int_{-\infty}^{+\infty} \text{sign}(x)\,\text{sign}(y) \exp\left(-\frac{x^2+y^2}{2(1-r_g^2)} + \frac{r_g xy}{(1-r_g^2)}\right) dx\,dy. \tag{6.11}$$

Putting $p = xy$ and $q = x/y$ and integrating over p and then q it is found that

$$r_b = \frac{2}{\pi} \arcsin(r_g). \tag{6.12}$$

From this result it may be noted that r_b can take values over the whole range from -1 to $+1$. In addition, except for the three points $r_b = r_g = -1, 0, +1$, $|r_b| < |r_g|$. This second property affects the way in which correlations decay with distance and the model will no longer have the geometric form characteristic of simple binary or Gaussian growth-disorder models.

Similar results can be obtained for values of $c \neq 0$, but an analytical expression is not available. A conversion to a required value of m_A may be effected numerically using an iterative procedure.

7

INTERACTIONS BETWEEN OCCUPANCIES AND DISPLACEMENTS

7.1 General intensity expressions

A general description of diffuse scattering that allows for both short-range compositional (chemical) order and local atomic distortions can be obtained by expanding the exponential in the kinematic scattering equation in powers of displacement:

$$
\begin{aligned}
I(\mathbf{k}) &= \sum_{n=1}^{N} \sum_{m=1}^{M} f_m f_n \exp\left(i\mathbf{k}.(\mathbf{R}_m + \mathbf{u}_m - \mathbf{R}_n - \mathbf{u}_n)\right) \\
&\simeq \sum_{n=1}^{N} \sum_{m=1}^{M} f_m f_n \exp\left(i\mathbf{k}.(\mathbf{R}_m - \mathbf{R}_n)\right) \times \Big(1 + i\mathbf{k}.(\mathbf{u}_n - \mathbf{u}_m) \\
&\quad - \frac{1}{2}\left(\mathbf{k}.(\mathbf{u}_n - \mathbf{u}_m)\right)^2 - \frac{i}{6}\left(\mathbf{k}.(\mathbf{u}_n - \mathbf{u}_m)\right)^3 + \dots\Big).
\end{aligned}
\tag{7.1}
$$

Here I is the scattered intensity and f_m is the scattering factor of the atom m associated with the lattice site at the location $\mathbf{R}_m$ and which is displaced from its site by a small amount $\mathbf{u}_m$. Equation (7.1) may be compared with eqn (2.8). The scattering vector, $\mathbf{k} = 2\pi \mathbf{S}$, is defined as

$$
\mathbf{k} = h_1 \mathbf{a}^* + h_2 \mathbf{b}^* + h_3 \mathbf{c}^*.
\tag{7.2}
$$

The intensity may be written as the sum of component intensities: the first term is independent of the displacements, the second term is dependent on the first moment of displacements, the third term on the second moment etc. At this point the expression includes both Bragg peaks and diffuse scattering. However, when the Bragg peaks are removed the diffuse part of the intensity may similarly be written as the sum of component intensities:

$$
I_{\text{Diffuse}} = I_0 + I_1 + I_2 + I_3 + \dots.
\tag{7.3}
$$

The individual diffuse components may be written

$$
I_0 = -N \sum_{ij} \sum_{lmn} c_i c_j f_i f_j^* \alpha_{lmn}^{ij} \cos\left(2\pi(h_1 l + h_2 m + h_3 n)\right)
\tag{7.4a}
$$

$$
I_1 = -2\pi N \sum_{ij} \sum_{lmn} c_i c_j f_i f_j^* (1 - \alpha_{lmn}^{ij}) \sin\left(2\pi(h_1 l + h_2 m + h_3 n)\right)
$$
$$
\times \left(h_1 \left\langle X_{lmn}^{ij} \right\rangle + h_2 \left\langle Y_{lmn}^{ij} \right\rangle + h_3 \left\langle Z_{lmn}^{ij} \right\rangle\right)
\tag{7.4b}
$$

$$I_2 = -2\pi^2 N \sum_{ij} \sum_{lmn} c_i c_j f_i f_j^* (1 - \alpha_{lmn}^{ij}) \cos\left(2\pi(h_1 l + h_2 m + h_3 n)\right)$$

$$\times \left(h_1^2 \left(\left\langle (X_{lmn}^{ij})^2 \right\rangle - (1 - \alpha_{lmn}^{ij})^{-1} \left\langle (X_\infty^{ij})^2 \right\rangle \right) \right.$$

$$+ h_2^2 \left(\left\langle (Y_{lmn}^{ij})^2 \right\rangle - (1 - \alpha_{lmn}^{ij})^{-1} \left\langle (Y_\infty^{ij})^2 \right\rangle \right) \tag{7.4c}$$

$$+ h_3^2 \left(\left\langle (Z_{lmn}^{ij})^2 \right\rangle - (1 - \alpha_{lmn}^{ij})^{-1} \left\langle (Z_\infty^{ij})^2 \right\rangle \right)$$

$$\left. + 2h_1 h_2 \left\langle X_{lmn}^{ij} Y_{lmn}^{ij} \right\rangle + 2h_1 h_3 \left\langle X_{lmn}^{ij} Z_{lmn}^{ij} \right\rangle + 2h_2 h_3 \left\langle Y_{lmn}^{ij} Z_{lma}^{ij} \right\rangle \right)$$

$$I_3 = \frac{4}{3} \pi^3 N \sum_{ij} \sum_{lmn} c_i c_j f_i f_j^* (1 - \alpha_{lmn}^{ij}) \sin\left(2\pi(h_1 l + h_2 m + h_3 n)\right)$$

$$\times \left(h_1^3 \left\langle (X_{lmn}^{ij})^3 \right\rangle + h_2^3 \left\langle (Y_{lmn}^{ij})^3 \right\rangle + h_3^3 \left\langle (Z_{lmn}^{ij})^3 \right\rangle \right.$$

$$+ 3h_1^2 h_2 \left\langle (X_{lmn}^{ij})^2 Y_{lmn}^{ij} \right\rangle + 3h_1^2 h_3 \left\langle (X_{lmn}^{ij})^2 Z_{lmn}^{ij} \right\rangle \tag{7.4d}$$

$$+ 3h_2^2 h_1 \left\langle (Y_{lmn}^{ij})^2 X_{lmn}^{ij} \right\rangle + 3h_2^2 h_3 \left\langle (Y_{lmn}^{ij})^2 Z_{lmn}^{ij} \right\rangle$$

$$+ 3h_3^2 h_1 \left\langle (Z_{lmn}^{ij})^2 X_{lmn}^{ij} \right\rangle + 3h_3^2 h_2 \left\langle (Z_{lmn}^{ij})^2 Y_{lmn}^{ij} \right\rangle$$

$$\left. + 6h_1 h_2 h_3 \left\langle X_{lmn}^{ij} Y_{lmn}^{ij} Z_{lmn}^{ij} \right\rangle \right).$$

The summation ij represents all atom species and sublattices, and the summation lmn represents all interatomic vectors. N is the number of unit cells. The concentration and scattering factors of the species are given by c, and f respectively. The SRO parameters, α, are those defined by Cowley (1950), that is,

$$\alpha^{ij} = 1 - \frac{P_{lmn}^{ij}}{c_j}, \tag{7.5}$$

where P_{lmn}^{ij} is the conditional probability of finding an atom with label j at the end of an interatomic vector lmn from the origin which contains an atom with label i. This definition of SRO parameters is somewhat more general than the C_{lmn} correlation values used earlier since it may be used in cases where there are more than two species present. For two component systems α^{ij} and C_{lmn} are equivalent. The displacement components, X_{lmn}^{ij}, Y_{lmn}^{ij} and Z_{lmn}^{ij}, are defined by

$$X_{lmn}^{ij} = u_{lmn}^{xj} - u_0^{xi} \quad \text{etc.,} \tag{7.6}$$

where u_{lmn}^{xj} is the displacement in the x direction of an atom with label j situated at the end of the vector $\mathbf{r}_{lmn}$ from the origin, where an atom with label i is displaced u_0^{xi}, also in the x direction.

I_0 is commonly known as the SRO component, I_1 as the 'size-effect' component and I_2 as the Huang scattering and thermal diffuse scattering (TDS) component (see Section 7.4).

7.2   A possible Ising-like model for occupations and displacements

In earlier sections models which were either purely occupational or purely displacive have been discussed. It is clear from the last section that it is necessary to include the possibility that the occupational and displacement variables interact with each other. This might be done using an Ising-like interaction potential of the form

$$E = \sum_{i,j} x_{i,j} \sum_{m,n} K_{m,n} x_{i-m,j-n} + \sum_{i,j} \sigma_{i,j} \sum_{m,n} L_{m,n} \sigma_{i-m,j-n}$$
$$+ \sum_{i,j} x_{i,j} \sum_{m,n} M_{m,n} \sigma_{i-m,j-n}. \tag{7.7}$$

The first term defines interactions between the displacement variable $x_{i,j}$ and neighbouring displacement variables within the range defined by the indices m and n. The second term defines the corresponding interactions between occupancy variables $\sigma_{i,j}$. The third term defines the interaction between the displacement variable $x_{i,j}$ and neighbouring occupancy variables, $\sigma_{i,j}$.

This form should be capable of producing lattice realisations containing disorder of considerable complexity, but when it is realised that a particular problem might involve a number of quite distinct displacement variables (x, y, z and orientation) the number of possible terms becomes rather unwieldy. In addition the form, though quite general, does not take into account the particular features of a physical problem that actually lead to the correlation between occupancy and displacement variables.

7.3   Use of force models and Monte Carlo simulation

A more direct approach to the treatment of the interaction between occupancy and displacement variables is to use a force model. It is imagined that the structure is held together by harmonic springs, the equilibrium lengths and force constants of which depend on the occupational variables defining the two atomic (or molecular) sites joined by the spring. The method is illustrated by a simple 2D model in which there are occupancy variables $\sigma_{i,j}$ distributed on a square lattice, say, and displacement variables $x_{i,j}$ and $y_{i,j}$ describing the displacements away from the average positions. If the distance between two interacting sites m and n is taken as d_{ave} then there will be an energy term of the form,

$$E = \sum_{n,m} K_{n,m} \left(d_{n,m} - d_{\text{ave}} (1 + \varepsilon_{n,m}) \right)^2, \tag{7.8}$$

where $\varepsilon_{n,m}$ is a 'size-effect' parameter which takes different values according to whether the two sites joined by the spring are $(0,0)$, $(0,1)$, $(1,0)$ or $(1,1)$. For a simple square lattice, with a cell repeat of unity, in which the interactions are along the sides of the unit cell $d_{n,m}$ can be replaced by the variables $x_{i,j}$ and $y_{i,j}$, so

$$d_{1,0} = \sqrt{(x_{i,j} - x_{i-1,j})^2 + (y_{i,j} - y_{i-1,j})^2}$$

and

$$d_{0,1} = \sqrt{(x_{i,j} - x_{i,j-1})^2 + (y_{i,j} - y_{i,j-1})^2}.$$

In Fig. 7.1 $K_{1,0}$ and $K_{0,1}$ have been set to the same value, K_1. With only these two springs this 2D system is unstable with respect to a shear force and so a pair of stabilising springs, $K_{1,1}$ and $K_{1,\bar{1}}$, have been introduced with a force constant K_2 (much smaller than K_1).

Realisations of this simple force model may be obtain via Monte Carlo simulation using eqn (7.8). Examples are shown in the next section.

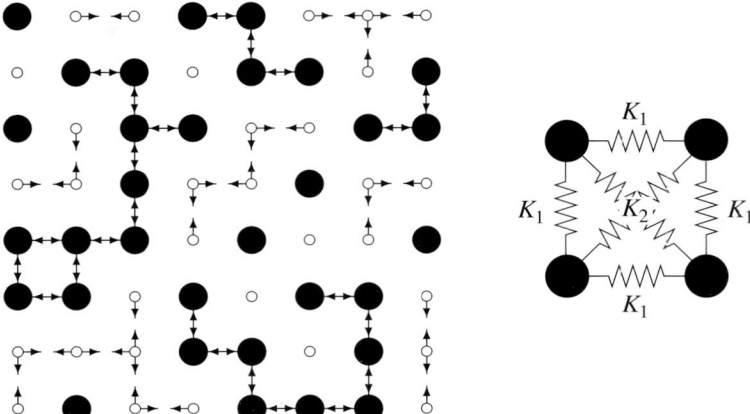

Fig. 7.1 Schematic diagram illustrating the size-effect force models. In the left figure arrows indicate the tendency for two large (black) atoms to move apart and for two small atoms to move together. The right figure indicates the springs used to induce the size-effect relaxations.

7.4   Illustration of the meaning of the different intensity components

The diffraction patterns shown in Fig. 7.2, which were all calculated from the simple model described above, illustrate the meaning of the different components of the intensity, I_0, I_1, I_2, etc. Figure 7.2(a) corresponds to a purely random distribution of the occupancy variables, that is, all correlations are zero, but all scatterers are perfectly on the lattice sites. Then the only term contributing to the diffuse intensity is the zero'th order correlation term of I_0. The diffraction pattern is a uniform distribution of intensity and is known as the Laue monotonic scattering. Figure 7.2(b) shows an example where there are also no displacements but now the short-range order parameters or correlations are non-zero. In this case the nearest-neighbours are negatively correlated in both axial directions producing a diffuse peak at the centre of the reciprocal unit cell. For this I_0 diffraction term, commonly known as the SRO component, the intensity is the sum of cosines and hence produces symmetric diffraction profiles.

Figure 7.2(c) and (d) show examples illustrating the component I_1, commonly referred to as the size-effect component. The Monte Carlo algorithm has been used to

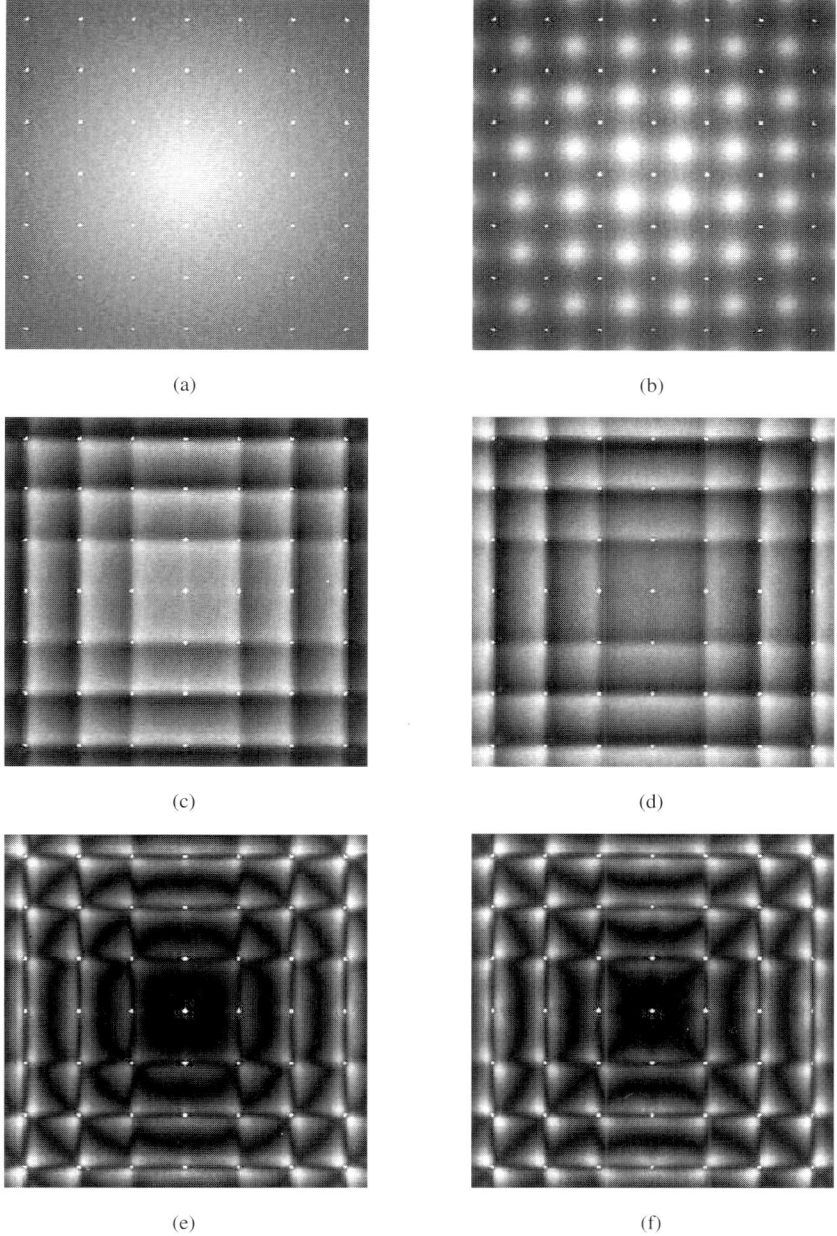

Fig. 7.2 Diffuse intensity components. (a) Laue Monotonic scattering. (b) SRO term, $2 \times -$ve correlation. (c) Size-effect $f_{\text{big}} > f_{\text{small}}$. (d) Size-effect $f_{\text{small}} > f_{\text{big}}$. (e) Second-order term $f_{\text{big}} = f_{\text{small}}$ (dilation). (f) Second-order term $f_{\text{big}} = f_{\text{small}}$ (shear).

relax the disordered structure so that individual atoms move off the average lattice positions. The same completely random occupancy distribution used for Fig. 7.2(a) was used here. In Fig. 7.2(c) the interactions are such that neighbouring atoms with large scattering factor tend to move away from each other and neighbouring atoms with small scattering factor move towards each other. In Fig. 7.2(d) the reverse is true and neighbouring atoms with large scattering factor tend to move towards each other while atoms with small scattering factor move away from each other. For this I_1 diffraction term the intensity is the sum of sines and hence produces asymmetric diffraction profiles.

Figure 7.2(e) and (f) illustrate the component I_2, commonly referred to as the TDS and Huang scattering component. Observation of this term is possible when the two different scatterers are given the same scattering factor. Then both I_0 and I_1 become zero and what is left is I_2 and higher terms. Figure 7.2(e) corresponds to the situation where the distortions are carried out exactly as for Fig. 7.2(c), that is, big atoms move apart; small atoms together, but the two types of atom have the same scattering factor. This occurs in both axial directions and so corresponds to a *dilation* of the structure (see Fig. 7.3(a)). Figure 7.2(f) was produced very similarly but here the two axial directions were treated differently. In one direction, say vertically, the treatment was the same as for Figure 7.2(e) but in the other direction (horizontally) the distortion was of the opposite sense, that is, small atoms move apart; big atoms together. This type of distortion corresponds to a local *shearing* of the structure (see Fig. 7.3(b)). For this I_2 diffraction term the intensity is the sum of cosines and hence produces symmetric diffraction profiles.

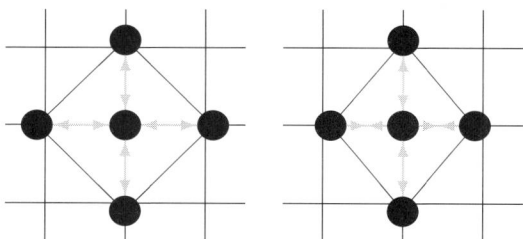

Fig. 7.3   Ilustrating dilation and shear. (a) dilation, (b) shear.

7.5   Size-effect and multi-site correlations

It was seen in earlier sections that a number of quite different distributions of occupancy variables could give rise to indistinguishable diffraction patterns. In particular there were examples of distributions with large three-body and four-body correlations which gave diffraction patterns that were indistinguishable from that of the purely random distribution. It is interesting to consider whether the presence of local relaxations using the Monte Carlo algorithm described above will show up any differences between these distributions. Figure 7.4 shows a sample of three different distributions together with diffraction patterns obtained from each. The three columns of examples are: column 1

Four-body correlations   Three-body correlations
Small   Large

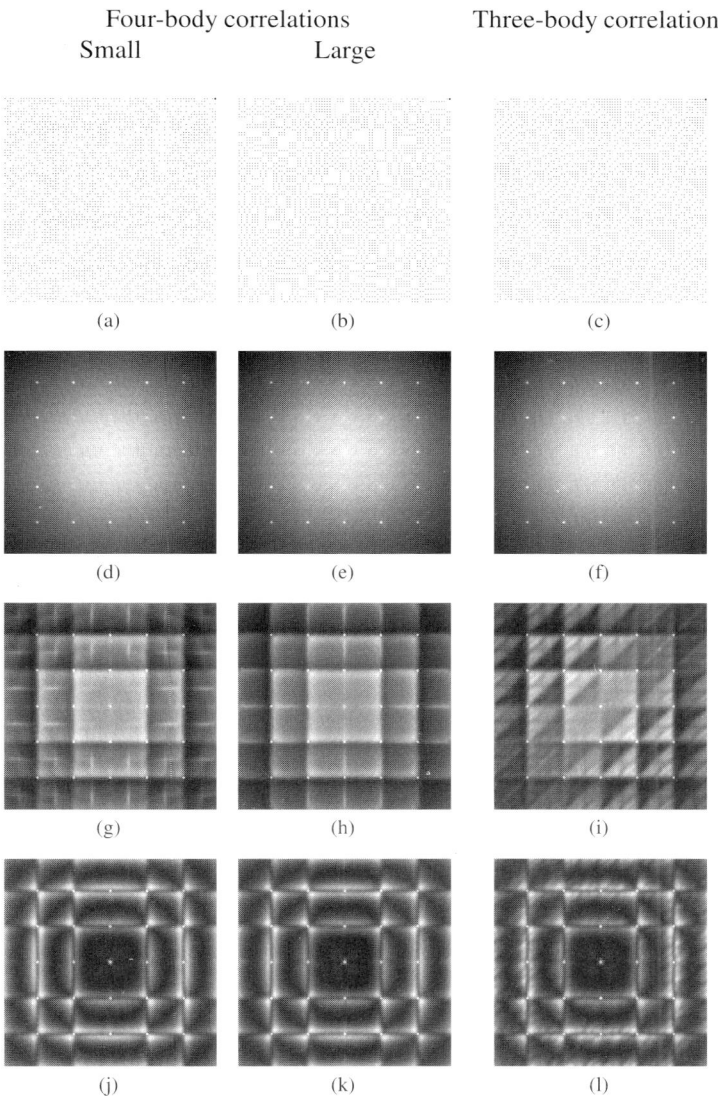

Fig. 7.4   Size-effect and multi-site correlations. (a), (b) and (c) show small representative regions of model distributions. (d), (e) and (f) show corresponding diffraction patterns with no size-effect. (g), (h) and (i) show corresponding diffraction patterns with size-effect for $f_{\text{big}} > f_{\text{small}}$. (j), (k) and (l) show corresponding diffraction patterns with size-effect for $f_{\text{big}} = f_{\text{small}}$.

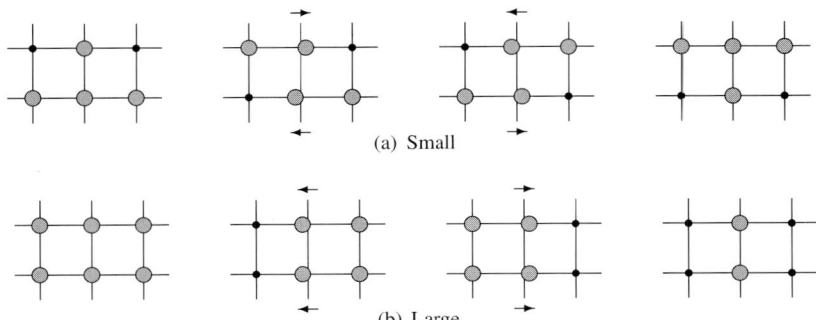

Fig. 7.5 Showing how size-effect relaxation induces transverse two-body displacement correlations when four-body correlations are present in a distribution.

has a small four-site correlation; column 2 has a large four-site correlation; column 3 has a large three-site correlation. For each of these examples diffraction patterns are shown with: no relaxation (I_0); with relaxation and different scattering factors (I_1); and with relaxation and the same scattering factors (I_2).

Although the patterns look broadly similar to those for the random distribution discussed above there are some distinct differences which appear in the I_1 term. For the small four-site correlation example some extra elongated diffuse peaks appear midway between the rows of Bragg peaks, while for the large four-site correlation example similar extra peaks appear along the rows of Bragg peaks. These may be understood by considering the diagram shown in Fig. 7.5. This figure shows groups of six points which are the four most frequently occurring local configurations for the two different four-site correlation examples. In each of the figures the central point in each of the two horizontal rows of sites is displaced according to its neighbours to the left and right. If the two neighbours are the same then there is no displacement, but if they are different then the site is displaced towards the smaller of the two neighbours. For Fig. 7.5(a), which corresponds to the low four-site case, it is seen that in two of the four cases this has the effect that atoms in successive rows are displaced in opposite directions. That is to say a negative transverse displacement correlation has been induced. In contrast, for Fig. 7.5(b), which corresponds to the high four-site case, it is seen that in two of the four cases atoms in successive rows are displaced in the same direction. This corresponds to a positive transverse correlation. Thus the presence of four-site correlations has induced extra two-site correlations in the displacements and their presence can be seen in the I_1 term of the diffraction pattern.

For the three-site correlation example the effect of introducing size-effect relaxation is more marked. The extra features that are seen in the I_1 term of the diffraction pattern are quite unlike those for the four-site examples and the new features show the asymmetry of intensity across rows of Bragg peaks which is characteristic of the size-effect. These new features appear in the diagonal direction indicating that a size-effect has been induced in the diagonal direction even though the actual forces are in the axial directions.

For the bottom row of figures which correspond to the I_2 diffraction term, the two four-site examples show virtually no sign of the four-site correlation, but the three-site example shows some texture that is oriented in the diagonal direction.

Part III

Examples of real disordered systems

8

1,3-dibromo-2,5-diethyl-4,6-dimethylbenzene (Bemb2)

8.1 Introduction

Crystals of the compound 1,3-dibromo-2,5-diethyl-4,6-dimethylbenzene (Bemb2) are
disordered because of the similarity in size of the bromo- and methyl-groups. In each
molecular site the molecule may take up one of two different orientations, effectively
resulting in the interchange of these substitutional groups on the sites of the average unit
cell. The average structure of Bemb2 viewed in projection down the **a**-axis is shown in
Fig. 8.1(a). The shaded atoms represent a site containing 50% bromo and 50% methyl.
The diffuse diffraction pattern, two sections of which are shown in Fig. 8.2, clearly indicates short-range ordering of this orientational disorder, but the rather broadly diffuse
nature of the scattering indicates that the correlations can extend only over a few unit
cells. In Fig. 8.2(b) it may be seen, however, that this scattering is interrupted close to
some Bragg positions by a diminution of intensity, giving the appearance that there is
a 'hole' in the scattering. This is particularly noticeable at the 0 1 0 position where the
Bragg peak is absent. The same point in Fig. 8.2(a) shows a marked narrowing of the
diffuse bands close to 0 1 0. The dimensions of this 'hole' and the narrow part of the
diffuse peak are such that correlations extending over many unit cells are indicated.

This system provides an excellent example of how the use of the modulation-wave
description of disorder described in Section 5.8 can provide insights into the origins of
these features. The system has been described in some detail in Welberry and Withers
(1990) and Welberry and Butler (1994).

8.2 Symmetry considerations

8.2.1 $\mathbf{q} = 0$ modulations

Modulations must be consistent with the space group symmetry of the average structure, and this imposes restrictions on the allowed values for the modulation amplitudes
$a_\mu(\mathbf{q})$ for occupancy and $e_\mu(\mathbf{q})$ for displacement, when the modulation wave-vector is
along certain symmetry directions.

The average structure of Bemb2 conforms to the space group $P2_1$, but in fact is
very close to $P2_1/c$ (Wood *et al.*, 1984). In Fig. 8.1(a) the shaded substituents have
a composition approximately $(0.5Br + 0.5CH_3)$. Only a small number of Bragg peaks
of the type $h\,0\,l$, with l odd, reveal the breaking of the c-glide symmetry. A $\mathbf{q} = 0$
modulation affects all unit cells within the crystal equally and thus corresponds to a
change in the average structure. The structure can therefore be considered as a $P2_1/c$
structure containing a weak $\mathbf{q} = 0$ modulation which destroys the c-glide but maintains
the 2_1-screw axis.

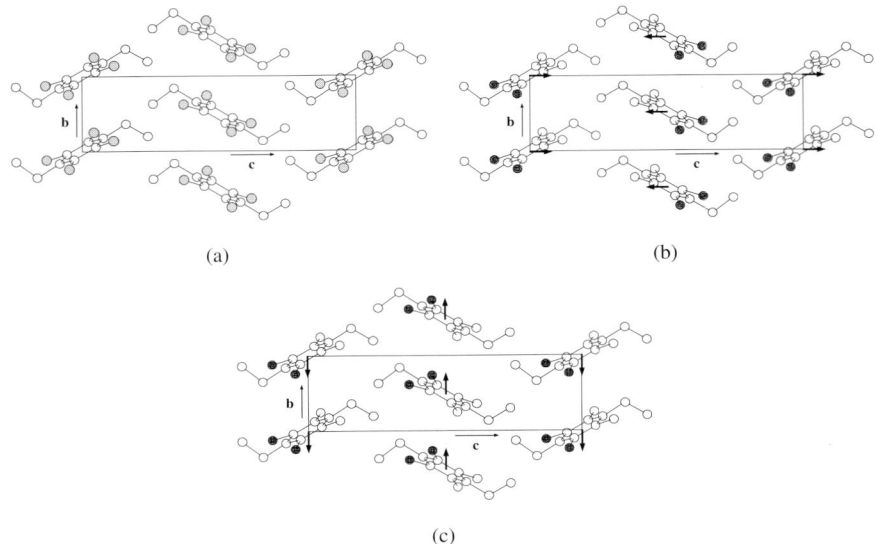

(a)   (b)

(c)

Fig. 8.1   The structure of Bemb2 viewed in projection down **a**.

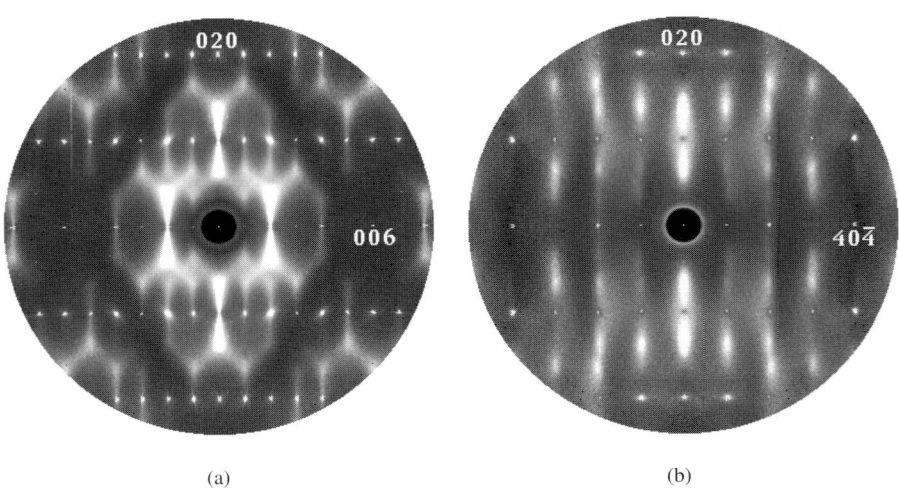

(a)   (b)

Fig. 8.2   Observed X-ray diffuse scattering patterns of Bemb2. (a) the $(0\,k\,l)$ section. (b) the $(h\,k\,\bar{h})$ section.

In order to see the effect of modulations with different symmetries it is necessary to consider the isogonal point group, F, of the space group (i.e. the point group obtained by taking just the point group operators from amongst the symmetry operators of the space group). For $P2_1/c$ this is $2/m$ or $\{E, i, C_{2y}, \sigma_y\}$ in Schoenflies notation. For waves of a particular $\mathbf{q}$ it is only necessary to consider the little co-group of $\mathbf{q}$, that is, the subgroup of F which leaves $\mathbf{q}$ invariant. For a general wavevector $\mathbf{q}$ this is simply $\{E\}$ but for wavevectors along $\mathbf{b}^*$ it is $\{E, C_{2y}\}$ while for $\mathbf{q} = 0$ the little co-group has the full symmetry of the isogonal point group, that is, $\{E, i, C_{2y}, \sigma_y\}$. The table of irreducible representations for this point group is

$$
\begin{array}{ccccc}
 & E & i & C_{2y} & \sigma_y \\
\Gamma_1 & 1 & 1 & 1 & 1 \\
\Gamma_2 & 1 & 1 & -1 & -1 \\
\Gamma_3 & 1 & -1 & 1 & -1 \\
\Gamma_4 & 1 & -1 & -1 & 1 .
\end{array}
\tag{8.1}
$$

A character of 1 under a particular symmetry operation R in the above table implies that the corresponding space group symmetry operation $\{R|v\}$ is preserved in the resultant structure. Thus a modulation which has a character of 1 in the third column will preserve the 2_1-screw, while one having a character of 1 in the fourth column will preserve the c-glide. It is clear that for compositional modulations in which individual sites must contain the non-centrosymmetric Bemb2 molecule in one or other of its two possible orientations, a modulation of Γ_1 or Γ_2 symmetry is not possible, since these require that the four substituent sites in one molecular site maintain the same composition. A modulation of Γ_3 symmetry (for which $a_1 = a_2$) is possible, however, and produces a structure with the appearance of Fig. 8.1(b), which preserves the 2_1-screw axis. A modulation of Γ_4 symmetry (for which $a_1 = -a_2$) is also possible, however, and produces a structure with the appearance of Fig. 8.1(c). This destroys the 2_1-screw axis and preserves the c-glide symmetry. Consequently the observed symmetry of the average structure implies that it must be describable as a $\mathbf{q} = 0$, Γ_3 modulation of the idealised $P2_1/c$ structure.

8.2.2 $\mathbf{q} \neq 0$ modulations

For a general wave-vector $\mathbf{q}$ there are no symmetry restrictions constraining the choice of the two permissible types of concentration modulation, but for modulation wavevectors along $\mathbf{b}^*$ there are restrictions and the amplitudes of the modulations must conform to

$$
a_2 = \pm a_1 \exp(2\pi \mathbf{q} . (\mathbf{b} + \mathbf{c})/2).
\tag{8.2}
$$

Clearly at $\mathbf{q} = 0$ the modulation merges with Γ_3 if the $+$ sign is used and Γ_4 if the $-$ sign is used. For the present purposes the two types of modulation at a general $\mathbf{q}$ are referred to as Γ_3 and Γ_4 although strictly these terms apply to $\mathbf{q} = 0$ only.

In general the lowest energy modulation at a particular $\mathbf{q}$ might be expected to be a linear combination of the two types of modulation. The eigenvector (a_1, a_2) defining this mixture depends on the energy of the system and not on the symmetry.

8.3   Calculated diffraction patterns

8.3.1   *The $(0\,k\,l)$ section*

It is interesting to compare the effects of these two different types of modulation, Γ_3 and Γ_4, on the diffuse scattering distribution. Use is made of the method of synthesising real-space distributions of scattering points described in Section 5.8. In this method random variables, $\sigma_{i,j}$ say for a 2D example, are used to represent the property of the site (i, j) in the lattice that is being modulated. Initially, $\sigma_{i,j}$ is assumed to be a real variable. It is then assumed that each $\sigma_{i,j}$ is the sum of a large number of simple sinusoidal perturbations of different wave-vectors $\mathbf{q}$. All possible wave-vectors $\mathbf{q}$ on a finely spaced grid in the first Brillouin zone are used. Each perturbing wave is given a random phase $\phi_{\mathbf{q}}$, and an amplitude $A_{\mathbf{q}}$ which is chosen to reflect the variation in intensity with $\mathbf{q}$ that is observed in the experimental diffraction pattern. The values of different $\sigma_{i,j}$ result from the sum of a large number of perturbations and so according to the central limit theorem they will be normally distributed. Sets of such real random variables, suitably normalised, may be used *directly* to represent atomic displacements, or *indirectly* by conversion to an analogous set of binary variables to represent site occupancies.

In the first Brillouin zone of the $(0\,k\,l)$ section of Bemb2 (Fig. 8.2(a)) the diffuse intensity is seen to be in the shape of a 'Y'—the intense vertical part of the 'Y' extending along $\mathbf{b}^*$ then splitting into two weaker arms which end on the zone boundary at $(0.5\mathbf{b}^* + 0.5\mathbf{c}^*)$. Real-space distributions of binary random variables $\sigma_{i,j,m}$ were constructed using modulation-wave amplitudes which qualitatively matched this 'Y'-shaped diffuse intensity distribution. Here i, j are indices defining the unit cell and m an index defining the two molecular sites at the corner and the centre of the projected unit cell shown in Fig. 8.1. In Fig. 8.3 some results of this synthesis are shown in the two cases when each of the two different types of modulation Γ_3 or Γ_4 were employed. Figure 8.3(a) and (c) show small portions of the real-space distribution in which a single 'atom' is placed at the corners and centre of the unit cell. A white dot represents a value of $\sigma_{i,j,m} = 1$ and black a value of $\sigma_{i,j,m} = -1$.

Figures 8.3(b) and (d) are corresponding diffuse diffraction patterns computed from these realisations, using a carbon scattering factor f_C for $\sigma_{i,j,m} = 1$ and zero for $\sigma_{i,j,m} = -1$. This figure clearly shows the effect of the two different types of concentration modulation, both in real-space and reciprocal-space. In Fig. 8.3(a) there are numerous occurrences of unit cells in which the corners are occupied and the centres are not (or vice versa), while in Fig. 8.3(c) there are numerous occurrences of cells in which corners and centres are both occupied. Some examples have been outlined in the figure. In the diffraction patterns the 'Y'-shaped scattering is seen to occur near the reciprocal points $(h + k) = $ odd for Fig. 8.3(b) and $(h + k) = $ even for Fig. 8.3(d). Apart from the origin shift, these two diffraction patterns are clearly identical.

Figure 8.4 shows diffraction patterns computed using the same distributions of $\sigma_{i,j,m}$ as used for Fig. 8.3. Now $\sigma_{i,j,m} = 1$ is taken to represent a whole Bemb2 molecule in orientation A and $\sigma_{i,j,m} = -1$ as a molecule in orientation B. For Fig. 8.4(a) the modulations (Γ_3) tend locally to maintain the 2_1-screw axis, while for Fig. 8.4(b) the modulations (Γ_4) tend locally to maintain the c-glide plane. Clearly the calculated pattern, Fig. 8.4(b), corresponds very closely to the observed pattern (Fig. 8.2(a)). This means

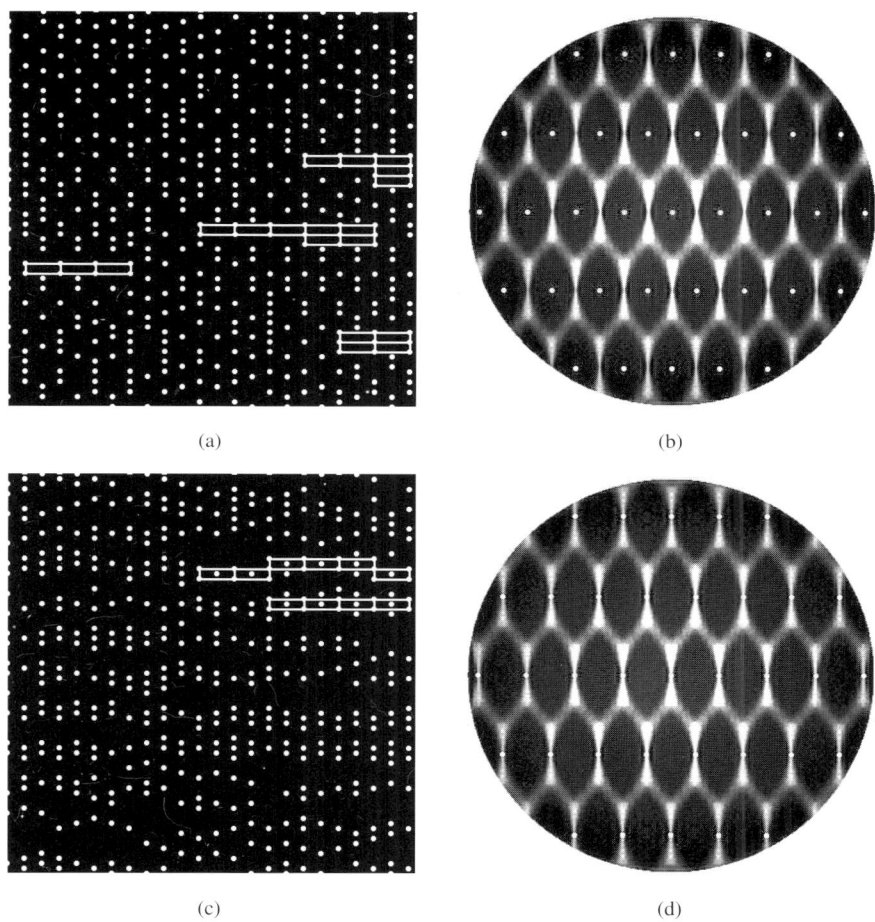

(a)

(b)

(c)

(d)

Fig. 8.3 Synthesised distributions and their corresponding calculated diffraction patterns corresponding to the $(0\,k\,l)$ section of Bemb2.

that, although there are no symmetry constraints demanding that only one of the types of concentration mode exists at a general wave-vector $\mathbf{q}$, the energetics are such that, for virtually the whole of $\mathbf{q}$-space, modulations of the second type (Γ_4) are the only ones that actually exist.

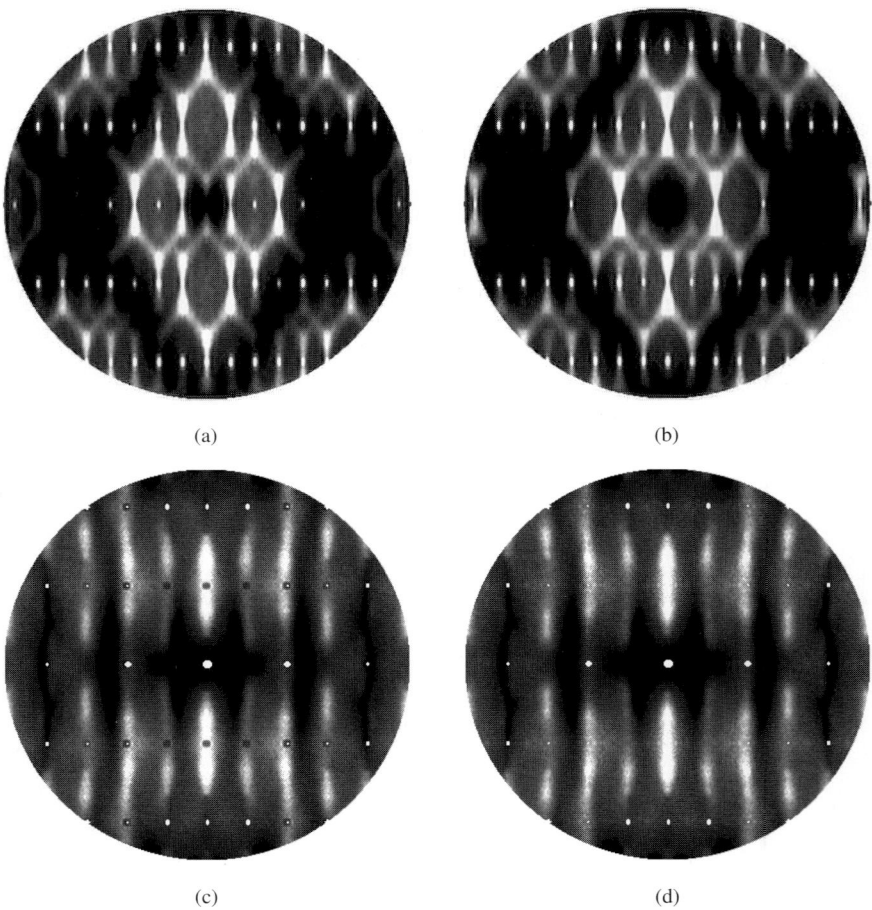

Fig. 8.4 Calculated diffraction patterns for Bemb2. See text for details.

It is thus seen from the $(0kl)$ patterns that there is seemingly a paradoxical situation in Bemb2, since the compositional modulations of the structure for all wave-vectors other than $\mathbf{q} = 0$ are of the Γ_4 symmetry (which tend to preserve the c-glide symmetry but destroy the 2_1-screw axis), while for $\mathbf{q} = 0$ the modulation is of Γ_3 symmetry (which preserves the 2_1-screw axis but destroys the c-glide symmetry).

8.3.2   The $(h\,k\,\bar{h})$ section

The 'hole' in the diffuse scattering which is seen clearly in the $(h\,k\,\bar{h})$ section, Fig. 8.2(b), is evidently a manifestation of this incompatibility between what the short-range and long-range forces are attempting to do with the structure. The diffuse scattering is explained in terms of short-range forces that favour modulations which preserve the c-glide symmetry, but for wave-vectors close to $\mathbf{q} = 0$, corresponding to large distances in real-space, such modulations are energetically unfavourable because over a long range the structure prefers to maintain the screw axis.

Figure 8.4(c) shows a diffraction pattern calculated from a synthesised real-space distribution corresponding to the $(h\,k\,\bar{h})$ section which again was obtained using only modulations of the Γ_4 symmetry. However, wave-vectors close to $\mathbf{q} = 0$ have been omitted to produce a 'hole' in the scattering. For comparison Fig. 8.4(d) shows a calculation from a similar synthesised distribution but, in this, modulations close to $\mathbf{q} = 0$ have been included.

8.4   Displacement modulations

The calculated diffraction patterns shown here were obtained from synthesised models using only compositional modulation waves. That these calculated patterns agree very well with the observed patterns is an indication that displacement modulations are unimportant at these relatively low diffraction angles. However it is usual for compositional modulations of a given symmetry to be accompanied by corresponding displacement modulations of the same symmetry (however small they may be). In Figs. 8.1(b) and (c) the displacement patterns are indicated for molecules undergoing rigid-body translational motion with the Γ_3 or Γ_4 symmetry. The presence of such accompanying displacement modulations may provide an explanation for the change of symmetry between long and short range, noted above. The short-range order may be explained in terms of the forces between the molecular dipoles, which provide a motive force tending to produce a structure with Pc symmetry. However, it is well established that for close-packing of non-centrosymmetrically shaped molecules like Bemb2 the space group $P2_1$ is more efficient than Pc (see Kitaigorodsky, 1973; Wilson, 1988, 1990). That is, there is a competition between the crystal packing forces and the local dipole–dipole interactions. When the molecules try to achieve the c-glide locally, there will be accompanying strains away from the preferred $P2_1$ packing, and these strains accumulate as the range of the Pc domain increases until beyond a certain range the c-glide type of packing cannot be sustained. This is discussed further in Section 17.5.

8.5   Comparison with correlation description

It is interesting to consider how the above modulation-wave description of Bemb2 corresponds to what would be learnt from an analysis in terms of correlations. The 'Y'-shape diffuse scattering distribution may be considered to reflect the correlations that exist between the molecules of each sublattice (e.g. the molecules at the corners of the unit cell). Given this correlation within a sublattice, then the two modes, Γ_3, Γ_4, correspond to whether there is positive or negative correlation between the two sublattices. Pure

Γ_4 modulation corresponds to when there is maximal positive correlation between the two sublattices and Γ_3 to when there is maximal negative correlation between them. While it is perfectly clear from the modulation-wave approach presented that for all but $\mathbf{q}$-values close to zero only Γ_4 are present, such a special result would not be so easily identifiable in the correlation method. This is because the maximum possible value for the correlation between the two sublattices depends on the correlations within each sublattice.

p-chloro-*N*-(*p*-methyl-benzylidene)aniline (MeCl)

9.1 Introduction

The compound *p*-chloro-*N*-(*p*-methyl-benzylidene)aniline (MeCl), $C_{14}H_{12}ClN$, is one of a number of para disubstituted benzylideneanilines (BAs) that have been of interest to theoretical chemists studying the way in which molecular shape and conformation influences the crystal structure of a compound. The average molecular site in this compound contains four basically different molecular orientations resulting from end-to-end disorder and side-to-side disorder (see Fig. 9.1). Which of these four possible orientations is contained in a given molecular site can be conveniently specified by using two sets of binary $(-1, +1)$ random variables ρ and σ defined in the figure; σ specifies the end-to-end disorder, while ρ specifies the side-to side disorder. In addition to the binary variables ρ and σ, additional (continuous) random variables U, V are used to represent local displacements of molecules away from their mean positions. These may in principle include pure rigid-body translations and librations but (a possibility for MeCl) also intramolecular conformational variations.

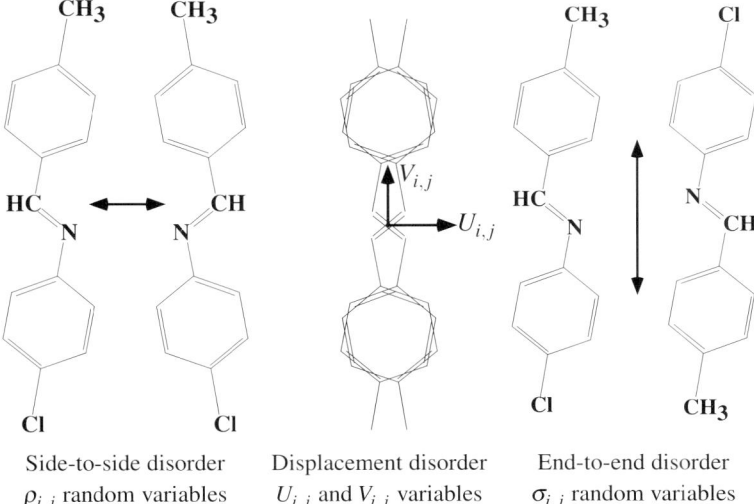

Side-to-side disorder	Displacement disorder	End-to-end disorder
$\rho_{i,j}$ random variables	$U_{i,j}$ and $V_{i,j}$ variables	$\sigma_{i,j}$ random variables

Fig. 9.1 The variables used to model the disorder in MeCl.

9.2   Cell data

Original crystal structural work for MeCl was published by Bar and Bernstein (1982, 1983). They found the space group to be monoclinic $P2_1/a$. Some weak additional peaks have since been seen (Welberry *et al.*, 1993a) indicating that the true unit cell has a doubled **c**-axis and reduced (triclinic) symmetry $P1$ or $P\bar{1}$ (see Table 9.1). However, for the purposes of investigation of the diffuse scattering, the smaller monoclinic cell given by Bar and Bernstein was used. It was considered that the small differences from this more symmetric cell were simply minor perturbations which could be ignored (see also Welberry and Butler, 1994).

Table 9.1   Cell data for MeCl.

Space Group	$P2_1/a$	$P1$ or $P\bar{1}$
a	5.960 Å	5.960 Å
b	7.410 Å	7.410 Å
c	13.696 Å	27.386 Å
β	99.2°	99.2°

9.3   X-ray diffuse scattering data

Diffuse scattering in MeCl was recorded using the position-sensitive detector (PSD) system described in Section 1.2.1. Figure 9.2 shows a plot of the $(h0l)$ section of data.

Fig. 9.2   The $(h0l)$ diffuse X-ray diffraction pattern of MeCl.

It is clear from this that the diffuse scattering is very rich in detail with a whole variety of different diffraction features to be seen. The aim in this section is to try to understand the origins of the various features that are visible. These are listed here for reference.

1. There are closely spaced diffuse fringes running across the Fig. 9.2. These fringes run approximately parallel to the $[2\,0\,1]^*$ reciprocal direction. It is clear that in some regions the fringes have good visibility (mark A) while in others (such as B) the visibility is poor.

2. There are strong diffuse peaks (C) at positions $h\ 0\ l+0.5$ where h = odd only. These are somewhat elliptical in shape with the long axis of the ellipse approximately parallel to the $[2\,0\,1]^*$ reciprocal direction.

3. There is diffuse scattering around Bragg peaks of the parent structure. These appear to be generally of an intensity proportional to the Bragg peak intensity so that a few appear very prominent. Some of these (D) have a fairly rounded shape, while others (E) are elongated, again in the $[2\,0\,1]^*$ direction. These peaks have all of the characteristics of thermal diffuse scattering but comparing data recorded at $120\,\mathrm{K}$ with corresponding room temperature data revealed very little change in the pattern, so that it may be assumed that they are largely not of thermal origin.

4. There is a diffuse streak (F) to be seen near the reciprocal point $\bar{1}\,0\,6$ which again is extended approximately in the $[2\,0\,1]^*$ reciprocal direction. In the range of scattering angle covered by this experiment this is the only such peak not associated with a strong Bragg peak.

5. There are a number of other fairly localised diffuse peaks (G) which do not seem to be associated with either the strong diffuse peaks (2 above) or the thermal like diffuse peaks (3 above), occurring as they do at quite general reciprocal positions.

9.4 The average structure and model for the disorder

The determination of the average structure of MeCl reported by Bar and Bernstein revealed each molecular site to contain molecules in one of four basically different orientations. The methyl and chlorine sites at the extremities of the molecule were each partially occupied by Cl and Me, implying an end-to-end type of disorder. Assumption of the space group $P2_1/a$ implies the average molecule has a centre of symmetry and hence each of these atomic sites has an occupancy, $m_A = 0.5$. If the variable σ is used to represent the end-to-end disorder of the molecules, then this requires $P(\sigma = +1) = 0.5$ and $P(\sigma = -1) = 0.5$, where $P(\sigma = +1)$ means the probability that $\sigma = +1$.

The second type of disorder was also detected in which the central –CH=N– bridge is flipped by rotating the molecule $180°$ about an axis approximately passing through the Cl and Me extremities of the molecule. For this disorder it is not necessary to have equal proportions of the two orientations. Using the variable ρ to represent this side-to-side type of disorder it was found that the proportions corresponded to the probability that $P(\rho = +1) = 0.7$ and $P(\rho = -1) = 0.3$ (see Fig. 9.1).

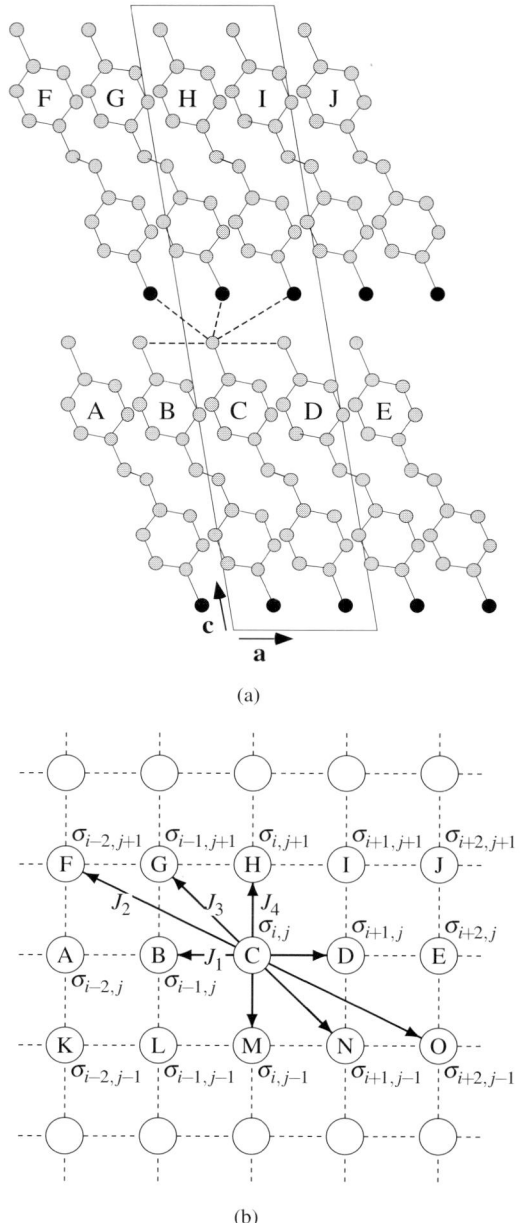

Fig. 9.3 Molecular interactions in MeCl.

9.5 Molecular interactions and MC simulation

To model the diffuse scattering to be found in the $(h\,0\,l)$ section the structure need be considered in projection only, and Fig. 9.3 shows a plot of this projection with the molecules labelled (A, B, ..., etc) for reference. It should be noted that the unit cell outlined in the figure corresponds to the doubled ($c = 27.386$ Å) cell, and all the molecules are plotted in the same orientation. The aim in performing Monte Carlo simulations is to arrive at distributions of random variables which correctly reproduce the various disordered features of the structure. First consider the random variables $\sigma_{i,j}$ that are used to define the end-to-end disorder in MeCl. The set of $\sigma_{i,j}$, may be considered to be defined on a simple 2D square lattice as shown in Fig. 9.3(b). Each point on this lattice is associated with a molecular site in the projection of the MeCl structure (Fig. 9.3(a)). That is, there is a one-to-one mapping of the $\sigma_{i,j}$ variables in Fig. 9.3(b) and the molecules in Fig. 9.3(a).

An interaction energy which depends on pair interactions only may then be specified. For example,

$$E = \sigma_{i,j}\Big(J_1(\sigma_{i-1,j} + \sigma_{i+1,j}) + J_2(\sigma_{i-2,j+1} + \sigma_{i+2,j-1})$$
$$+ J_3(\sigma_{i-1,j+1} + \sigma_{i+1,j-1}) + J_4(\sigma_{i,j+1} + \sigma_{i,j-1})\Big) \qquad (9.1)$$

In Fig. 9.3(b) the solid lines indicate with which neighbours the variable $\sigma_{i,j}$ interacts. The constants J_2, J_4 refer to interactions between molecules in the **ac** plane which have direct Cl/Me contacts. J_1 and J_3 refer to interactions between molecules in one plane with ones in the plane below.

Given the energy of interaction, E, the Monte Carlo procedure used to obtain lattice realisations is as follows. A site (i, j) on the lattice is chosen at random and the contribution to the total lattice energy of all interactions involving the variable $\sigma_{i,j}$ is computed. The value of $\sigma_{i,j}$ is then changed (e.g. 1 to -1 or vice versa) and the energy computed again. The new value is accepted if the difference in energy, $\Delta E = E_{\text{new}} - E_{\text{old}} < 0$. If $\Delta E \geq 0$ the new value is accepted with probability

$$P = \frac{\exp(-\Delta E/kT)}{1 + \exp(-\Delta E/kT)}. \qquad (9.2)$$

The process is repeated for many Monte Carlo cycles, where a cycle is defined as the number of steps required to visit all sites once on average. For the MeCl study a lattice of 256×256 sites was used and ~ 100 cycles was found sufficient to attain a reasonable steady-state.

The above procedure is appropriate for producing a lattice realisation corresponding to a given choice of J_1, J_2, J_3 and J_4. It is, however, often useful to produce realisations in which specific correlation values between neighbours in given directions are defined *a priori*. To do this an iterative procedure can be used in which the values of J_1, J_2, J_3 and J_4 are continually adjusted during the Monte Carlo process so that the desired correlation structure is achieved. After each cycle of iteration, the lattice averages $\langle \sigma_{i,j}\sigma_{i-1,j} \rangle$, $\langle \sigma_{i,j}\sigma_{i-2,j+1} \rangle$, $\langle \sigma_{i,j}\sigma_{i-1,j+1} \rangle$, and $\langle \sigma_{i,j}\sigma_{i,j+1} \rangle$ are computed and according to whether these are greater or less than the desired correlation values J_1, J_2, J_3 and J_4 can be adjusted to compensate for the discrepancies. Eventually, provided the system is not in a multi-phase region, the values of J_1, J_2, J_3 and J_4 settle down to constant values and the Monte Carlo process can proceed normally.

This same basic method has been used not only for the binary variables σ and ρ representing the molecular orientations of MeCl, but also for the continuous variables U, V etc. The only difference when dealing with continuous variables is that at each Monte Carlo step the $U_{i,j}$ variable is given a small random shift (+ve or −ve) instead of the $-1/+1$ flip. In addition, to maintain the mean and variance of the variables (it is convenient to use zero mean and unit variance variables) the $U_{i,j}$ are renormalised after each Monte Carlo cycle.

The particular choice of interactions depicted in the figure was used to generate suitable distributions for the σ variables. For other variables, different choices of neighbour may be important. For example, the interactions influencing molecular displacements may not be the same as those influencing the end-to-end disorder. Going further than this, it may also be appropriate to have a Monte Carlo energy in which there are cross-terms linking different kinds of variables, for example, $L_1 \sigma_{i,j} U_{i,j}$, $L_2 \sigma_{i,j} U_{i-1,j}$, etc., since displacements of a molecule may be strongly coupled to its orientation or that of its neighbours. To arrive at a suitable distribution in this case, simultaneous iteration on the different types of variable would be necessary.

In the present case, however, only four *independent* sets of random variables were used:

1. $\sigma_{i,j}$ binary variables used to define the end-to-end disorder. The interactions used in the Monte Carlo simulation were the four terms J_1, J_2, J_3 and J_4 shown in the figure and defined in the equation above.
2. $\rho_{i,j}$ binary variables used to define the side-to-side disorder. Initially interactions K_1 and K_3 along the same vectors as J_1, J_3 were used to generate realisations of these variables, but no evidence for the existence of short-range order amongst these variables could be detected in the X-ray pattern and it was subsequently assumed that the $\rho_{i,j}$ were completely random.
3. $U_{i,j}$ Gaussian variables were used to define rigid-body molecular displacements in a direction perpendicular to the length of the molecule.
4. $V_{i,j}$ Gaussian variables were used to define rigid-body molecular displacements in a direction parallel to the length of the molecule.

9.6 Results

Figure 9.4(a) shows the diffraction pattern calculated for the final model of MeCl which was considered to give a qualitatively good fit to the observed data. Comparing this to the observed data it is seen that most of the features of the experimental pattern have been reproduced reasonably well. It should be noted that this calculated pattern is of the *diffuse* part of the intensity only. Scattering due to the average unit cell (which gives rise to Bragg peaks) has been subtracted. The positions of the Bragg peaks have been marked artificially by inserting small white dots. This final model was arrived at after numerous

Table 9.2 Correlations between the molecule C in Fig. 9.3 and the neighbouring molecules A, B, F, G, H, I and J which are present in the example simulation corresponding to the diffraction pattern in Fig. 9.4(a)

	A	B	F	G	H	I	J
$\sigma_{i,j}$ [a]	0.39	−0.60	−0.29	0.31	−0.30	0.25	−0.20
$\rho_{i,j}$ [b]	0.0	0.0	0.0	0.0	0.0	0.0	0.0
$V_{i,j}$ [c]	0.47	0.58	0.20	0.20	0.20	0.19	0.17
$U_{i,j}$ [d]	0.28	0.20	0.40	0.20	0.40	0.16	0.19

[a] binary variables defining the end-to-end disorder.
[b] binary variables defining the side-to-side disorder.
[c] variables defining displacements *parallel* to the length of the molecule.
[d] variables defining displacements *normal* to the length of the molecule.

trial runs in which different choices were made both of the particular interactions that were to be employed in generating the distributions of the random variables $\sigma_{i,j}$, $\rho_{i,j}$, $U_{i,j}$ and $V_{i,j}$, and of the correlation values themselves. Table 9.2 lists the correlation values that have been introduced along each of the principal intermolecular vectors for this model. It is not claimed that these values provide the 'best-possible' fit to the experiment but rather that they represent a set of parameter values defining a simple model that reproduces moderately well the main features of the observed pattern.

A description is given below of how each of the observed features arises and how it is related to the correlation values given in Table 9.2. The effect of changing the various parameter values is discussed and the other calculated diffraction patterns shown in Fig. 9.4 are used to illustrate the effects.

1. For the realisation corresponding to Fig. 9.4(a) the $\sigma_{i,j}$ variables have a strong negative correlation (−0.6) corresponding to molecular pairs such as CB. In addition there is a (weaker) negative correlation (−0.3) between both types of pair such as CF and CH, with a corresponding positive correlation (+0.3) between pairs such as CG. These together give the strong diffuse spots that occur at the points $h\ 0\ l+0.5$ where $h =$ odd only and which correspond to the peaks labelled C in the observed data. These correlations reflect the tendency of the structure to form a superlattice in which molecules pack alternately head-to-tail along the a-direction.

2. Figures 9.4(c) and (e) show the effect of changing the values for these correlations. In Fig. 9.4(c) the negative values for the CF and CH correlations are replaced by corresponding positive ones (i.e. +0.3). Comparison with Fig. 9.4(a) shows that this has a twofold effect: while the diffuse peaks are similar in shape they now occur at $h\ 0\ l+0.0$, $h =$ odd only, but in addition the distribution of strong and weak peaks has also changed. Figure 9.4(e) shows the effect of reducing the value of the CH correlation (now −0.1) relative to the CF correlation (still

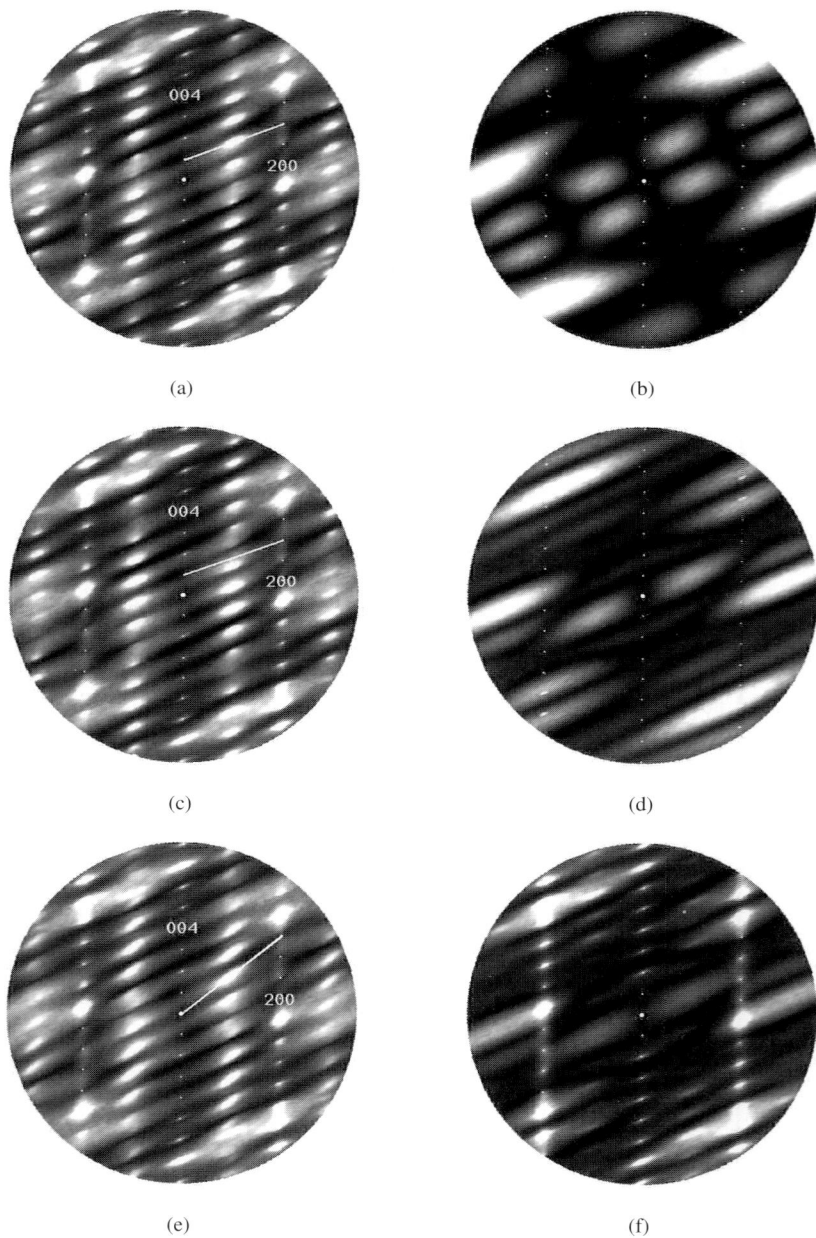

Fig. 9.4 Calculated diffraction patterns.

−0.3). The diffuse peak is now seen to have a different shape, with its longer axis oriented more towards the normal of the CF vector.

3. The fact that the $\rho_{i,j}$ correlations are all zero in the table implies that the side-to-side disorder is random. In order to see the effect of this disorder in the total pattern Fig. 9.4(b) shows a pattern calculated from a simulation in which all the $\sigma_{i,j}$ were set to $+1$ and $U_{i,j}$ and $V_{i,j}$ set to zero, so that the end-to-end disorder scattering and that due to molecular displacements has been removed. It is clear that this scattering contributes to the loss of visibility in the basic fringing pattern in Fig. 9.4(a), (c) and (e) and the similar loss of visibility in the experimental pattern can be attributed to this effect.

4. The effect on the diffraction pattern of the molecular displacements described by the $U_{i,j}$ and $V_{i,j}$ variables is illustrated in Fig. 9.4(d) and (f). For both of these figures the effects of the end-to-end and side-to-side disorder have been removed by setting all $\sigma_{i,j}$ and $\rho_{i,j}$ to $+1$ so that the scattering seen is purely displacive in origin. In Fig. 9.4(d) both the U and V displacements are completely uncorrelated and this figure corresponds to the well known difference Fourier transform (DFT) (see Amorós and Amorós (1968)). Introduction of correlations between the displacements results in a bunching-up of the intensity around Bragg-peak positions as shown in Fig. 9.4(f). Note how the strong peaks arise in the strongest regions of the DFT.

5. The $V_{i,j}$ variables which describe rigid-body displacements parallel to the long axis of the molecules are seen to be strongly correlated from molecule to molecule within an **ab** layer (CA and CB pairs; 0.47 and 0.58 respectively) but less correlated between layers (CF, CG, CH pairs; all 0.2). However, since the intermolecular distances involved are much greater between layers than within layers such correlation values correspond to a fairly isotropic falling-off of correlation with distance. The $V_{i,j}$ displacements give rise to the thermal diffuse scattering (TDS)-like peaks around the strong Bragg peaks labelled D in the observed data. The scattering due to this type of displacement is strongest in directions normal to the displacements, that is, normal to the long axis of the molecule.

6. The $U_{i,j}$ variables which describe rigid-body displacements perpendicular to the long axis of the molecules are seen to be correlated quite differently. The displacements within each layer are much more weakly correlated than for the $V_{i,j}$ displacements and this results in the much more elongated diffuse spots such as E in the observed data. The scattering due to this type of displacement is again strongest in directions normal to the displacements, that is, parallel to the long axis of the molecule. The correlation of the $U_{i,j}$ displacements between **ab** layers (i.e. CF and CH; 0.4) is stronger than for the $V_{i,j}$ displacements. In particular it is necessary that CF and CH are stronger than CG in order to produce the elongated diffuse peak F visible in the observed data.

7. It is the displacement correlations $U_{i,j}$ and $V_{i,j}$ that give rise to the shape of the diffuse TDS like peaks described above. To a first approximation the magnitude of these displacements does not affect the shape but only the intensity of these peaks. In the simulation corresponding to Fig. 9.4(a) the magnitude of the displacements

was chosen to have an r.m.s value of ~ 0.18 Å. The intensity of the peaks D and E (in the observed data) relative to the peaks C can be increased by increasing this value.

10

UREA INCLUSION COMPOUNDS

10.1   Introduction

In these compounds the 'host' substructure consists of an open framework of hydrogen-bonded urea molecules, in which exist parallel one-dimensional tunnels of diameter about 5.5–6 Å, see Fig. 10.1(a). These tunnels can accommodate 'guest' molecules such as the n-alkanes or similar long-chain molecules with limited side group substitution. Here a description is given of how diffuse X-ray scattering in conjunction with Monte Carlo (MC) computer simulation can be used to investigate guest-host interactions, with particular reference to n-hexadecane-urea (HD-urea) and 1,10-dibromo-n-decane-urea (DBD-urea), see Fig. 10.1(b) (see Welberry and Mayo, 1996).

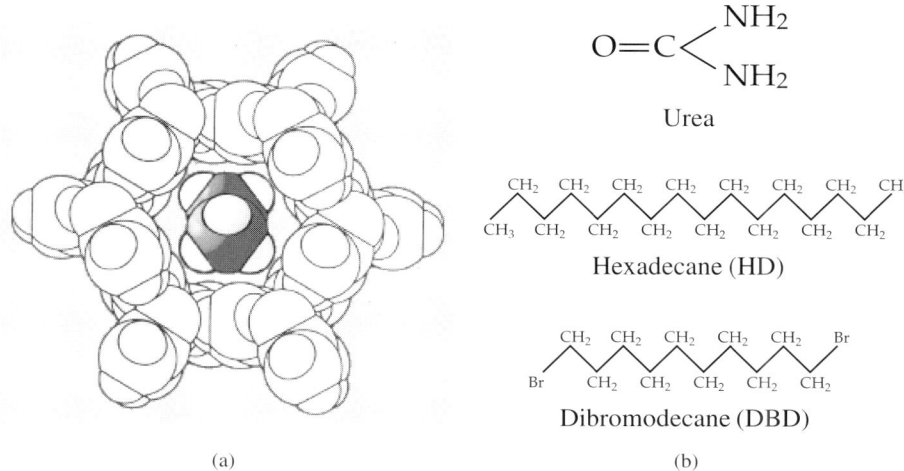

(a)   (b)

Fig. 10.1   Building blocks of urea inclusions compounds.

X-ray diffraction patterns of urea inclusion compounds contain scattering of two types—sharp Bragg reflections and various diffuse diffraction features which are observed at general positions in reciprocal-space. Studies using Bragg reflection data reveal only information of the average structure. In most cases such studies have revealed details of the urea framework, but have been able to model the disordered 'guest' only relatively poorly. Diffuse X-ray scattering gives information concerning departures from the average and is thus potentially a powerful probe of such local structural details. A number of studies have described the different diffuse features that typically occur in diffraction patterns of urea inclusion compounds (e.g. see Forst *et al.*, 1987).

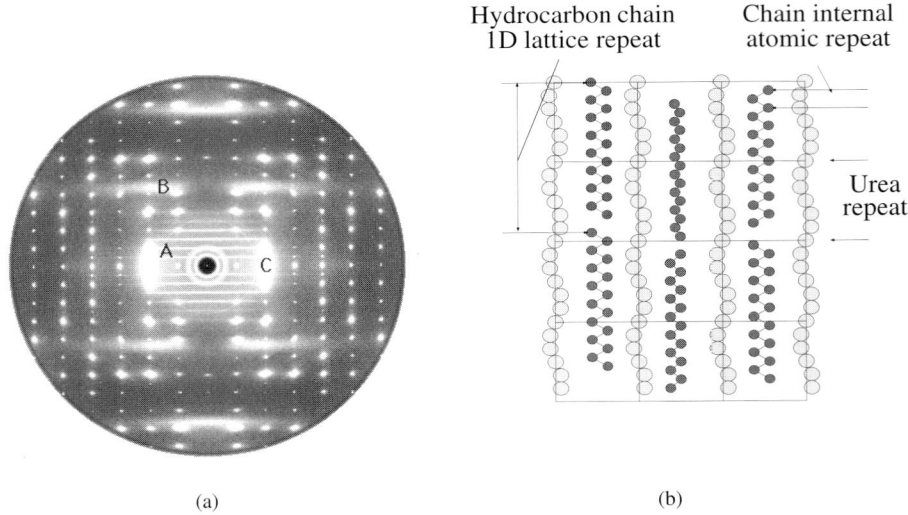

(a)   (b)

Fig. 10.2   HD-urea. (a) Diffuse scattering of the $(0\,k\,l)$ section. (b) Diagram illustrating the important lattice repeat distances.

These features, examples of which can be seen in the observed $(0\,k\,l)$ diffraction pattern of HD-urea, (Fig. 10.2(a)), can be briefly summarised as follows. Refer also to Fig. 10.2(b).

1. There are thin diffuse layers normal to **c**, labelled (A), occurring between the rows of Bragg peaks. These have a spacing which is incommensurate with the spacing of the Bragg peaks. For HD-urea the first of these layers occurs at a position only slightly less than $0.5\mathbf{c}^*$, so that the incommensuration is only apparent at higher-order levels. For DBD-urea the first layer occurs at about $0.6\mathbf{c}^*$. These incommensurate layers can be attributed to the guest molecules which form a 1D crystal within the urea channel with a repeat distance different from the urea repeat (see also Section 4.4).

2. A second type of diffuse layer of scattering (very broad), the first order of which is labelled (B) in Fig. 10.2(a), corresponds to a real-space distance of 2.56 Å and can be attributed to the repeat period of the nearly periodic alkane molecule. This scattering occurs as a result of random displacements of the molecules within the channels.

3. There is also strong diffuse scattering (C) within each Bragg layer. In addition to

the diffuse peaks around the Bragg positions visible in Fig. 10.2(a) the scattering is found, in each reciprocal layer normal to **c**, to extend as a series of streaks linking the Bragg positions (see later).

4. There is an overall continuous background scattering distributed nearly continuously in reciprocal-space (monotonic Laue scattering). This scattering has been ascribed to uncorrelated fluctuations in atomic positions.

Here the aim is to try to understand the details of the type of scattering described in 3 above.

10.2   X-ray diffuse scattering data

Diffuse X-ray scattering patterns were recorded for HD-urea and DBD-urea using the PSD system described in Section 1.2.1. The data for Fig. 10.2(a) were obtained using two settings of the PSD, but for all other data presented here only a single low-angle setting was used. Cu$K\alpha$ radiation was used throughout. The needle-shaped morphology of the crystals of both compounds is ideally suited to the Weissenberg geometry when sections normal to **c**, the needle axis, are required, and such crystals were used without any reshaping. For recording the section normal to **a**, shown in Fig. 10.2(a), a small length cut from a needle was reshaped into a crystal of cylindrical cross-section. Low-temperature data were collected using an Oxford Cryosystems Cryostream cooling device.

Figure 10.3 shows for comparison the $(h\,k\,0)$ section of data for both HD-urea and DBD-urea at three different temperatures. The marked similarity of the two systems is evident from this comparison. One major difference between the two systems was observed. The phase transition in HD-urea occurs abruptly at ~ 147K and on close inspection the superlattice peaks are found to be split, indicating a slightly different geometry for the different orthorhombic domains. In DBD-urea the pattern at 150 K appears very similar to that of HD-urea but the change to the low temperature structure appears to be rather gradual with changes apparent over a 20 K, or more, range. The superlattice peaks in the low-temperature phase of DBD-urea do not appear to be split. At temperatures just above the phase transition the diffuse streaking, which extends away from the strong diffuse peaks surrounding the Bragg positions, can be seen to contain peaks which are evidently precursors of the eventual superlattice peaks. These gradually disappear as the temperature is raised towards ambient.

10.3   Monte Carlo simulation

Evidence from previous work seemed to suggest that the phase transition which occurs at about 147 K in HD-urea involves the ordering of the alkane orientations within neighbouring tunnels to form orthorhombic domains of a herring-bone type of pattern (see Fig. 10.4(a)). Above the transition it was thought that the alkanes are free to rotate between the three symmetry-equivalent orientations within the urea cavity. To take this possibility into account, and also to allow for the interaction between the alkane and the host lattice, the following 2D model was set up to represent the system.

HD-urea   DBD-urea

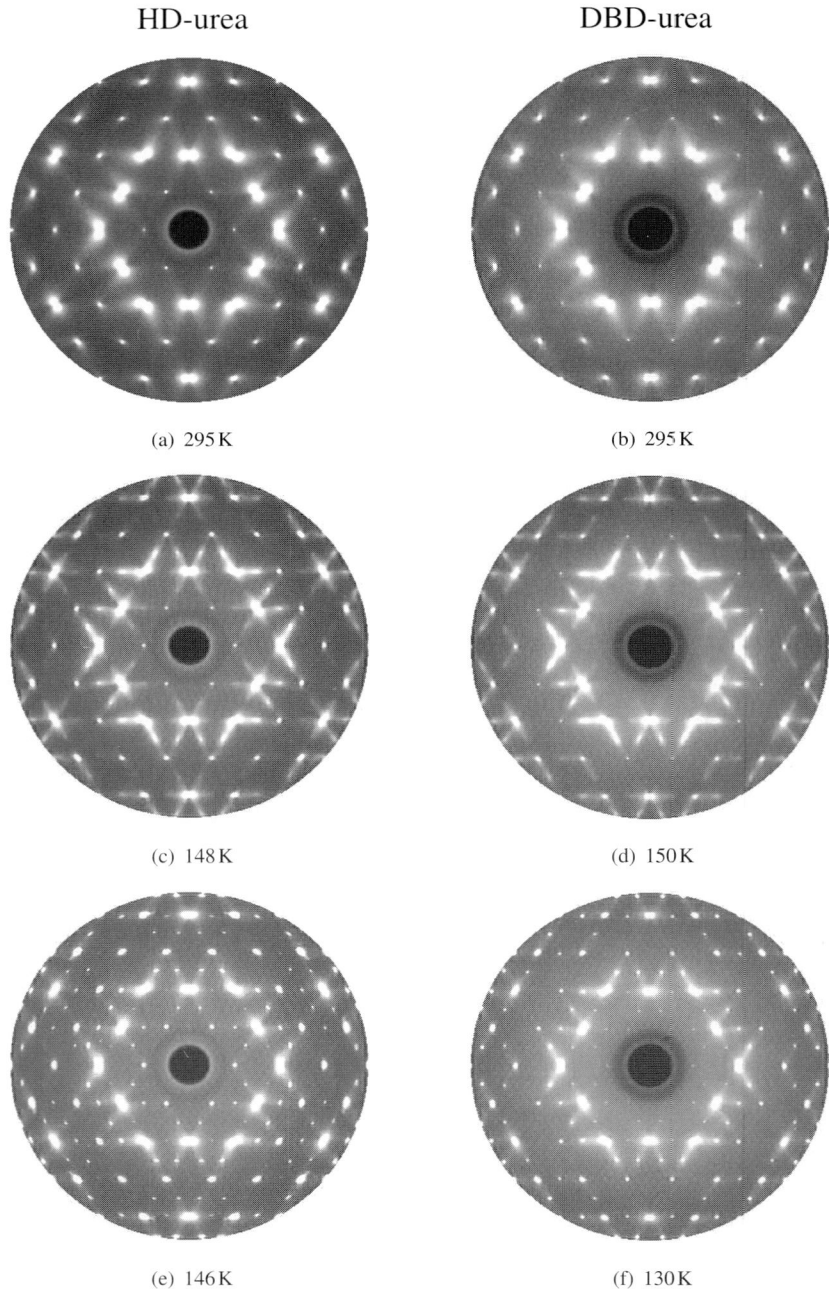

(a) 295 K   (b) 295 K

(c) 148 K   (d) 150 K

(e) 146 K   (f) 130 K

Fig. 10.3   Observed $(hk0)$ X-ray diffraction patterns.

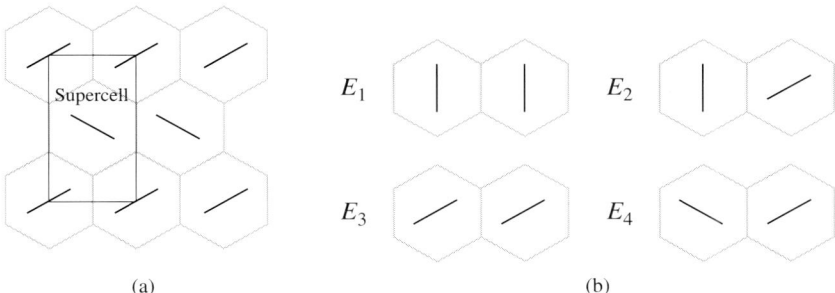

Fig. 10.4   Herring-bone arrangement of urea compounds. (a) Illustrating the supercell arrangement. (b) Definition of the four different possible nearest-neighbour interactions energies.

The model consisted of two stages; first a scheme for ordering the alkane orientations and second a scheme for allowing the surrounding urea framework to relax around the disordered alkane molecules.

10.3.1   *Ordering of alkanes*

In the 2D model described here the alkane molecules were represented in projection by a pair of atomic sites symmetrically disposed about the centre of each hexagonal cavity of the urea framework. It is convenient in drawing the structures to represent the orientation of these dimers by a line through their centres which represents the plane of the alkane backbone (see Fig. 10.4). It is assumed that the lines are directed towards the corners of the hexagon. Then interactions between neighbouring alkanes can be one of four types (as shown in Fig. 10.4(b)). The quantities E_1 to E_4 are used both to identify the different types of nearest-neighbour alkane–alkane configuration (and their symmetry equivalents) and to represent the energy associated with each of these configurations. Monte Carlo simulation was carried out on an array of 128×128 unit cells of the model lattice. Initially the lattice was set-up with the orientation of each alkane arbitrarily assigned to one of the three different possible orientations. In order to keep the numbers of alkanes in each of the three orientations constant, a form of Monte Carlo iteration was employed in which each step involved interchanging the alkanes between two sites chosen at random. After a given number of Monte Carlo steps intermediate states of order are achieved. In the present case configurations obtained after only 1 cycle of iteration (a cycle is defined as that number of individual Monte Carlo steps required to visit each site once on average) and after 100 cycles of iteration were used.

The envisaged low temperature herring-bone pattern superstructure (Fig. 10.4(a)) consists of nearest-neighbour pairs of types E_2 and E_3 in the ratio 2:1, while the pairs of types E_1 and E_4 are absent. In order to generate realisations containing ordered domains of this structure, simulations were carried out with the energies E_2 and E_3 set to zero and E_1 and E_4 set to positive values. When domains of the superstructure exist, the configurations E_1 and E_4 would be expected to occur on the domain boundaries. The values of the energies E_1 and E_4 were chosen after some initial experimentation to

Table 10.1   Energy values and resulting frequencies

Example	Cycles	E_1	E_2	E_3	E_4	E_1	E_2	E_3	E_4
			Energies				Resulting frequencies		
1	0	1.0	0.0	0.0	3.0	0.11	0.44	0.22	0.22
1	1	1.0	0.0	0.0	3.0	0.08	0.50	0.28	0.14
1	100	1.0	0.0	0.0	3.0	0.04	0.59	0.32	0.05
2	100	1.0	0.0	0.0	3.0	0.00	0.67	0.19	0.14
3	100	0.0	0.0	0.0	3.0	0.13	0.40	0.44	0.03

have values of 1.0 and 3.0 on an arbitrary scale. This combination was chosen as it reduced the two unwanted configurations E_1 and E_4 to a more or less equal extent. The relative frequencies with which the different pairs occurred after different numbers of iterations are given in Table 10.1, as example 1. This particular set of realisations for the alkane distribution proved to be very effective in reproducing features of the observed pattern and so was adopted in most subsequent work. To demonstrate the effects of changing the energy parameters other examples were investigated, and data for two of these (examples 2 and 3) are also given in Table 10.1.

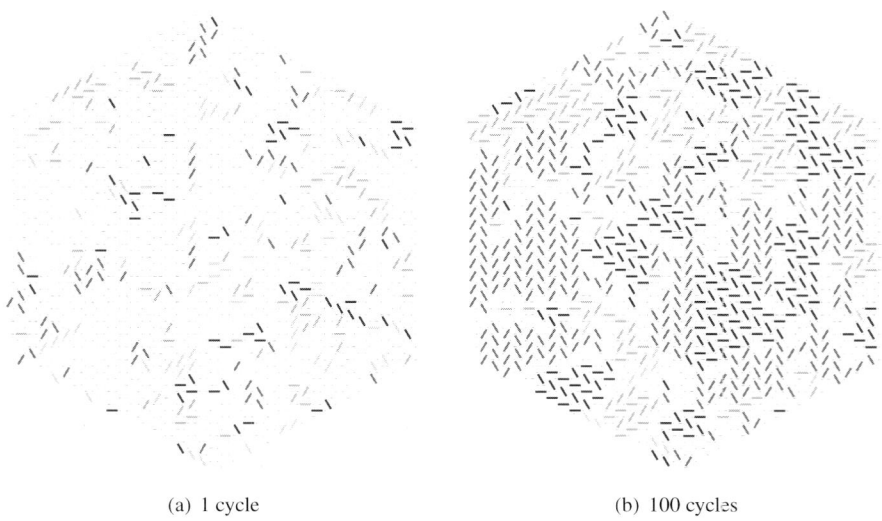

(a) 1 cycle   (b) 100 cycles

Fig. 10.5   Plots of part of a 2D realisation of the structure of example 1.

In Fig. 10.5(a) a plot is shown of part of a 2D realisation of the structure of example

1 after 1 cycle and after 100 cycles of iteration. The line segments drawn within each hexagon formed by the urea framework represent schematically the orientation of the alkanes. In Fig. 10.5(b) three differently oriented types of domain of the herring-bone pattern can be clearly seen, though in Fig. 10.5(a) only very small such regions can be identified. The domains have been printed in different shades of grey for emphasis. In drawing this figure a given alkane is considered to be part of a particular domain if together with at least four of its six neighbours it conforms to one orientation of the superlattice structure (Fig. 10.4(a)). Alkanes which could not be so assigned are indicative of more disordered boundary regions between domains and these are printed in a lighter pen.

10.3.2   *Modelling interactions with the urea framework*

In the second part of the model consideration is given to how the distribution of alkane orientations affects the regularity of the urea framework. Account needs to be taken of the fact that the anisotropic shape of the inclusion will tend to locally expand the framework in directions parallel to the plane of the alkane backbone and similarly locally contract the framework in the perpendicular direction.

The urea framework is held together by a network of hydrogen bonds. In three dimensions each hexagonal tunnel is surrounded by a spiral of linked urea molecules. In the projection down **c**, corresponding to the 2D model, each edge of the hexagonal tunnel consists of two urea molecules viewed almost edge-on so that the two N's are almost superposed. The hydrogen bond linkages between the molecules allows that the hexagon is able to flex both at the corners, where the O's from three molecules almost overlap, and in the middle of the sides, where the N's from two molecules face each other. In order to keep the computation as simple as possible but to allow the structure to undergo reasonably realistic motions the system was approximated in 2D as follows.

1. Each Urea 'molecule' consisted of a rigid group of three atomic sites $(O-C-N')$ where N' indicates an atomic site containing two N atoms. The centre of mass $(x, y$ coordinates) and the orientation (ϕ angle) of each group were variables in the simulation.

2. The flexible interactions mimicking the hydrogen bonding network of the ureas were provided by Hooke's law springs with force constants k_{OO} and k_{NN} between the three O's at the corners of the hexagons and between the N's in the middle of the sides, respectively (see Fig. 10.6). For example, if d_0 is the average (projected) interatomic distance between the O sites obtained from the average structure analysis, the contribution to the energy when the interaction distance at any instant is d is $\Delta E = k_{OO}(d - d_0)^2$.

3. The projected alkane 'molecule' consisted of a pair of sites each representing (in the case of HD-urea) 4 C atoms (the 16 C alkane extends over two unit cells of the urea framework structure). The orientations of these dimers were determined as described above and were treated as fixed throughout the relaxation, but their centres of mass were allowed to vary.

4. Interaction of the alkane with the surrounding ureas was via Hooke's law springs of force constants k_{AO} and k_{AN} from the alkane centre of mass to the O– and

N′–end of the urea 'molecule'. That is, $3 \times k_{AO}$ springs to the three O's near the corners of the hexagon and $2 \times k_{AN}$ springs to the two N's near the centres of the sides, making 30 interactions in all.

5. The anisotropy of the alkane/urea interaction was introduced by allowing the equilibrium length of the Hooke's law springs to vary with the angle, ϕ (see Fig. 10.6), between the vector to the particular O or N′ site on the ureas and the alkane orientation vector. If d_{AN}, d_{AO} are the (projected) interatomic distances observed from the average crystal structure, then the equilibrium distances assumed for the Hooke's law springs were $d_0 = d_{AN}(1 + \varepsilon)$, $d_0 = d_{AO}(1 + \varepsilon)$ respectively. Here ε is obtained from the dot product of a unit vector, $\mathbf{v}$, in the direction of the interaction and a vector, $\mathbf{u}$, along the line of the projected alkane backbone. That is, $\varepsilon = (\cos^2 \phi - 0.5) \times 0.04p$, where p is a parameter defining the percentage variation of the spring length. The Hooke's law contribution to the energy is then, as before, $\Delta E = k_{AN}(d - d_0)^2$ and $\Delta E = k_{AO}(d - d_0)^2$.

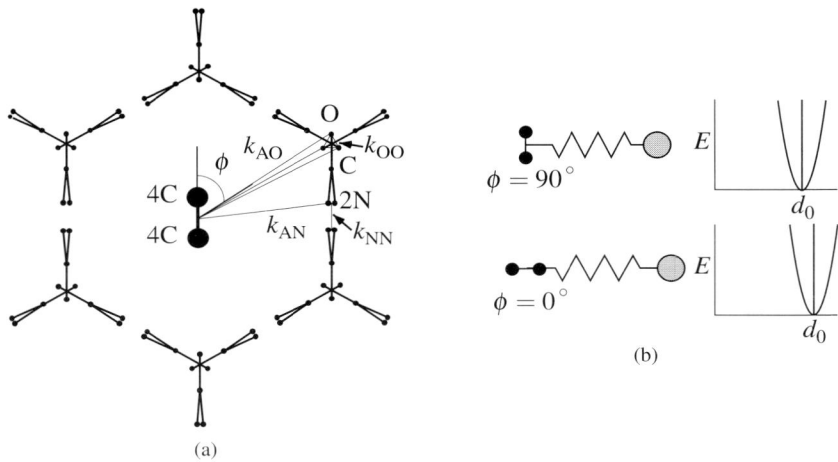

Fig. 10.6 Spring model description.

10.4 Relaxation simulation

With the relaxation model specified as above, simulations were carried out to explore the effects of varying the strengths of different force constants and the anisotropy parameter, ε. All simulations were carried out with a Boltzmann temperature $kT = 1.0$ so that the values adopted for the force constants are measured relative to this. Each of the 128×128 unit cells contains six ureas and one alkane. For each Monte Carlo run 800 cycles of iteration were carried out. As usual a cycle is defined as that number of individual Monte Carlo steps required for each urea and each alkane to be visited once on average. The magnitude of the random shifts in centre of mass and angular variables used, were chosen to allow acceptance/rejection ratios in the region of 25–50% to be achieved. The normal Metropolis algorithm involving the Boltzman probability $P = \exp(-\Delta E/kT)$ was used. The choice of 800 cycles was dictated by the total computation time available. Some initial tests showed that only small changes resulted from substantially longer runs. After considerable experimentation with different parameters, the set given in Table 10.2 was chosen as it produced a good qualitative fit to the observed patterns. Diffraction patterns obtained from simulations in which these parameters were used are shown in Fig. 10.7.

Table 10.2 Values of parameters used in the Monte Carlo relaxation simulation.

Force constants	alkane-urea	k_{AN}	25
		k_{AO}	25
	urea-urea	k_{NN}	200
		k_{OO}	200
Anisotropy parameter, ε			2%
Temperature, kT			1.0

10.5 Results

The diffraction patterns shown in Figs. 10.7(a) and (b) were obtained from the alkane ordering model, example 1, and used the same relaxation parameters given in Table 10.2. Figure 10.7(a) corresponds to 100 cycles of alkane ordering while Fig. 10.7(b) corresponds to only one cycle. These two patterns show good qualitative agreement with the observed patterns shown in Fig. 10.3. The pattern obtained for the purely random distribution of alkane orientations, Fig. 10.7(c), does not show the characteristic 'butterfly' shape seen in the observed pattern, indicating that at room temperature the alkane orientations must still be far from random.

Of particular significance is the asymmetry of the scattering across the line joining $0\,2\,0$ and $2\,0\,0$ and the line joining $0\,4\,0$ and $4\,0\,0$, indicated by the direction of the arrows in Fig. 10.7(a). This intensity transfer is characteristic of the 'size-effect' (see Section 7.4). It may be seen from Fig. 10.3 that the effect is more pronounced in the DBD-urea example. The sign of the intensity transfer depends on the difference in scattering factor of the scatterers. Here the relevant scatterers are the differently oriented alkanes which have highly anisotropic molecular scattering factors: large in directions parallel to the plane of the backbone and small in directions normal to this. As expected, reversing the sign of ε results in this transfer of intensity being in the opposite direction. Figure 10.7(d) shows the diffraction pattern of the same simulation used for Fig. 10.7(a) but instead of having two atomic sites to represent the alkanes, these have now been coalesced so that all of the alkane scattering originates from a single site near the centre of

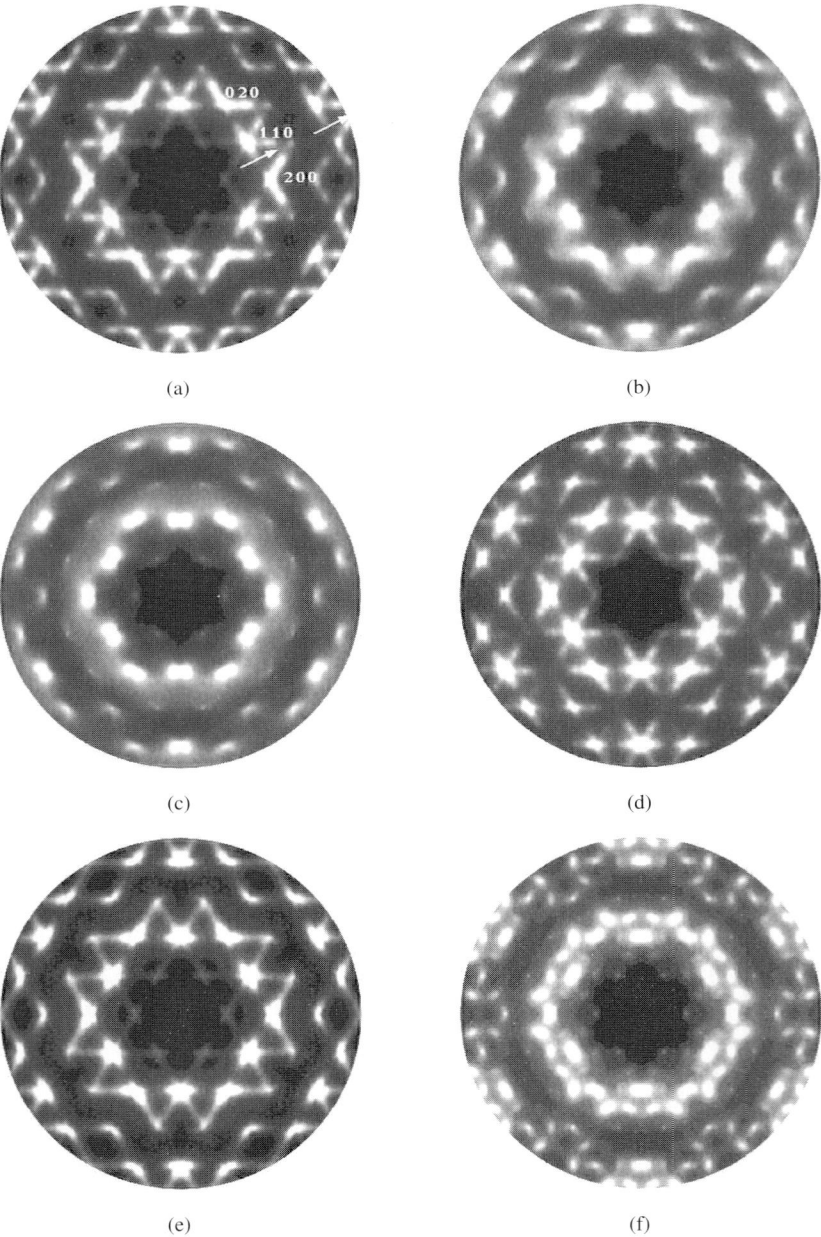

(a)   (b)

(c)   (d)

(e)   (f)

Fig. 10.7   Modelled diffuse scattering patterns for urea inclusion compounds.

the hexagon tunnel. Now the size-effect transfer has disappeared, but so also have the embryonic superlattice peaks, leaving only the linear streaking.

That the general form of the calculated scattering in Figs. 10.7(a) and (b) is in broad agreement with observation provides strong evidence that the model for the way in which the alkanes distort the surrounding urea framework is reasonable.

Figures 10.7(e) and (f) show two examples, example 2 and example 3, which illustrate the effect of changing the alkane ordering energies. In both cases 100 cycles of iteration were used. The frequencies with which the different combinations of nearest-neighbour orientations occur are given in Table 10.1. It is seen that for Fig. 10.7(e) the configuration of type E_1 has been completely eliminated, but for Fig. 10.7(f) the number of E_4 type neighbours has only been reduced to $\sim 3\%$. Both of these examples reflect a considerable departure from the $E_2{:}E_3$ ratio of 2:1 characteristic of the herring-bone superlattice, but in opposite directions. These two patterns show substantial differences from that of Fig. 10.7(a), although the same relaxation of the structures was carried out with the parameters given in Table 10.1. In Fig. 10.7(f) diffuse superlattice peaks can be seen, indicating that there is a tendency to form the herring-bone domains, but these have a different shape from those of Fig. 10.7(a) and are not connected by diffuse streaking.

11

MULLITE

11.1 Introduction

The mineral mullite, $Al_2(Al_{2+2x}Si_{2-2x})O_{10-x}\otimes_x$, where $\otimes$ represents an oxygen vacancy and x ranges from 0.17 to 0.59, has a stability field intermediate in composition between sillimanite Al_2SiO_5 and alumina Al_2O_3. Mullite is an important material since it is a major component of all alumino-silicate ceramics. In addition it has attracted much interest because the structure develops an incommensurate modulation in which the occupancies and positions of the tetrahedral cation sites and some oxygen sites vary through the structure in a periodic way (see McConnell and Heine, 1985). In diffraction space this modulation produces incommensurate diffraction peaks at wave vectors $G \pm q$, where G is a reciprocal lattice vector and $q = 0.5c^* + \gamma a^*$ and γ varies with composition, but in addition to this, specimens of mullite also exhibit strong diffuse scattering. Details of the average structure of mullite of composition $Al_2(Al_{2+2x}Si_{2-2x})O_{10-x}$ where $x = 0.4$, are shown in Fig. 11.1 (see Angel and Prewitt, 1987).

Chains of edge-sharing AlO_6 octahedra, running parallel to the crystallographic **c**-axis, are cross-linked by $(Al, Si)O_4$ tetrahedra. In sillimanite the tetrahedrally coordinated cations are bridged by a single oxygen atom and are ordered in such a way that the bridging anion always connects a Si containing tetrahedron with an Al containing tetrahedron. In mullite some fraction, x, of these bridging oxygen sites are vacant. To preserve charge balance each of the O^{2-} vacancies are accommodated by the exchange of $2Al^{3+}$ for $2Si^{4+}$. Because the tetrahedrally coordinated Al and Si are no longer present in the ratio of 1:1 these sites necessarily become chemically disordered. Furthermore, crystal structure analysis has identified the presence of a second tetrahedral site—displaced from the original by approximately 1.3 Å—that is presumably occupied by cations which have lost bridging oxygen atoms. The bridging oxygen atom was also found to occasionally occupy a site displaced 0.5 Å from its normal position and appears to be associated with the displaced tetrahedral site. In Fig. 11.1 the (Al, Si) tetrahedral sites are labelled T and T^* and the bridging anion sites are labelled O_c and O_c^* where * indicates the displaced positions mentioned above. It should be stressed that it is only *presumed* that the T^* sites are occupied by cations which have lost bridging oxygen atoms and that the displaced O_c^* sites are directly associated with them. The Bragg experiment only reveals the overall average occupancies.

11.2 X-ray diffuse scattering data

Sections of diffuse X-ray scattering data were recorded using the dedicated PSD diffractometer system described in Section 1.2.1. Fig. 11.2 shows plots of the data from six

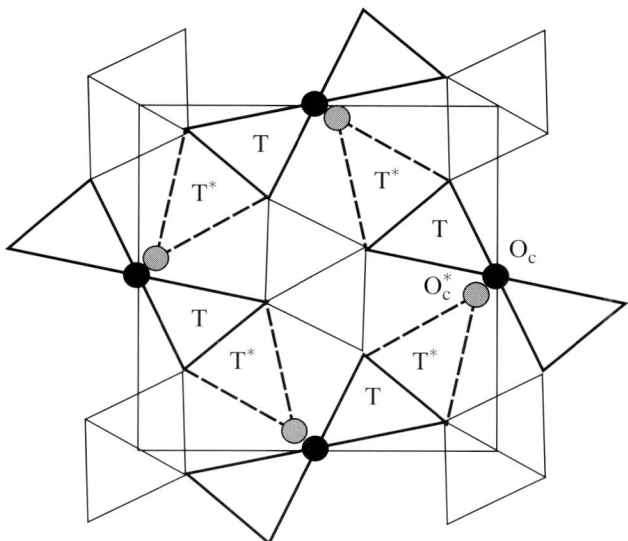

Fig. 11.1 The average structure of mullite seen in projection down the crystallographic **c**-axis. The sites labelled O_c and O_c^* contain oxygen atoms which bridge two or more tetrahedra. In an $x = 0.4$ mullite the T, T^* sites are disordered such that 80% of the T sites and 20% of the T^* sites are occupied by Al or Si. 40% of the O_c sites and 20% of the O_c^* sites are occupied.

different reciprocal sections. Each pattern consists of $\sim 180 \times 180$ pixels. Each pixel corresponds to an increment in the Miller indices $\Delta h_1, \Delta h_2 = 0.05$. This resolution is rather poorer than the resolution that the instrument can provide, but the data were binned at this resolution to provide a manageable number of independent measurements. Eleven sections in $\Delta h_3 = 0.05$ increments from $h_3 = 0.5 \ldots 1.0$ were measured, yielding in excess of $130,000$ independent measurements.

The plots of data in Fig. 11.2, which are all made on the same scale, demonstrate the rich detail that is present in the diffuse data. Many different features may be seen, varying from the relatively sharp incommensurate peaks visible in the $0.5\mathbf{c}^*$ section to the more diffuse circular features in the $0.7\mathbf{c}^*$ section and the narrow lines of the $0.9\mathbf{c}^*$ section. Although in the past much interest has centred on the incommensurate peaks, these patterns reveal that these are just an integral part of the continuous diffuse distribution. The total integrated intensity in each of the different recorded sections is fairly constant.

11.3 A simple model

In order to carry out an analysis two simplifying assumptions were made. First it was assumed that only the distribution of oxygen vacancies is important, and that the shift of cations from the T to the T^* sites and the shift of the oxygens from O_c to O_c^* follow from these as a direct consequence (see Fig. 11.3(a)). Second, it was assumed that certain local configurations, which are chemically implausible, do not occur at all in the

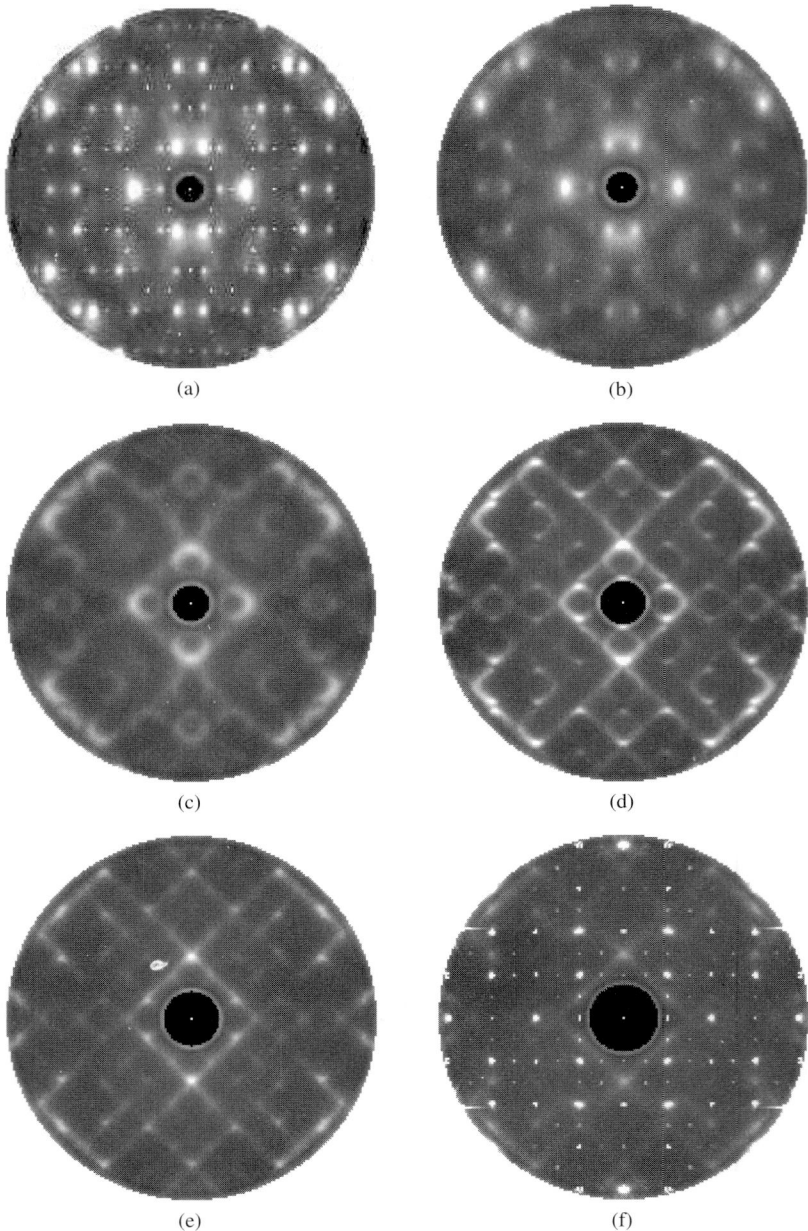

Fig. 11.2 Observed X-ray diffuse scattering data for mullite in six sections normal to **c**. (a) $0.5\mathbf{c}^*$; (b) $0.6\mathbf{c}^*$; (c) $0.7\mathbf{c}^*$; (d) $0.8\mathbf{c}^*$; (e) $0.9\mathbf{c}^*$; (f) $1.0\mathbf{c}^*$.

structure. These are shown in Figs. 11.3(b) and (c). In Fig. 11.3(b) the presence of two vacancies separated by $\frac{1}{2}\langle 1\,1\,0\rangle$ would cause the cation 'M' to have no bridging oxygen atom to bond with. In Fig. 11.3(c) the presence of two vacancies separated by $[1\,1\,0]$ would cause four tetrahedral cations to share the same bridging oxygen.

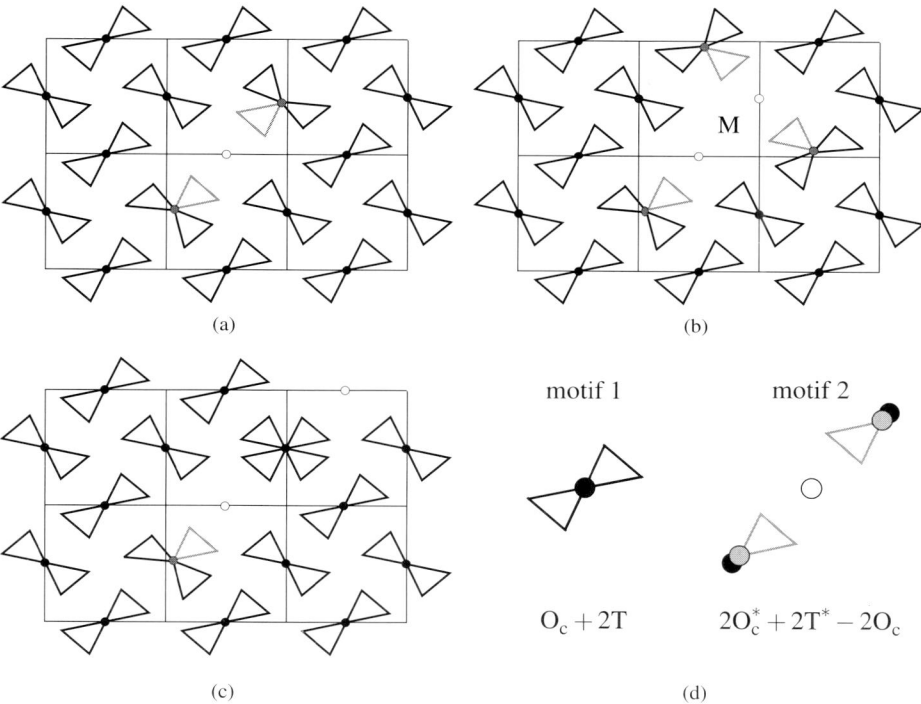

Fig. 11.3   Description of the simple model used for Mullite.

A description of the mullite structure based on the occupancy of the bridging oxygen sites alone can then be formulated as follows. When a bridging site is occupied by an oxygen atom that site is 'decorated' with one O_c oxygen atom and two T cations (motif 1; see Fig. 11.3(d)). When the site is vacant it is 'decorated' instead with two T^* cations, two neighbouring O_c^* oxygen atoms and two neighbouring O_c oxygen atoms are subtracted (motif 2). Note that as long as the ordering rules stated above are obeyed and all the bridging sites have been 'filled' with either an oxygen atom or a vacancy the positions of all tetrahedrally coordinated cations and all of the bridging oxygen atoms will have been specified. The 'negative' O_c oxygen atoms that are included with an oxygen vacancy are used as a means to displace O_c atoms from their normal positions to positions labelled O_c^* while still preserving the total number of oxygen atoms in the crystal. Once the entire lattice has been decorated in this way there will be no remaining 'negative' oxygen atoms.

With this formulation the intensity components, I_1 and I_2 can be discarded, and use made of only the short-range order (SRO) term, I_0. In this, though, the atomic scattering factors are replaced by structure factors for the chemical motifs of an occupied or a vacant site. This simplification merely incorporates the displacements which would normally appear in I_1 and I_2 into I_0, but in doing so automatically dictates that the displacements when they occur are always of exactly the same magnitude. The scattering power of an occupied (O) and vacant (V) O_c site can be calculated as a simple structure factor of the chemical motifs described above:

$$F_O^1 = f_O + 2f_{Al}\cos\left(2\pi(x_T h_1 + y_T h_2)\right) \tag{11.1}$$

$$F_V^1 = 2f_O\cos\left(2\pi(x_{O_c^*}h_1 + y_{O_c^*}h_2)\right) + 2f_{Al}\cos\left(2\pi(x_{T^*}h_1 + y_{T^*}h_2)\right) \\ - 2f_O\cos\left(2\pi(x_{O_c}h_1 + y_{O_c}h_2)\right). \tag{11.2}$$

Here, f_O, f_{Al} are the atomic scattering factors of O and Al; x_T, y_T, etc. are the fractional coordinates of the T, T^*, O_c, and O_c^* sites measured relative to the O_c site in question; h_1, h_2 are continuous reciprocal-space coordinates; and the superscript on the structure factors indicates the sublattice upon which the oxygen atom or vacancy resides. Sublattice 1 refers to the O_c site at $(\frac{1}{2}, 0, \frac{1}{2})$ and translationally equivalent sites, while sublattice 2 refers to the O_c site at $(0, \frac{1}{2}, \frac{1}{2})$. The above equations give the expressions for motifs centred at sites on sublattice 1. Similar, symmetry-related, expressions may be written for motifs centred on sublattice 2. Replacing the atomic scattering factors in I_0 by these chemical motif structure factors, after some manipulation the following expression for the SRO diffuse scattering in Mullite may be obtained. (for further details see Butler and Welberry, 1994).

$$I_{SRO}^{mullite} = \mu_1 I_1 + \mu_2 I_2 + \mu_{12} I_{12} \tag{11.3}$$

where,

$$I_1 = \sum_{\substack{lmn \\ l,m,n \text{ integer}}} \alpha_{lmn}^{O_1 V_1} \cos(h_1 l + h_2 m + h_3 n) \tag{11.4}$$

$$I_2 = \sum_{\substack{lmn \\ l,m,n \text{ integer}}} \alpha_{lmn}^{O_2 V_2} \cos(h_1 l + h_2 m + h_3 n) \tag{11.5}$$

$$I_{12} = \sum_{\substack{lmn \\ l = \text{int}+1/2 \\ m = \text{int}+1/2 \\ n = \text{integer}}} \alpha_{lmn}^{O_1 V_2} \cos(h_1 l + h_2 m + h_3 n) \tag{11.6}$$

and

$$\mu_1 = c_O c_V \left(F_{O_1} - F_{V_1}\right)^2 \tag{11.7}$$

$$\mu_2 = c_O c_V \left(F_{O_2} - F_{V_2}\right)^2 \tag{11.8}$$

$$\mu_{12} = 2c_O c_V \Re\left(\left(F_{O_1} - F_{V_1}\right)\left(F_{O_2} - F_{V_2}\right)^*\right). \tag{11.9}$$

The three intensities, I_1, I_2 and I_{12}, are periodic functions in reciprocal-space with a repeat that is commensurate with the first Brillouin zone and the coefficients, μ_1, μ_2, μ_{12}, are continuous functions in reciprocal-space. I_1 and I_2 involve only correlations between sites on the same sublattice (1 and 2 respectively), while μ_{12} involves cross-correlations between the two different sublattices. Figure 11.4 shows plots of the functions μ_1, μ_2, μ_{12}, which are calculated from the fractional coordinates of the average structure obtained from the Bragg reflections. It should also be noted that since T, T^*, O_c, and O_c^* all have the same z-coordinate in the structure the functions μ_1, μ_2, μ_{12} are virtually independent of h_3, the reciprocal coordinate in the $\mathbf{c}^*$ direction. The variation of the intensity distribution with reciprocal section thus results entirely from the Warren–Cowley SRO parameters, α_{lmn}, appearing in the three intensity components I_1, I_2, and I_{12}.

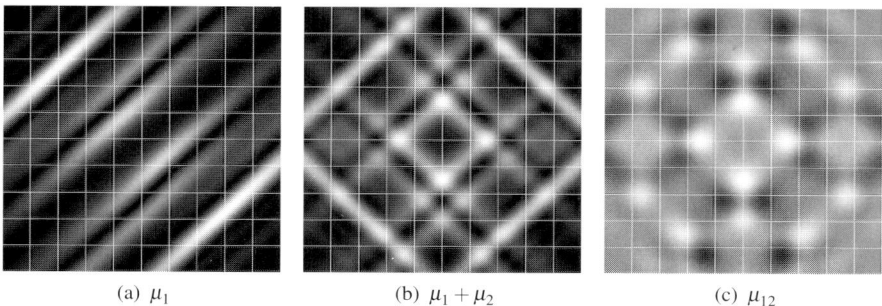

(a) μ_1 (b) $\mu_1 + \mu_2$ (c) μ_{12}

Fig. 11.4 The intensity coefficients. White lines show the integral h_1 and h_2 positions.

11.4 Results

A value for each of the three intensity components can be obtained for a given point inside the first Brillouin zone by gathering together several measurements of the diffuse intensity from points in the reciprocal-space that are related by symmetry with this point. The set of measurements related by symmetry in this way is commonly referred to as an *associated set* and the point inside the minimum repeat volume is referred to as a *minimum volume point* (see Fig. 11.5). For example, if the point (h_1, h_2, h_3), lies inside the first Brillouin zone then the points $(1 - h_1, h_2, h_3)$, $(h_1, 2 + h_2, h_3)$, $(2 + h_1, 3 - h_2, 1 + h_3)$, etc. form the associated set of the minimum volume point (h_1, h_2, h_3). Each of the intensity components I_1, I_2 and I_{12} have the same numerical magnitude (but not necessarily sign) for all points inside a particular associated set but the coefficients, μ_1, μ_2, μ_{12}, vary predictably in a way that does not repeat commensurately with the underlying reciprocal lattice. Any collection of three or more measurements in an associated set can therefore be used to solve for the three intensity components at one minimum volume point. In the present case upwards of 30–40 measurements were available in each associated set and with only three intensity components to solve for the problem is highly overdetermined. Standard least-squares procedures were used to solve for the three intensity components at each of 1331 points in the minimum volume.

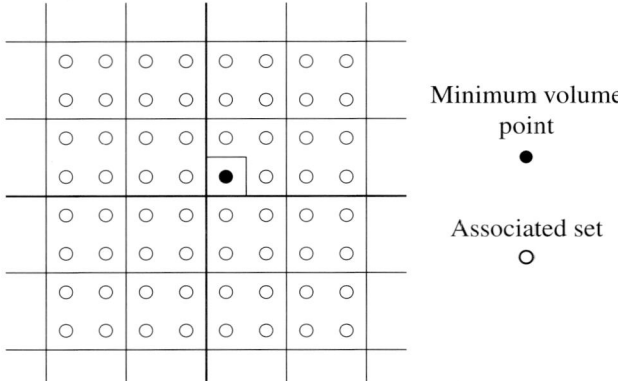

Fig. 11.5   Definition of *associated set* and *minimum volume point*.

When this had been done for every point in the minimum volume, the Warren–Cowley SRO parameters, α_{lmn}, were extracted from each of the three component intensities by simple Fourier inversion of eqns (11.3)–(11.6). The most significant of these are given in Table 11.1. Assuming that the three SRO intensities derived this way are the only significant contributions to the total diffuse intensity in mullite then the oxygen-vacancy SRO parameters obtained by this procedure will contain all the information that the diffuse part of the diffraction pattern can provide.

The intensity R factor, $R = \Sigma (I_{\text{obs}} - I_{\text{calc}})^2 / \Sigma I_{\text{calc}}^2$, for a typical least-squares solution of one of the 1331 minimum volume points was approximately 35%. This obvious quantitative discrepancy may seem very high by conventional crystallographical standards but the resulting least-squares fit is not as bad as this figure implies. When all 1331 independent solutions are used, together with the μ_1, μ_2, μ_{12} coefficients, to obtain a 'best-fit' calculated diffraction pattern for visual comparison with the observed patterns, the results are really rather convincing (see Fig. 11.6).

Both Fig. 11.2 and Fig. 11.6 are reproduced on the same scale so that the shades of grey can be compared to see how the two agree. There are a few obvious differences. For instance in the $0.7c^*$ reciprocal layer the measured intensities near the origin and at large scattering angles are noticeably higher than in the calculated pattern whereas at intermediate angles this situation is reversed. There are also distinct diagonal dark bands visible in the $0.5c^*$ layer of the calculated pattern that are much less distinct in the data. These differences may indicate minor short-comings in the model used to derive eqn (11.3) which does not allow for atom displacements other than those included in the simple chemical motifs described.

The most significant values for the Warren-Cowley SRO parameters, α_{lmn}, obtained from the analysis are shown in Table 11.1. For a vacancy concentration of 0.2 the most

Table 11.1 Local order parameters for mullite determined from the least-squares analysis

Interatomic vector	SRO param, α_{lmn}	Interatomic vector	SRO param, α_{lmn}
$\frac{1}{2}\langle 1\,1\,0\rangle$	−0.250	$\langle 0\,1\,1\rangle$	−0.122
$[1\,1\,0]$	−0.250	$[1\,1\,1]$	−0.079
$[1\,\bar{1}\,0]$	−0.081	$\langle 0\,\bar{1}\,1\rangle$	+0.000
$\langle 0\,0\,1\rangle$	−0.227	$\frac{1}{2}\langle 1\,1\,2\rangle$	+0.109
$\langle 1\,0\,0\rangle$	−0.004	$\frac{1}{2}\langle 3\,1\,0\rangle$	+0.141
$\langle 0\,1\,0\rangle$	+0.015	$\frac{1}{2}\langle 1\,3\,0\rangle$	+0.077
$\langle 0\,0\,2\rangle$	−0.019	$\frac{1}{2}\langle 3\,1\,2\rangle$	−0.092
$\langle 0\,0\,3\rangle$	−0.124	$\frac{1}{2}\langle 1\,3\,2\rangle$	+0.058
$\langle 0\,0\,4\rangle$	+0.047	$\langle 0\,1\,2\rangle$	+0.088
$\langle 1\,0\,1\rangle$	+0.113	$\langle 0\,2\,2\rangle$	+0.051

negative value of α_{lmn} feasible is $\alpha_{lmn} = -0.25$, which corresponds to complete avoidance of defect pairs separated by the vector $[l\,m\,n]$. Thus it is seen that, in addition to the absence of defects separated by the vectors $\frac{1}{2}\langle 1\,1\,0\rangle$ and $[1\,1\,0]$ (which results from the simplifying assumptions that were made), there is also a strong tendency to avoid defects separated by $\langle 0\,0\,1\rangle$ and a moderately strong tendency to avoid pairs of defects separated by $\langle 0\,0\,3\rangle$, $\langle 0\,1\,1\rangle$, $[1\,\bar{1}\,0]$ and $\frac{1}{2}\langle 3\,1\,2\rangle$. The most common inter-defect vectors found were $\langle 1\,0\,1\rangle$, $\frac{1}{2}\langle 1\,1\,2\rangle$ and $\frac{1}{2}\langle 3\,1\,0\rangle$.

11.5 Conclusions

Comparison of Fig. 11.2 and Fig. 11.6 demonstrates convincingly that, for this seemingly complicated mineral, the diffuse scattering can be described simply as the sum of three component SRO diffuse intensities. Two of these components involve correlations between the basic defects on each of the two different sublattices (symmetry-related) and the third involves correlations between the two sublattices. Although there are quantitative discrepancies indicating some short-comings in the fine detail of the model, there is no doubt that the model is essentially correct.

The basic defect consists of a vacant O_c site together with a transfer of its two neighbouring tetrahedral cations from a T to a T^* site. Each T^* cation then shares a further bridging oxygen which shifts from O_c to O_c^* (see Fig. 11.3(a)). An integral part of the model was the necessity to assume that two chemically implausible configurations (shown in Figs. 11.3(b) and (c)) were forbidden. If two defects were separated by $\frac{1}{2}[1\,1\,0]$ the intervening cation would be left without a bridging O_c atom, while if two defects were separated by $[1\,1\,0]$ it would be necessary for four cations to share the same bridging O_c atom. The analysis revealed that defect pairs separated by $\langle 0\,0\,1\rangle$

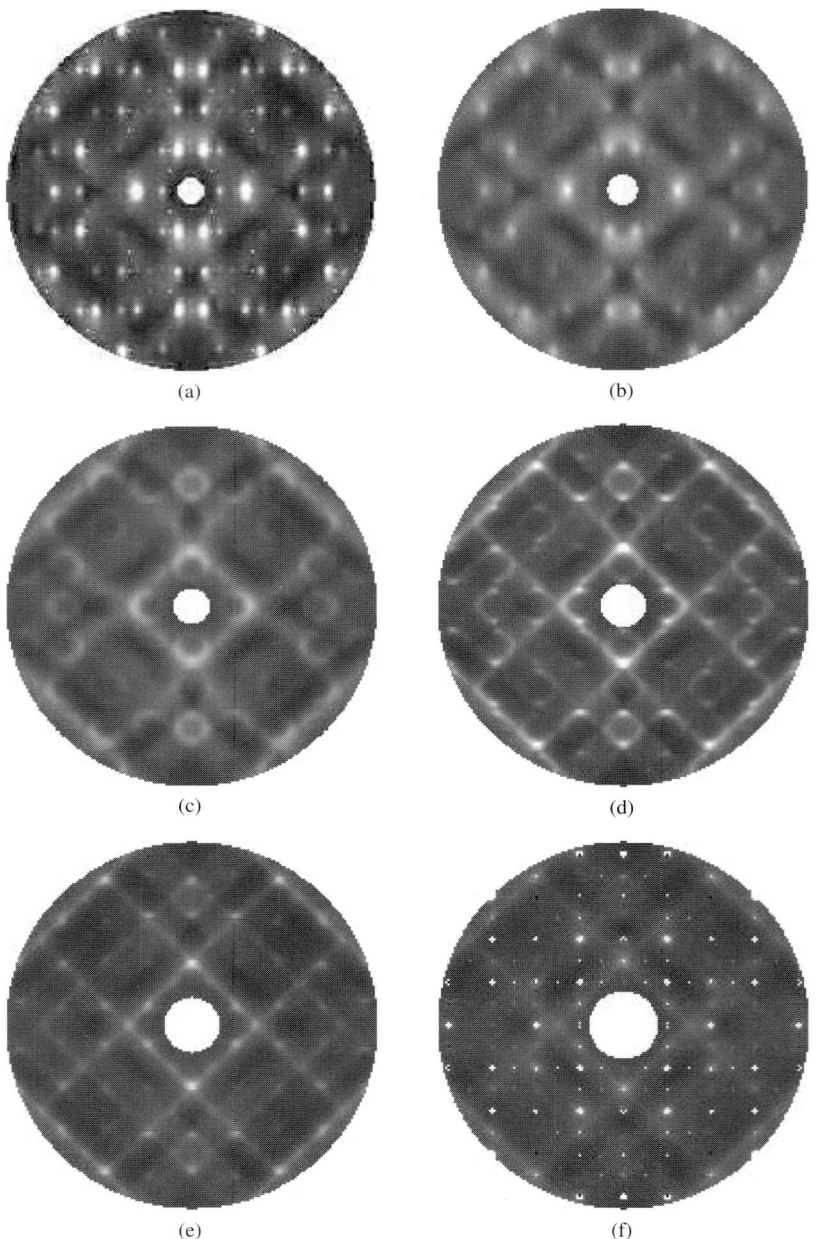

Fig. 11.6   Intensities calculated for mullite from final fitted values of SRO parameters. (a) $0.5c^*$; (b) $0.6c^*$; (c) $0.7c^*$; (d) $0.8c^*$; (e) $0.9c^*$; (f) $1.0c^*$. Cf. Fig. 11.2.

were strongly avoided, and that defects separated by $\langle 0\,0\,3 \rangle$, $\langle 0\,1\,1 \rangle$, $[1\,\bar{1}\,0]$ or $\frac{1}{2}\langle 3\,1\,2 \rangle$ also tended to be avoided, although rather less strongly. The most common inter-defect vectors found were $\frac{1}{2}\langle 3\,1\,0 \rangle$, $\langle 1\,0\,1 \rangle$, $\frac{1}{2}\langle 1\,1\,2 \rangle$, $\langle 0\,1\,2 \rangle$, $\frac{1}{2}\langle 1\,3\,0 \rangle$, $\frac{1}{2}\langle 1\,3\,2 \rangle$ and $\langle 0\,2\,2 \rangle$.

WÜSTITE

12.1 Introduction

The structure of the non-stoichiometric oxide wüstite, $Fe_{1-x}O$, has been the subject of much debate over many years (e.g. see Roth, 1960; Koch and Cohen, 1969; Garstein and Cohen, 1980; Schweika *et al.*, 1995) The material is of considerable importance as it is thought to be a major constituent of the Earth's lower mantle. There is general agreement that the average structure is that of the rock-salt type with both O^{2-} and Fe^{2+} ions forming interpenetrating f.c.c. sublattices. The O sublattice is considered to be complete while the Fe sublattice contains vacancies, but in addition there are Fe^{3+} ions present on interstitial sites. There also appears to be agreement that the Fe^{2+} vacancies and Fe^{3+} interstitials form clusters, although quite different defect clusters have been proposed. The smallest such cluster is the V_4T cluster consisting of 4 vacancies (V) forming a tetrahedron on neighbouring Fe^{2+} sites, together with one Fe^{3+} interstial (T) at the centre of the tetrahedron (see Fig. 12.1). Larger clusters are envisaged to contain aggregations of two or more tetrahedral clusters in various corner-, edge- or face-sharing arrangements. Figure 12.2 shows three such defects which have cubic symmetry that are referred to later.

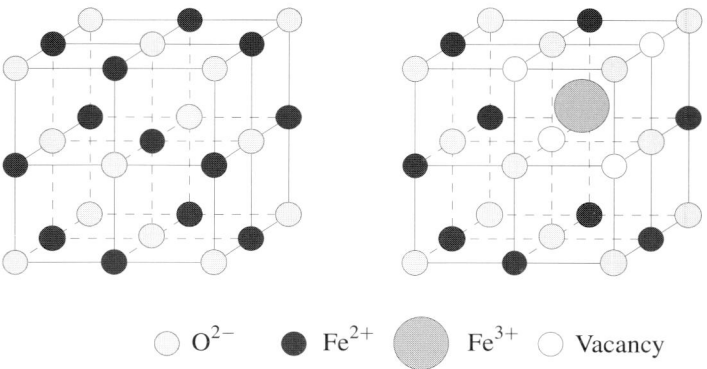

$\bigcirc\ O^{2-}$ $\bullet\ Fe^{2+}$ ⬤ Fe^{3+} $\bigcirc$ Vacancy

Fig. 12.1 Rock-salt structure and formation of a tetrahedral defect.

Among the experimental methods that have been used to investigate the wüstite structure, X-ray (or neutron) diffraction has featured prominently, though relatively few of these studies have involved the measurement of diffuse X-ray scattering. In this chapter a step-by-step description is presented of how the complex diffuse diffraction patterns of wüstite arise and are influenced by various possible real-space variables—for example, defect distribution, defect cluster size, number of interstitials and lattice strain. For further details see Welberry and Christy (1997).

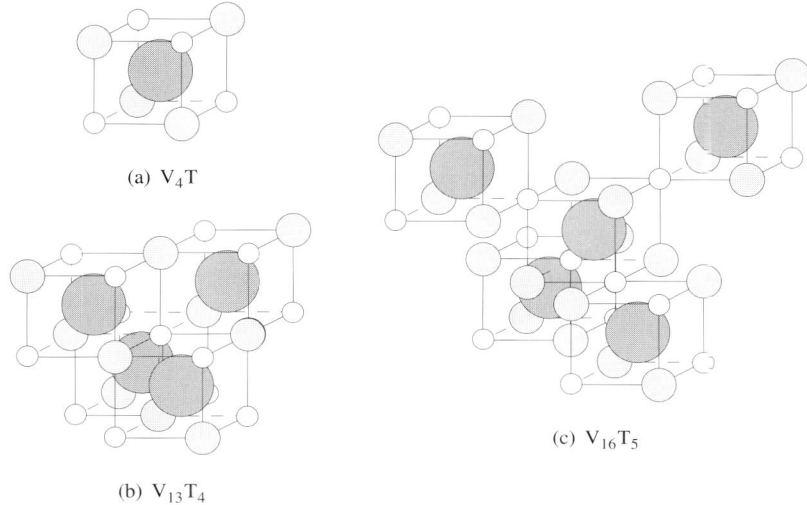

(a) V_4T

(c) $V_{16}T_5$

(b) $V_{13}T_4$

Fig. 12.2   Three defect clusters with cubic symmetry.

12.2   X-ray diffuse scattering data

Figure 12.3 shows X-ray diffraction patterns of various reciprocal sections for wüstite. These were all recorded using the PSD diffractometer system (see Section 1.2.1). Figures 12.3(a) and (b) were recorded using Mo$K\alpha$. For these data aluminium foil was used to attenuate the strong fluorescence coming from the iron. For Figs 12.3(c) and (d) Co$K\alpha$ was used which avoids the fluorescence problem and also gives much greater spatial resolution. In broadest terms the form of the patterns may be understood in terms of a model for the distribution of defects in the basic rock-salt type FeO lattice. In this description the defects are assumed to be arranged on a primitive lattice of average spacing of $\sim 2.7\mathbf{a}_0 \times 2.7\mathbf{a}_0 \times 2.7\mathbf{a}_0$ where $\mathbf{a}_0$ is the basic rocksalt repeat. If the real-space distribution of wüstite is considered as a perfect lattice of the ideal wüstite *multiplied* by this second larger-scale paracrystal distribution function describing the position of the defects, the diffraction pattern then consists of the ideal wüstite diffraction pattern *convoluted* with the diffraction pattern of the paracrystal distribution. That is, around each parent Bragg peak appears a motif of scattering which is the Fourier Transform of the paracrystal distribution. See Fig. 12.4.

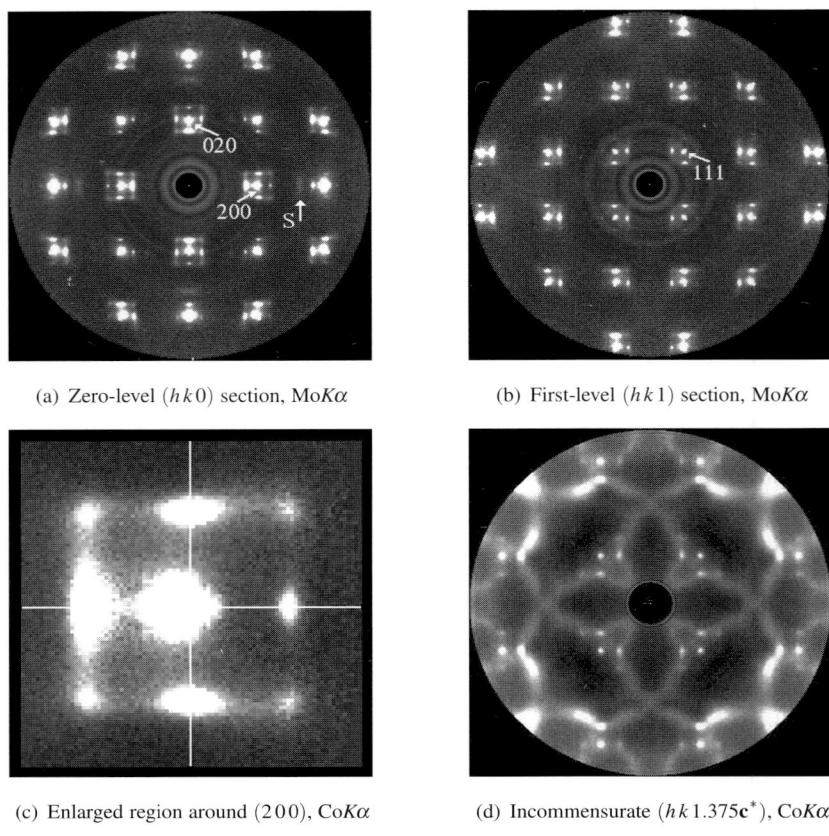

(a) Zero-level $(hk0)$ section, MoKα   (b) First-level $(hk1)$ section, MoKα

(c) Enlarged region around (200), CoKα   (d) Incommensurate $(hk\,1.375\mathbf{c}^*)$, CoKα

Fig. 12.3   Observed X-ray diffraction patterns of wüstite.

12.3   Summary of X-ray diffraction features

For convenience a brief summary of the main diffraction features that may be seen in the diffraction patterns of Fig. 12.3 are presented here for later reference.

1. To a first approximation the diffraction motif surrounding the $(2,\ 0,\ 0)$ Bragg peak consists of a square of satellite peaks at positions $(2 \pm \delta,\ 0,\ 0), (2,\ \pm \delta,\ 0),$ $(2 \pm \delta,\ \pm \delta,\ 0)$. (N.B. this is a cube in 3D). The intensities of peaks on the low-angle side of the Bragg position are much stronger than those on the high-angle side.

2. The satellite peaks are diffuse and, particularly those at $(2 \pm \delta,\ 0,\ 0)$ and $(2,\ \pm \delta,\ 0)$, are quite anisotropic. The width of these peaks along a vector directed towards the Bragg peak is quite small, while in the transverse direction the width is much larger.

3. In additition to the diffuse satellite peaks there is also strong diffuse scattering around the Bragg peak itself. This scattering has an intensity distribution that is

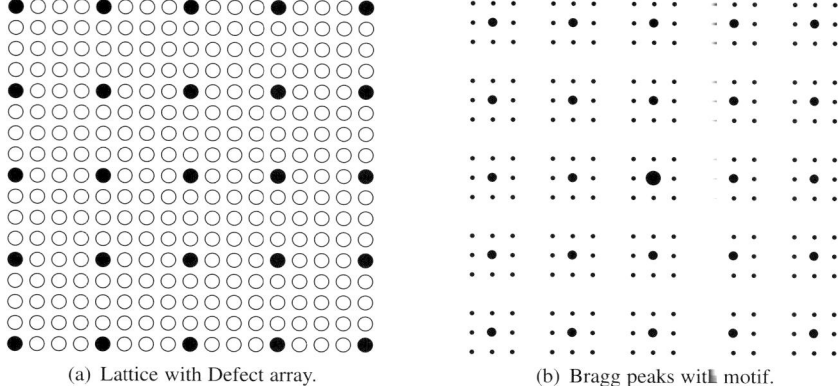

| (a) Lattice with Defect array. | (b) Bragg peaks with motif. |

Fig. 12.4   Real and reciprocal-space description of wüstite.

highly asymmetric with respect to the Bragg peak position, with the bulk of the intensity occurring on the low-angle side of the Bragg position.

4. Although the peaks at the corners of the square are quite diffuse the two on the left of Fig. 12.3(c) appear to be somewhat closer together than the two on the right, giving the square a distinctly trapezoidal appearance.

5. In addition to the diffraction motifs around the Bragg peaks there are weaker much broader diffuse features in the pattern, the most striking example of which is in the $1.375c^*$ section. So far these broad diffuse distributions are unexplained.

12.4   Paracrystal model to account for diffraction features

Point 1 and point 2 above may be explained by a simple paracrystalline description of the distribution of defects, characterised by the transverse and longitudinal correlations ρ_T and ρ_L and the variance σ. Instead of the defects occurring on a simple cubic lattice it is supposed that this lattice is highly distorted. In order to give rise to the shape of diffuse peaks observed, it is necessary that the paracrystal lattice has very strong longitudinal correlation and rather weaker transverse correlation. (cf. Fig. 6.5). The value of the variance must be sufficiently high to result in a fall-off in intensity that gives only first-order satellites.

For the examples that are shown here the same paracrystal distribution has been used throughout, in order that differences due to changes of other variables may be more readily distinguished. For this distribution $\rho_L = 0.935$, $\rho_T = 0.8$, and $\sigma = 1.7$ (i.e. 31.5% of the defect spacing). Defects were placed on a 2D lattice of dimensions 1024×1024, centred at sites n, m given by,

$$n = \text{nint}(5.4i + \sigma X_{i,j}),$$
$$m = \text{nint}(5.4j + \sigma Y_{i,j}),$$

where n, m are indices defining points on the basic square lattice which corresponds to the $\mathbf{a}_0 \times \mathbf{a}_0$ repeat of the projected structure. i, j are indices (in the range 1–189) referring to points on the paracrystal array. A small representative portion of a realisation of this paracrystal distribution is shown in Fig. 12.5(a). In this figure the underlying perfect rock-salt structure is shown as a white background and the black squares represent $V_{13}T_4$ defect clusters (see Fig. 12.2). It will become apparent below that these defects are both crystal-chemically reasonable and give a good fit to the observed diffraction behaviour. The effect of the large value of σ can be seen by noting how a given row of defects deviates considerably from the underlying straight lattice rows. The high value of the longitudinal correlation ensures that, despite the high σ, the spacing between neighbouring defects varies only a small amount. The variance of this spacing, $\sigma_d^2 = \sigma^2(1 - r)$, gives a value for σ_d of 0.43 for the values of σ, ρ_L given above. Consequently only a very few instances occur where the spacing between neighbouring defects takes a sufficiently low value that the defects touch or overlap, and the majority of spacings will be either 5 or 6 $\mathbf{a}_0$ units. Figures 12.5(b) and (c) show diffraction patterns calculated from this model. Figure 12.5(b) corresponds to the case when the interstitial cations have been included and Fig. 12.5(c) to the case when they have been omitted. Note the effect on the relative intensity of the (2 0 0) and (2 2 0) reflection. Note also that these patterns are of the diffuse intensity only. The positions of the Bragg peaks indicated by a white dot have been inserted for reference. It is also clear from this that a simple paracrystal model does not result in diffuse intensity around the central Bragg position.

12.5 Relaxation of structure around defects

The asymmetry in the diffraction motif described in point 1 of Section 12.3 is characteristic of size-effect distortion due to relaxation of the lattice around the defects. To carry out relaxation of the rock-salt lattice around the defect clusters a simple Monte Carlo algorithm was used. For simplicity interstitial atoms were not included in the simulation but were inserted afterwards. As before a simple interaction potential was used in which harmonic (Hooke's law) springs connect primary lattice sites in the the $\langle 1\,0 \rangle$ and $\langle 1\,1 \rangle$ directions. The Hamiltonian is given by

$$E = \sum_{a,b} K_{ab}(R_{ab} - d_{ab})^2.$$

Here the summation is over all nearest-neighbour pairs of sites in the $\langle 1\,0 \rangle$ and $\langle 1\,1 \rangle$ directions. d_{ab} is the instantaneous length of a given inter-site vector. The equilibrium length of the spring, R_{ab}, between a pair of sites ab was assumed to be equal to $\mathbf{a}_0$ for $\langle 1\,0 \rangle$ springs and $\sqrt{2}(1 + \varepsilon_{ab})\mathbf{a}_0$ for $\langle 1\,1 \rangle$ springs. The strengths of the force constants, K_{ab}, for the $\langle 1\,0 \rangle$ and $\langle 1\,1 \rangle$ springs were assumed to be 20 and 100 respectively in arbitrary units. For the $\langle 1\,0 \rangle$ springs the distortion parameter ε_{ab} was set to be -0.06 for vectors between a vacant site and an occupied site; $+0.02$ between two occupied sites; 0.0 between two vacancies. After distortion this choice of parameters resulted in the mean cation/vacancy vector being reduced by $\sim 4\%$. [This figure is consistent with the octahedral Fe—O bondlengths in the first cation shell around the defects, predicted by

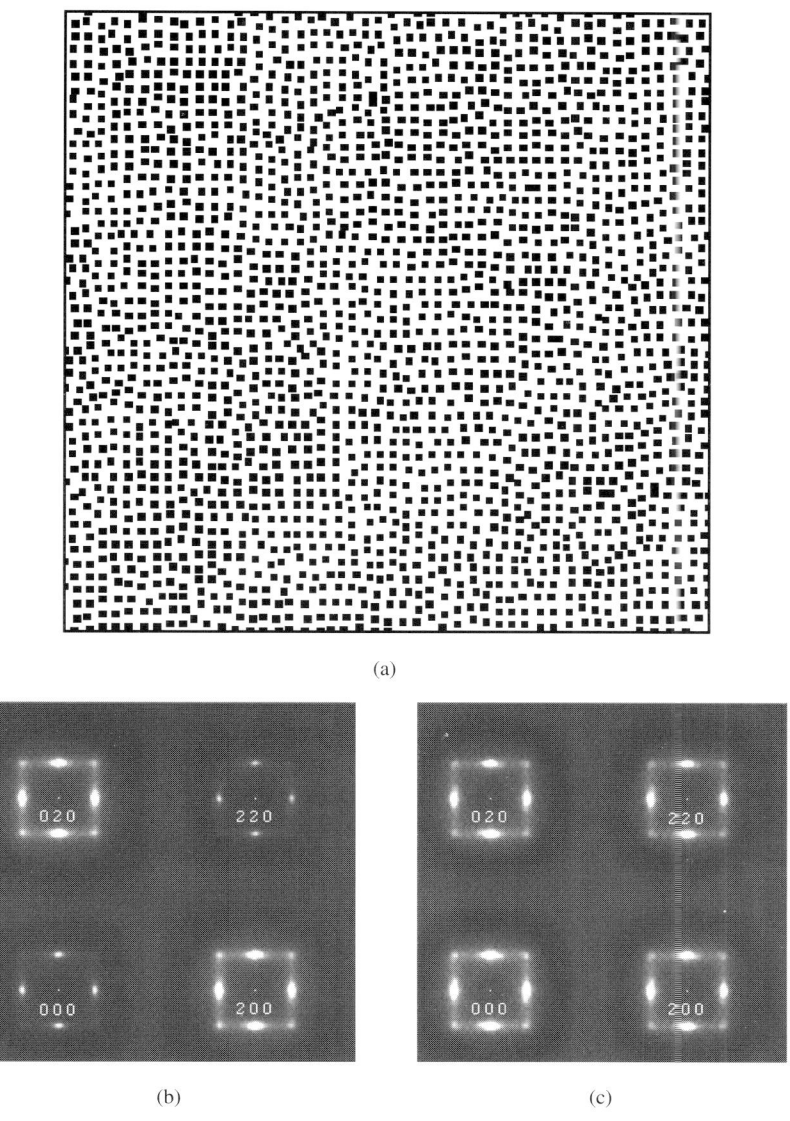

(a)

(b)   (c)

Fig. 12.5 The paracrystal distribution used in the examples.

calculations performed in 3D using the bond-length/bond-valence formalism (see Brese and O'Keeffe, 1991). If all the octahedral Fe^{3+} is concentrated in this shell then the mean valence there is 2.47 (i.e assuming 14 out of 30 next-nearest neighbours to the T cations are trivalent, to balance charges). The bond-length/bond- valence relation for Fe then predicts a corresponding mean bond shortening of 3.8%.] Monte Carlo simulation was carried out for 25 cycles of iteration. After each cycle, lattice averages of the different types of nearest-neighbour distances were computed to monitor the progress of the relaxation. After iteration the interstitial Fe^{3+} cations were placed at the centre of mass of the surrounding groups of four vacancies. An example diffraction pattern calculated after this kind of relaxation is given in Fig. 12.6(a) and a schematic diagram indicating the direction of the size-effect intensity transfer is given in Fig. 12.6(b). Note here, however, that the pattern shown in Fig. 12.6(a) corresponds to a model which does have intensity around the central Bragg peak positions.

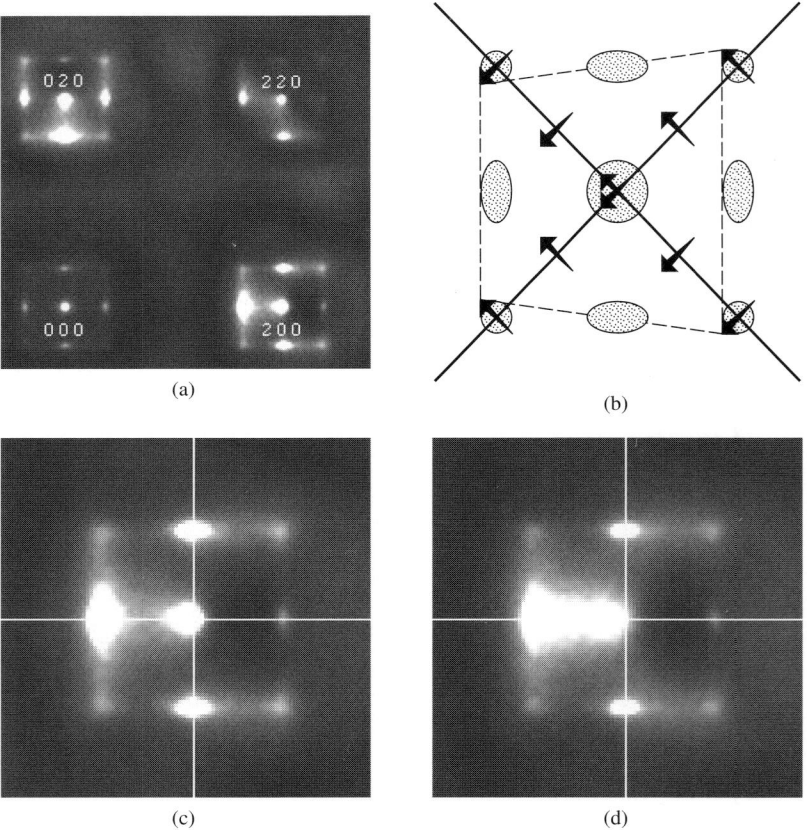

Fig. 12.6   The effect of lattice relaxation and inhomogenity.

One possible explanation for the presence of this diffuse scattering around the Bragg position is thermal diffuse scattering or Huang scattering. However, though these types of scattering undoubtedly will exist, they arise from second-order terms of the diffraction equation and are symmetric about the Bragg peak position. The strong asymmetry exhibited by the peaks in the observed patterns clearly points to an effect involving atomic displacements which are directly coupled to an occupancy factor (i.e. deriving from a first-order term in the diffraction equation, eqn (7.4b)). This suggests that in order to explain the presence of this scattering around the Bragg positions the distribution of defects must be inhomogeneous. For example, there might be regions in the structure where a paracrystalline distribution of defect clusters occurs, while other regions are defect free.

To investigate this possibility a simple modification to the basic paracrystal distribution of Fig. 12.5(a) is proposed. The same paracrystal distribution as before is assumed but for each site i, j the decision whether to place a defect cluster or not is decided with reference to a second independent distribution on the same lattice. For this second distribution a simple binary Ising model was used. The Ising spin variable $\sigma_{i,j}(= \pm 1)$ was used to designate which defect clusters on the paracrystal lattice were present ($\sigma_{i,j} = +1$) or absent ($\sigma_{i,j} = -1$). Ising model realisations were generated with varying degrees of local order (or domain size) and these were used in conjunction with the paracrystal lattice to obtain the final defect distributions. Two examples of these are shown in Figs. 12.6(c) and (d). In Fig. 12.6(c) the Ising lattice has a nearest-neighbour correlation (i.e. $C_{10} = \langle \sigma_{i,j}\sigma_{i-1,j} \rangle$) of 0.6, while for Fig. 12.6(d) the nearest-neighbour correlation is 0.2.

It is clear from Fig. 12.6 that a model of this kind satisfactorily accounts for the presence of the strong diffuse peak around the Bragg position. It is also clear that the domain size produced by a correlation $C_{10} = 0.6$ gives better agreement with the observed patterns than that for $C_{10} = 0.2$, the central peak in this latter pattern being far too diffuse. In both of these examples 50% of the original defect clusters have been removed. In addition the size-effect transfer of intensity is seen to operate on the central diffuse peak as well as the motif peaks.

It should be noted that if the paracrystal array of V_4T clusters is complete this leads to a total number of Fe vacancies of only 3.8%. For $V_{13}T_4$ clusters the number of Fe vacancies would be 11.4% and for the larger $V_{16}T_5$ clusters there would be 14%. Consequently an inhomogeneous array of $V_{13}T_4$ clusters in which 50% of the clusters are missing would give a composition corresponding closely to the observed composition in which there are 5.5% vacancies (i.e. $x = 0.055$).

12.6 Effect of cluster volume fraction

Despite the generally good qualitative agreement between Fig. 12.6(c) and the observed pattern, Fig. 12.3, it is clear that the relative magnitude of the central diffuse peak and that of the motif peaks is not correct, the central peak being rather too weak in comparison to the motif peaks. It is necessary to consider how the magnitude of these peaks changes as the volume fraction of the paracrystal array that is occupied by defects is changed. The motif peaks will simply have an intensity that is proportional to the amount of paracrystal that is present. In contrast the central peak due to the inhomogeneous distribution will have an intensity that is proportional to $m_A(1 - m_A)$ where m_A is the fraction of the paracrystal array occupied. For a 30:70 ratio of defect to defect-free domains the central peak would diminish to $0.21/0.25 = 84\%$ of its original value, but the paracrystal peaks would diminish to $0.3/0.5 = 60\%$ of the original value. For a 70:30 ratio, the central peak would again diminish to 84% but the paracrystal peaks would increase to $0.7/0.5 = 140\%$. Figure 12.7 shows parts of the diffraction patterns of examples containing regions of $V_{13}T_4$ clusters and regions of no defects in the ratios 30:70 and 70:30. In each case the lattice relaxation described above has been applied. Note how the central diffuse peak in the different motifs is relatively much stronger in (a) than in (c).

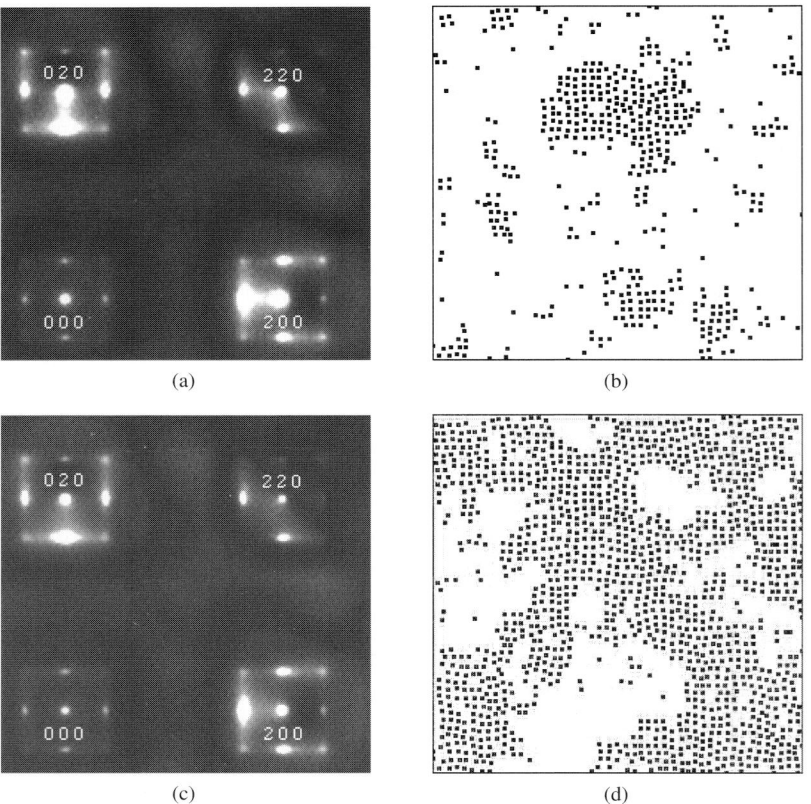

(a)   (b)

(c)   (d)

Fig. 12.7   The effect of cluster volume fraction.

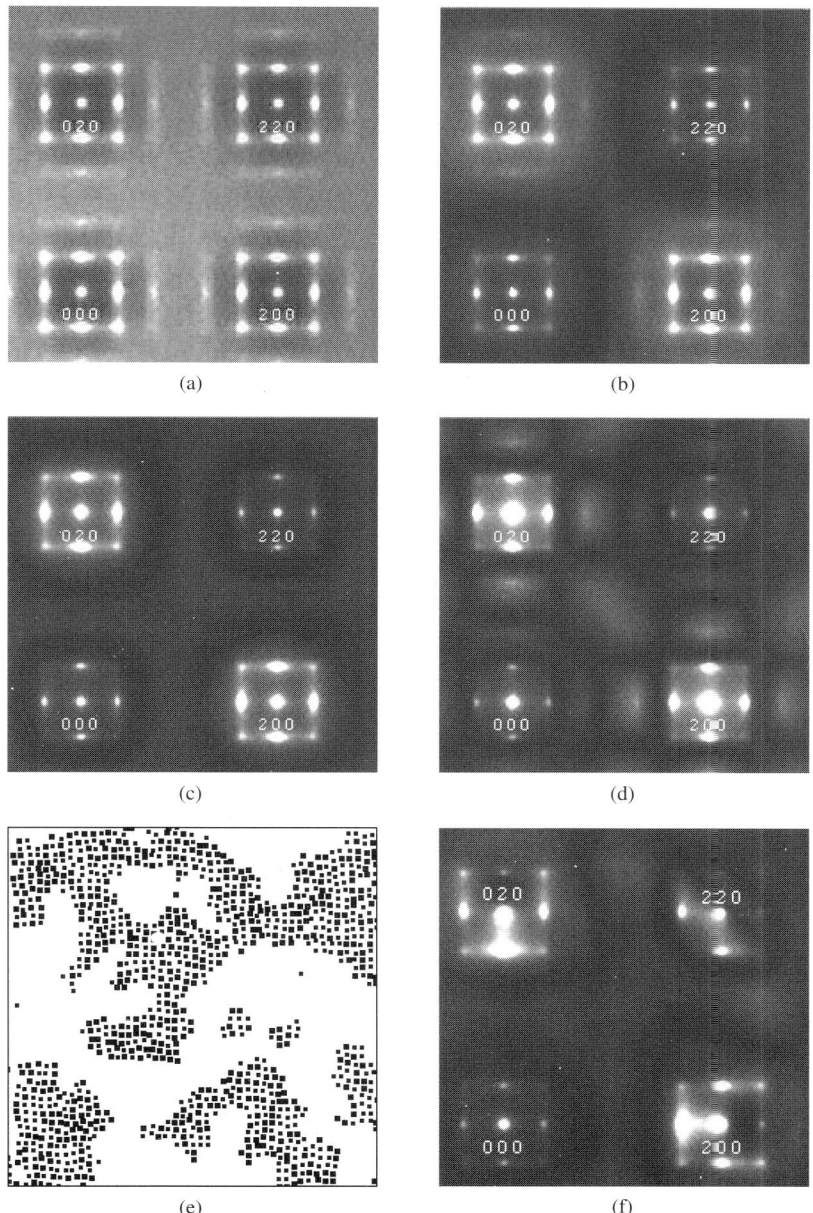

Fig. 12.8 The effect of cluster size.

12.7   Effect of cluster-size

Although, as seen above, reducing the volume fraction of the paracrystal lattice that is occupied by defects results in an increase in the intensity of the central diffuse peak relative to the motif peaks, the resulting distribution does not contain a sufficient number of Fe vacancies to satisfy the observed overall composition, if the clusters are of the $V_{13}T_4$ type. Therefore in this section the effect on the diffraction pattern of using different sized clusters while using the same paracrystal distribution is investigated. Figure 12.8 shows diffraction patterns calculated using the same inhomogeneous paracrystal distribution used for Fig. 12.6(c) with 50% volume fraction but in which (a) single Fe vacancy defects; (b) V_4T defects; (c) $V_{13}T_4$ defects; (d) $V_{16}T_5$ defects are used. For these figures the size-effect relaxation has not been applied. It is therefore possible to see that having $V_{16}T_5$ defects results in a much larger central peak. Since having only $V_{16}T_5$ clusters would result in too large an overall concentration of Fe vacancies it seems likely that a model in which a mixture of $V_{13}T_4$ and $V_{16}T_5$ clusters is present would allow both the constraint on the overall concentration of vacancies and the intensity of the central diffuse peak to be satisfied. Such a composite distribution together with its diffraction pattern is shown in Figs. 12.8(e) and (f).

13

CUBIC STABILISED ZIRCONIAS

13.1 Introduction

The class of compounds known as cubic stabilised zirconias (CSZs) have an extremely simple average structure (the fluorite CaF_2 structure, see Fig. 13.1) but they exhibit extremely complex diffuse X-ray diffraction patterns, see Fig. 13.3. Despite numerous attempts over many years to understand the disorder in these materials (Allpress and Rossell, 1975; Morinaga et al., 1980; Neder et al., 1990; Proffen et al., 1993) a completely statisfactory model for the local order remains elusive, although considerable progress has been made in recent years (Welberry et al., 1995; Welberry, 2001). In this chapter, the study of a CSZ system is used as a pedagogical example to show how various of the simulation methods together with a knowledge of the basic diffraction theory described in previous chapters can be used to give insight into a complex diffraction problem.

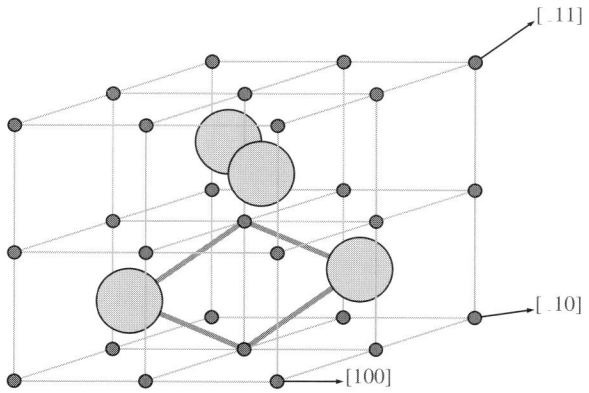

Fig. 13.1 Average structure of a CSZ. The thick lines indicate that between a pair of cations there are two bridging oxygen sites.

The fluorite structure has eight oxygen sites per cell and four cation sites. The oxygen array is primitive but in CSZs contains vacancies, while the cation array is f.c.c. and in CSZs, though complete, is disordered. Cubes of oxygens along the three cubic directions are alternately occupied by a cation, or unoccupied, while along $\langle 1\,1\,0\rangle$ there exist chains of cubes all occupied by a cation.

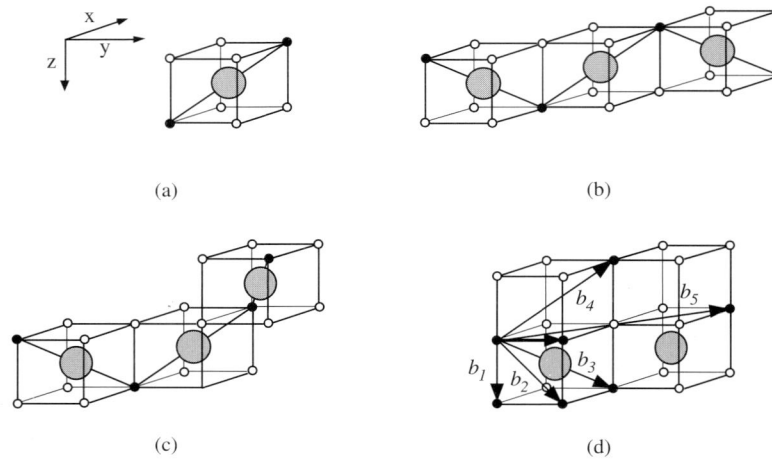

(a)   (b)

(c)   (d)

Fig. 13.2   Schematic representation of possible $\frac{1}{2}\langle 1\,1\,1\rangle$ structures.

In commencing a study of this system it may be noted that in all known fluorite-related superstructure phases (with the one exception of the C-type rare-earth oxide structures) all anion vacancies occur in pairs separated by $\frac{1}{2}\langle 1\,1\,1\rangle$ (in a cube con-taining a cation). Vacancy pairs separated by $\frac{1}{2}\langle 1\,0\,0\rangle$ and $\frac{1}{2}\langle 1\,1\,0\rangle$ are avoided. These $\frac{1}{2}\langle 1\,1\,1\rangle$ cation/vacancy-pair units (see Fig. 13.2(a)) may be isolated (as in M_7O_{13}), in linear chains (as in M_7O_{12}), linked into zig-zag chains (as in the pyrochlore struc-ture, see Fig. 13.2(b)), helical chains (as in $Ca_6Hf_{19}O_{44}$), or into helical clusters (as in $CaZr_4O_9$, see Fig. 13.2(c)).

From a chemical point of view it seems reasonable to suppose that the structure of a disordered CSZ of a given composition might consist of small domains of one or other of the known super-lattice phases which occur nearby in the phase diagram. A useful approach to try to understand the complex diffraction patterns that are ob-served, therefore, is to generate model structures in which such small domains of the various known superstructures occur, and compare the diffraction patterns with the ob-served patterns. For this purpose the example is used of an yttria-stabilised cubic zirco-nia, $Zr_{0.61}Y_{0.39}O_{1.805}$, for which numerous sections of X-ray diffraction data have been recorded (see Fig. 13.3). Before proceeding with this survey of possible models for the oxygen vacancy ordering, however, it is necessary to consider how the structure relaxes when an oxygen vacancy is introduced.

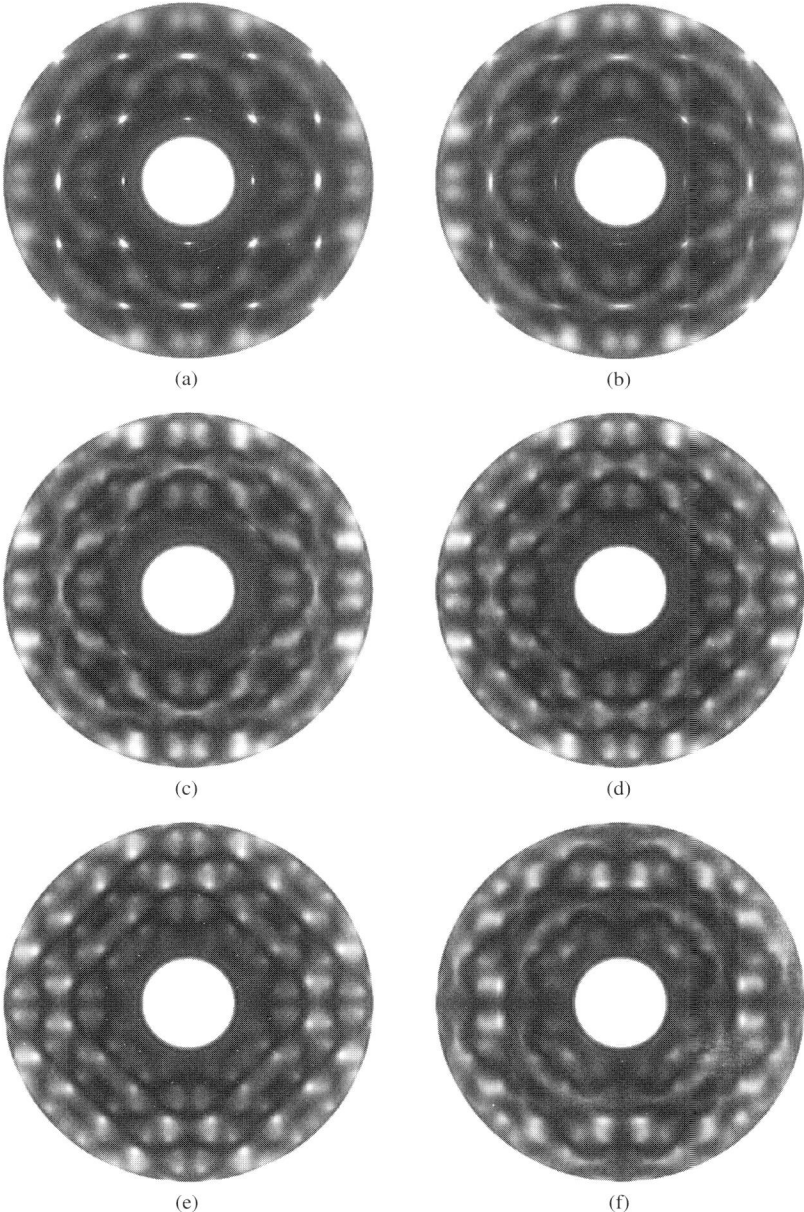

Fig. 13.3 Observed X-ray diffuse scattering patterns for yttria-stabilised cubic zirconia normal to **c**. (a) $0.1\mathbf{c}^*$; (b) $0.2\mathbf{c}^*$; (c) $0.3\mathbf{c}^*$; (d) $0.4\mathbf{c}^*$; (e) $0.5\mathbf{c}^*$; (f) $0.7\mathbf{c}^*$.

13.2   Model for relaxation

Since Zr and Y differ in atomic number by only one their X-ray scattering factors are practically identical. Consequently to a good approximation the terms that occur in the diffraction equation that depend on $(f_Y - f_{Zr})$, that is, I_0 and I_1, are absent. Although mean-square atomic displacements for O are somewhat higher than for the cations, their contribution to the intensity is relatively minor because of their much lower scattering factor. Consequently the observed diffraction patterns are dominated by the displacements of the cations and are described by diffraction terms I_2 and higher.

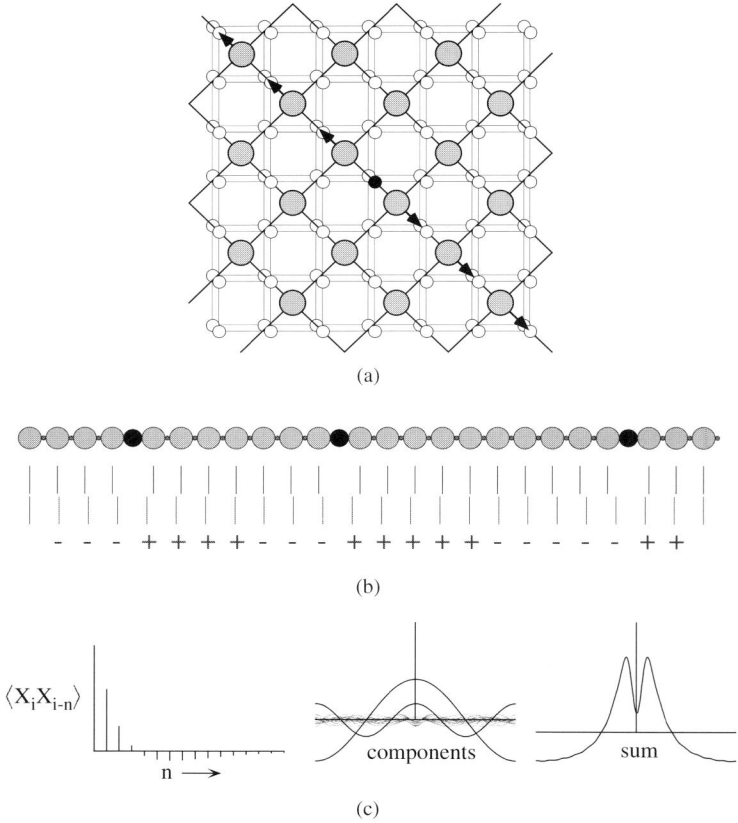

Fig. 13.4   The model for relaxation.

A simple relaxation model has been developed which describes the way in which cations are displaced away from their average position in response to the removal of an oxygen neighbour. Nearest-neighbour cation pairs are separated by $\frac{1}{2}\langle 1\,1\,0 \rangle$, with a pair of bridging oxygen sites mid-way between. The simple relaxation model supposes that if either of these oxygen sites is vacant then the two cations will move further away from each other, while if both are occupied they will tend to be slightly closer than average to compensate. This is shown schematically in Fig. 13.4(a). The vacancy (black circle) near the centre of the figure results in the cations (grey circles) being displaced as indicated by the arrows. The effect is transmitted along the $[1\,1\,0]$ row of cations. Figure 13.4(b) shows schematically the effect of this on the displacements away from the average positions in such a one-dimensional $[1\,1\,0]$ row. The top row of vertical lines indicates the actual position of the cations (large open circles) as a result of being displaced by the vacancies (black circles). Small circles indicate the position of bridging oxygens. The lower row of vertical lines indicates the positions of the average cation lattice, that is, a lattice of regular spacing with the same average spacing as the actual lattice. At the bottom of the figure the shift of the actual lattice relative to the average is indicated by the symbols "−" if the displacement is to the left and "−" if it is to the right. This clearly indicates that near-neighbour cations tend to be shifted in the same direction so that the lattice average $\langle X_i X_{i-n} \rangle$ is positive, while for rather longer vectors the shifts tend to be of opposite sign so that $\langle X_i X_{i-n} \rangle$ is negative. Figure 13.4(c) shows how the summation of Fourier terms resulting from these correlations gives a broad diffuse peak resulting from the near-neighbour positive correlations, with a 'dark line' at the centre resulting from the more distant negative correlations.

This same basic distortion model has been used for all subsequent calculations. Examples of calculated diffraction patterns for the $(h\,k\,\frac{1}{2})$ section are shown in Fig. 13.5 and the characteristic 'dark line' is seen to be common to all of them, as well as to the X-ray pattern (Fig. 13.5(a)). It should be stressed that this 'dark-line' phenomenon is clearly visible only because of the condition that $f_Y \approx f_{Zr}$ and the I_0 and I_1 diffraction terms are absent. In calcia stabilised zirconia, although the diffraction patterns have many similarities, the effect is less marked.

A second feature of all the diffraction patterns shown in Fig. 13.5 is of note. It may be seen that there is a strong asymmetry in the intensity between the high- and low-angle sides of the dark lines. Since the I_0 and I_1 diffraction terms are absent and I_2 involves only cosine modulations, this asymmetry must originate from odd-order terms I_3 and higher. This asymmetry only occurs when the magnitude of the distortion is sufficiently high for these higher order terms to be important. For the distortion model described above this effect requires a difference in the inter-cation spacing, with and without bridging oxygens, of $\sim 3\%$. The same patterns computed with a smaller distortion of $\sim 0.5\%$ show very little asymmetry.

13.3   Vacancy ordering via MC simulation of pair correlations

To test the effects of introducing short-range correlations between oxygen vacancies, a simple pair-interaction model was set up in which a given oxygen site interacts with the five near-neighbour types of site, as shown in Fig. 13.2(d). These different interactions are specified by the parameters $b_1, \ldots, b_5$. In addition to the five interactions shown, all symmetry related vectors are included. Note that b_3 and b_4 are both $\frac{1}{2}\langle 1\,1\,1 \rangle$ vectors but b_3 is the diagonal of a cube containing a cation whereas b_4 is the diagonal of an empty cube.

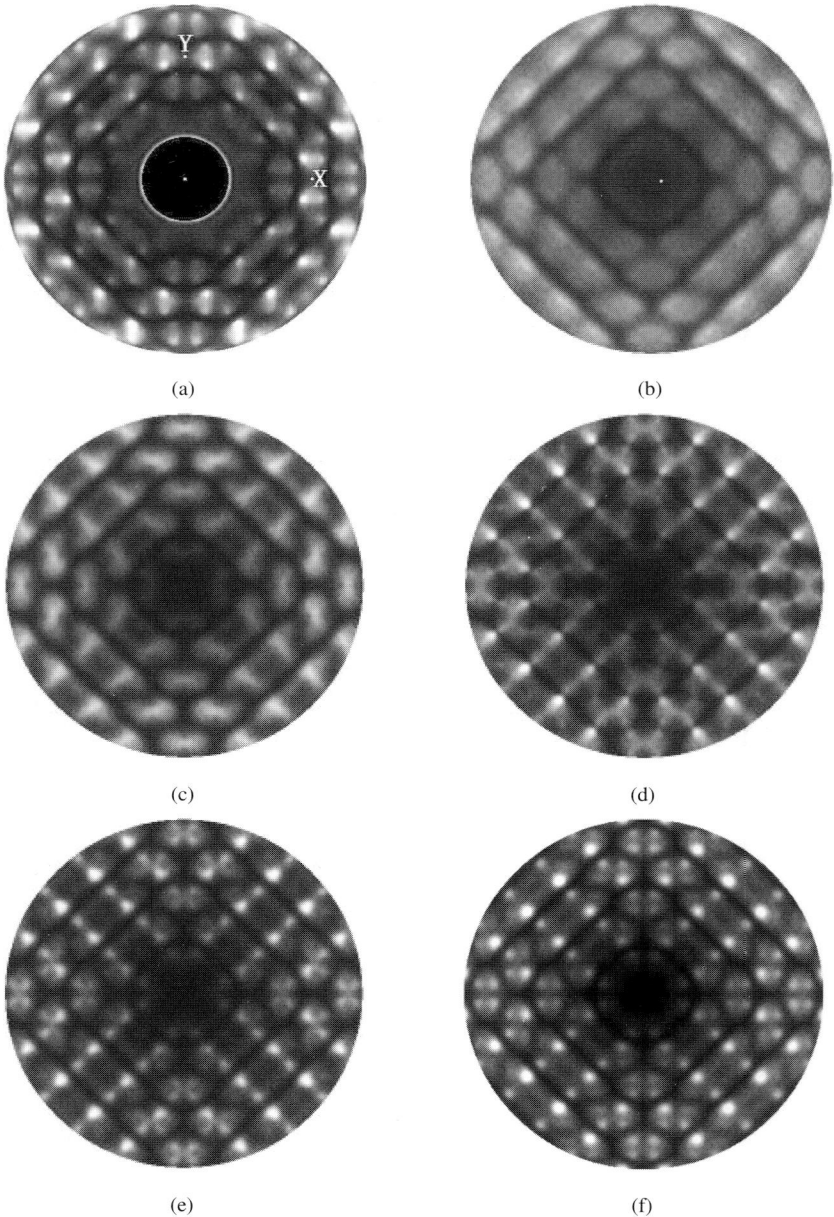

(a)

(b)

(c)

(d)

(e)

(f)

Fig. 13.5 Calculated diffraction patterns for the $(hk\frac{1}{2})$ section of Y-CSZ. (a) shows the observed data for comparison.

Starting with a random distribution of 10% vacancies the effect of adjusting the different b_i can be tested. Fig. 13.5 shows the diffraction pattern of the $(h\,k\,\frac{1}{2})$ section calculated from different realisations. Figure 13.5(a) shows the observed X-ray pattern for comparison. For Fig. 13.5(b) the distribution of oxygen vacancies was purely random. For Fig. 13.5(c) b_1 and b_2 were set to large positive values in order to induce large negative correlations along nearest-neighbour, $\frac{1}{2}\langle 1\,0\,0 \rangle$, next-nearest-neighbour, $\frac{1}{2}\langle 1\,1\,0 \rangle$ vectors. Note that for a concentration of vacancies of 10% the largest negative correlation that can be achieved is -0.11, which corresponds to the total avoidance of vacancy pairs. For Figs. 13.5(d) and (e) b_4 was additionally set to a large positive value in order to induce a large negative correlation along third-nearest-neighbour, $\frac{1}{2}\langle 1\,1\,1 \rangle^{\dagger}$ vectors. Here † is used to denote a vector across a cube of oxygens not occupied by a cation. For Fig. 13.5(d), in order to promote vacancy pairs along $\frac{1}{2}\langle 1\,1\,1 \rangle$ in occupied cubes, a target correlation of 0.35 was set and b_3 was adjusted during iteration to achieve this, but b_5 was zero. For Fig. 13.5(e), the same target correlation of 0.35 was set for $\frac{1}{2}\langle 1\,1\,1 \rangle$ in occupied cubes and b_3 adjusted accordingly, but in addition a target correlation of 0.0 was set for the $\frac{1}{2}\langle 1\,1\,0 \rangle$ vector and b_5 was adjusted to achieve this.

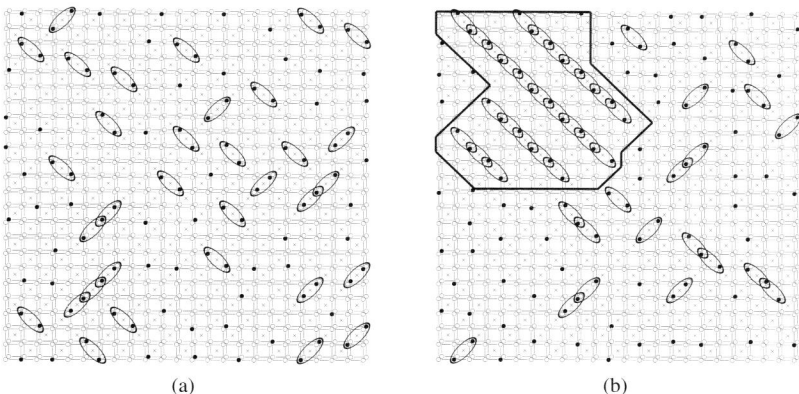

(a)   (b)

Fig. 13.6 Real-space representations of sections of realisations from which diffuse scattering was calculated. (a) Section of realisation used for Fig. 13.5(e). (b) Section of realisation used for Fig. 13.5(d).

Figure 13.6 shows parts of the two realisations corresponding to the examples of Figs. 13.5(d) and (e). In Fig. 13.6(a), corresponding to the diffraction pattern in Fig. 13.5(e), it is seen that the structure consists of mainly isolated vacancy pairs, while in Fig. 13.6(b), corresponding to the diffraction pattern in Fig. 13.5(d), the vacancy pairs tend to be linked into longer chains. In particular the region marked by the heavy line is a crystallite of the pyrochlore structure.

The diffraction pattern in Fig. 13.5(e) qualitatively reproduces all the features in the observed diffraction pattern, although clearly there are quantitative differences. The pattern of Fig. 13.5(d) clearly does not and so this model can be eliminated as a suitable model for the structure. Together, the calculated patterns in Fig. 13.5 serve to illustrate how a model can be progressively explored in order to arrive at a structure which gives a calculated diffraction pattern which agrees with the observed pattern. The model for Fig. 13.5(e) is still relatively simple and appears capable of further refinement. Comparison with the observed pattern reveals that the diffuse peaks which are visible either side of the dark lines differ in their detailed shape. In the X-ray pattern the peaks seem sharper in the middle but with rather broader wings whereas in the calculated pattern the peaks are rather less sharp but with most intensity near the peak. Such qualitative differences should provide guidance on how the model could be developed further, but attempts to do this using a more elaborate pair-interaction scheme have been unsuccessful.

13.4   Multi-site correlations

Since the diffraction pattern of any object is the Fourier transform of the pair-correlation function it might be argued that a model involving only pair-interactions should be all that is required to satisfactorily reproduce the observed diffraction pattern, and terms in the Hamiltonian such as the 3-site and 4-site terms serve no useful purpose. While it is true to say that multi-site correlations do not directly contribute to the diffraction patterns, their effects are felt indirectly, for example in the constraints that are imposed on the 2-site correlations and the way in which these decay with distance. Moreover it was shown in Section 7.5 that multi-site occupational correlations can result in distinctive diffraction effects when additional relaxation displacements are considered.

In the present case, in studying vacancy correlations in CSZs, it is important to point out that consideration of multi-site correlations is necessary in order to generate some of the known superstructures that occur. For example, the $CaZr_4O_9$ structure contains helical clusters, shown in Fig. 13.2(c), in an ordered arrangement in which both left-handed and right-handed helices occur. Since such left- and right-handed arrangements cannot be distinguished by 2-site correlation parameters it is not possible to generate an ordered crystal of this phase using only pair interactions. To define the right-handed helix shown in Fig. 13.2(c), use of the 4-site interaction, $s_i s_{i-n} s_{i-m} s_{i-l}$, where i, $i-n$, $i-m$, $i-l$ define the four vacant sites, would be an obvious choice for an interaction term.

Although the inclusion of such multi-site interactions may ultimately prove to be necessary for a full description of the CSZ systems it represents a level of complexity, both conceptually and in terms of implementation in a model, that it would be preferable to avoid if possible. The question may be asked, therefore, as to whether, by redefining the problem sufficiently, it may be possible to remain with a consideration of 2-site interactions only.

Considering again the helix in Fig. 13.2(c), it is possible to see that what defines the handedness of the helix is the dihedral angle between the $\frac{1}{2}\langle 1\,1\,1 \rangle$ vacancy pairs in the left-most and right-most cubes. Consequently if a vacancy pair is considered as the basic structural unit then each cube (occupied by a cation) can be considered as having a vacancy pair in one of four different orientations, together with the possibility of having no vacancy pair. Thus by replacing the simple binary variables representing single vacancies with 5-state variables representing the position and orientation of vacancy-pairs it is possible to again revert to considering only pair interactions, albeit with a considerable increase in complexity. For Monte Carlo simulation it is not clear that such a change of variables offers a distinct advantage. However, casting the problem in this new way does make it possible to make use of the modulation wave direct synthesis approach (described in Section 5.8) to generate suitable disordered distributions.

13.5 Modulation-wave direct synthesis of vacancy distributions

One of the seeming paradoxes of diffuse scattering in CSZs is that while the X-ray patterns appear complex, electron diffraction patterns appear relatively simple and moreover are basically much the same for a whole range of compositions in many different systems. A typical CSZ electron diffraction pattern is shown in Fig. 13.7(a). The pattern, corresponding to the $(1\,1\,\bar{2})$ zone axis, shows a series of diffuse circles. These are visible both as clearly defined circles but also as pairs of peaks resulting from the intersection of the Ewald sphere with circles which are inclined to the projection axis. All the circles can be shown to occur at positions centred on $\frac{1}{2}\{1\,1\,1\}$ and oriented normal to $\langle 1\,1\,1\rangle$ as shown in Fig. 13.7(b). The main reason that the electron diffraction patterns are so different from the X-ray patterns is that the scattering, being displacive in origin, has strong azimuthal variation. This azimuthal variation is largely removed in electron diffraction as multiple scattering results in intensity being transferred from one region of the pattern to another translated by a whole reciprocal lattice vector. Thus all reciprocal unit cells in Fig. 13.7(a) look essentially the same. On the other hand the pairs of peaks which straddle the dark lines in Fig. 13.5(a) may be recognised as being the same features as the pairs of peaks in Fig. 13.7(a).

As a result of these comparisons it might be supposed that the circular features observed in the electron diffraction pattern are a reflection of the basic compositional ordering of defects that occurs in CSZs and that the X-ray patterns give a detailed picture of how the atoms relax about these defects. To test this, therefore, the modulation wave direct synthesis method can be used to generate distributions of defects and then the same relaxation procedure as before can be applied to obtain the final calculated diffraction pattern.

A particularly simple model is to suppose that each of the four different orientations of diffuse ring shown in Fig. 13.7(b) is due to the distribution of defects ($\frac{1}{2}\langle 1\,1\,1\rangle$ vacancy pairs) in the particular $\{1\,1\,1\}$ plane normal to the $\langle 1\,1\,1\rangle$ vector which is bisected by the ring. That is, the plane is considered which consists of the triangular mesh shown in heavy lines in Fig. 13.7(c). This links all the vacancy pairs, which are drawn as filled circles. Wave-vectors $\mathbf{q}$ corresponding to points on one of the diffuse circles may then be used to carry out a direct synthesis of the distribution of vacancy pairs of the corresponding orientation. Since 10% vacancies are required in total only 2.5% are required for one orientation, and for the total synthesis the process is repeated using diffuse circles, planes and vacancy-pair orientations in each of the other three $\{1\,1\,1\}$ orientations. One plane from the resulting 3D synthesis is shown in Fig. 13.7(d). Here the black dots indicate the positions of the vacancy pairs and the larger circles drawn around most of

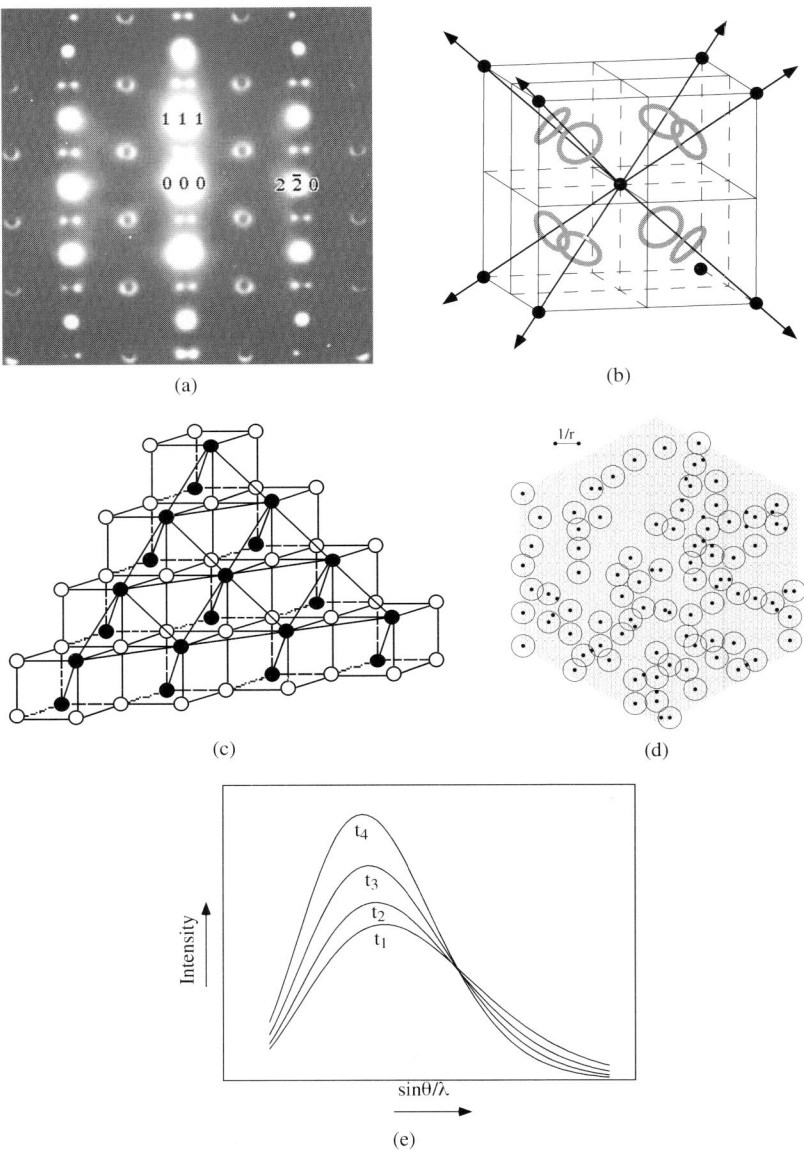

Fig. 13.7 Modulation-wave synthesis. (a) A typical CSZ electron diffraction photograph. (b) Schematic diagram showing the positions of the diffuse rings in reciprocal-space. (c) The (111) layer in the CSZ structure. (d) A realisation of one of the {1 1 1} layers from the synthesised distribution containing 2.5% vacancy pairs. The large circles have a diameter corresponding to the reciprocal of the radius of the diffuse circles in reciprocal space. (e) Schematic drawing of the development of radius of the diffuse peaks with time.

the vacancies have a diameter corresponding to the reciprocal of the radius of the diffuse circle.

For simplicity these four separate syntheses are quite independent, but with such a low concentration of defects, the chance of more than one vacancy pair occurring in a given cube of oxygen sites is very low ($\sim 0.06\%$), so that the coordination of the cation site is only very rarely less than a chemically plausible six. A calculated diffraction pattern for the $(h\,k\,\frac{1}{2})$ section obtain from this kind of synthesised distribution is shown in Fig. 13.5(f). The pattern (and ones calculated for other reciprocal sections, not shown here) is in excellent (qualitative) agreement with the observed pattern. By adjusting the detailed distribution of the modulation amplitudes as a function of the radius r it is likely that the agreement could be further improved, although this has not been attempted. The extreme simplicity of the model, as described, allows new insight into the origins of the disorder to be gained, and it is doubtful whether this would be enhanced by a more quantitative fit.

Inspection of the distribution of defects in the plane normal to the $\langle 1\,1\,1\rangle$ shown in Fig. 13.7(d) provides a clue to a possible mechanism for the disorder. Around each defect (except in a few places where two defects occur very close together which may be attributed to approximations inherent in the synthesis method) circles have been drawn with a diameter equal to the reciprocal of the radius of the diffuse circle used in the synthesis. It is seen that these circles tend to be predominantly in close contact with each other but with very little overlap. That is, the figure is very suggestive that around each defect there is a zone of exclusion where it is energetically unfavourable for another defect to occur, but that at a larger distance of, $\sim 1/r$, it becomes energetically favourable again. The whole plane appears to consist of two types of region, one defect free and the other consisting of defects closely packed with a mean separation of $\sim 1/r$. It should be noted that, in an equilibrium situation, if it were simply that defects tended to repel each other it might be expected that the distribution would be more like that of a liquid with a mean inter-defect spacing defined by the concentration. The tendency for defects to avoid each other may be understood in terms of the strain field that will exist around the defect, and which will require a certain distance to dissipate, but the tendency to cluster at a preferred distance of $\sim 1/r$ is not so easy to understand. These ideas are developed further in Chapter 17.

Such phenomena have been explained by Cahn (1967), in the case of alloys, in terms of spinodal decomposition. This occurs when a system is placed in a part of the phase diagram where a homogeneous single phase mixture is unstable and unmixing tries to occur. The description requires the solution of a diffusion equation involving thermodynamic, elastic and interface energy parameters. The outcome, however, is that compositional modulations develop and, as a function of time, the distribution of the wavelengths of these modulations shows a trend towards quite a narrow spread centred around a dominant wavelength (see Fig. 13.7(e)). Although in elastically anisotropic materials the wavelength may be orientation dependent, for cubic materials such as CSZs the wavelength may be expected to be orientation independent, hence producing the uniform 'ring' of scattering.

14

AUTOMATIC REFINEMENT OF A MONTE CARLO MODEL

14.1 Introduction

Monte Carlo (MC) simulation of a computer model has been used to aid in the interpretation of observed diffuse X-ray scattering patterns in numerous studies of disordered crystals, including a number of examples already discussed in earlier chapters of this book (see Welberry *et al.*, 1993*a,b*; Welberry and Glazer, 1994; Welberry and Mayo, 1996; Welberry and Christy, 1997). The same basic procedure is adopted in all cases. First a model is set up in the computer in terms of sets of random variables representing the atomic occupancies and positions or molecular orientations. A relatively small number of energy parameters are used to define the way in which these atoms or molecules interact. MC simulation is then carried out for a time sufficient to allow the system to reach (or at least closely approach) equilibrium. The final atomic coordinates of this model crystal realisation are used to calculate diffraction patterns, which may then be compared to the observed X-ray patterns. After assessing the points of agreement and disagreement from this comparison the model parameters are adjusted and the whole process is repeated iteratively until a satisfactory agreement between observed and calculated patterns is obtained.

Although convincing results have been obtained by this method for a variety of quite different systems, the crucial step of comparing the observed and calculated patterns has, until relatively recently, been performed visually and decisions regarding the adjustment of the system parameters relied heavily on an accumulation of experience, gained over a number of years. In this chapter a description is given of the first attempt to perform this iterative MC procedure solely by computer, using quantitative rather than visual comparison of observed and calculated diffraction patterns and with automatic updating of the model parameters, using a least-squares algorithm.

It should be stressed at the outset that this represents a formidable computational task. At each stage of iteration, complete MC simulations are carried out for the current set of model parameters, diffraction patterns are calculated and the goodness-of-fit parameter, χ^2, is obtained as a quantitative measure of the agreement with the observed data. In addition, complete MC simulations, together with accompanying calculations of their diffraction patterns, are carried out for sets of parameters in which each of the parameters has, in turn, been changed by a small amount. From these, numerical estimates of the differentials of χ^2 with respect to each of the system variables are obtained, and are used to form the least-squares matrix, $\mathbf{A}$ (see later). Although computer speeds have increased rapidly over recent years, it is still not feasible to perform such a calculation using extensive three dimensional diffuse scattering data nor using models which represent the real-space structure in as much detail as would ideally be desired. In order to make any progress a number of approximations have to be made, and these are discussed below.

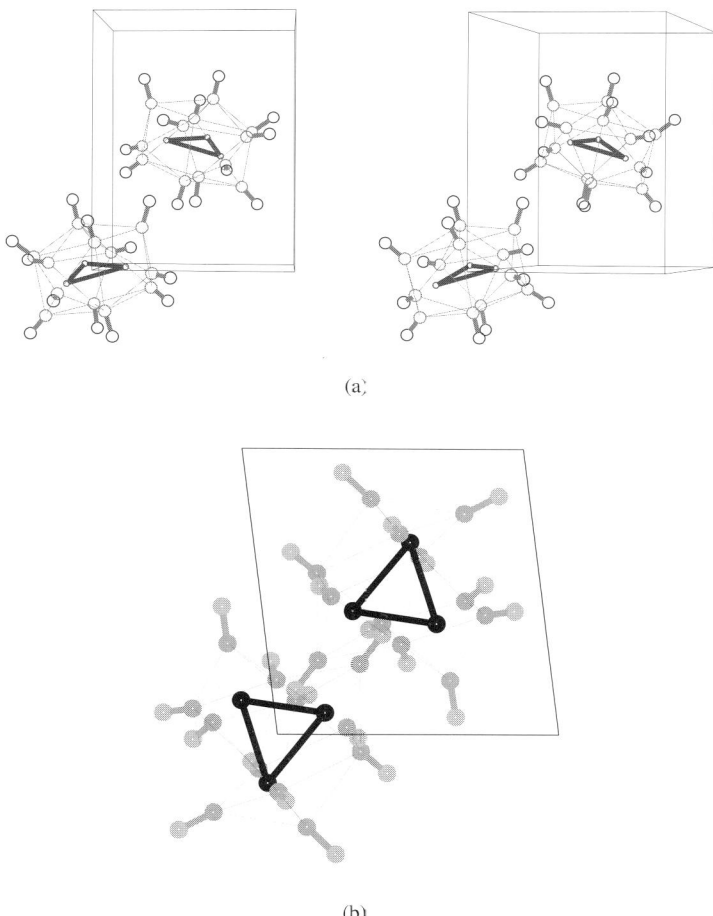

(a)

(b)

Fig. 14.1 Structure of $Fe_3(CO)_{12}$ (a) Stereoscopic view down **c** (b) Projection down **b**. Only one orientation of the Fe_3 triangles is shown.

The system chosen for this first study was triirondodecacarbonyl, $Fe_3(CO)_{12}$. The molecule consists of a triangle of iron atoms, surrounded by an approximately icosahedral cage of coordinating carbonyl (CO) groups (see Fig. 14.1). Crystal structure determination (Cotton and Troup, 1975; Braga *et al.*, 1994) revealed that the molecular sites are disordered but can be well modelled by assuming two different orientations of the same molecular geometry, related by a centre of symmetry. Cell data and atomic coordinates for the two orientations of the Fe_3 triangle are given in Table 14.1. Although not previously reported, the presence of strong diffuse scattering was established in initial experiments and the rather simple form of these diffraction patterns seemed to indicate that a relatively simple model should be capable of describing them.

Table 14.1 Fractional Fe coordinates for the two alternative orientations of the Fe_3 triangle in $Fe_3(CO)_{12}$. The space group is $P2_1/n$ with $a = 8.359$ Å, $b = 11.309$ Å, $c = 8.862$ Å, $\beta = 97.0°$ (see Cotton and Troup, 1975).

Atom	Orientation (+)			Orientation (−)		
	x	y	z	x	y	z
Fe(1)	−0.0544	−0.0204	−0.1684	0.0544	0.0204	0.1684
Fe(2)	0.1735	−0.0162	0.0716	−0.1735	0.0162	−0.0716
Fe(3)	−0.1123	0.0476	0.1100	0.1123	−0.0476	−0.1100

14.2   X-ray diffuse scattering data

Data for three reciprocal sections were recorded using the PSD diffractometer system (see Section 1.2.1). These were $(0\,k\,l)$, $(h\,k\,\bar{h})$ and $(h\,\frac{1}{2}\,l)$. Initial investigations were carried out using only the $(0\,k\,l)$, $(h\,k\,\bar{h})$ data but subsequently the $(h\,\frac{1}{2}\,l)$ data were added when it became apparent that two sections were insufficient to obtain an unambiguous solution. In the latter case the $(h\,\frac{1}{2}\,l)$ section was used instead of the $(h\,0\,l)$ section since most of the diffuse scattering for the $[0\,1\,0]$ zone axis lies between the Bragg layers. Figure 14.2 shows plots of the diffuse scattering data for the three sections used.

Complete reciprocal sections of data such as those displayed in Fig. 14.2 contain $\sim 400 \times 400$ pixels. This scale corresponds approximately to that at which an individual pixel matches the resolution of the observed X-ray data. However, this number of data far exceeds that which could feasibly be handled for the envisaged fitting process. One possible way of reducing the number of data would have been to re-bin the data to a lower resolution. However, it was decided instead to retain the spatial resolution and to use only selected subsections of the data in the actual fitting process. The areas chosen in each of the observed sections were those regions where the diffuse scattering was strongest and most clearly delineated. The regions of data actually used are indicated in Fig. 14.2 by the white rectangles. A small region around each Bragg peak was also excluded from the calculations.

14.3   Monte Carlo model

Of crucial importance in this process is the need to have a model system of sufficient size that statistical variations are small and the resulting calculated patterns are relatively noise-free and of a comparable quality to the observed data. Experience gained from numerous earlier studies is that a minimum system size of $\sim 32 \times 32 \times 32$ unit cells

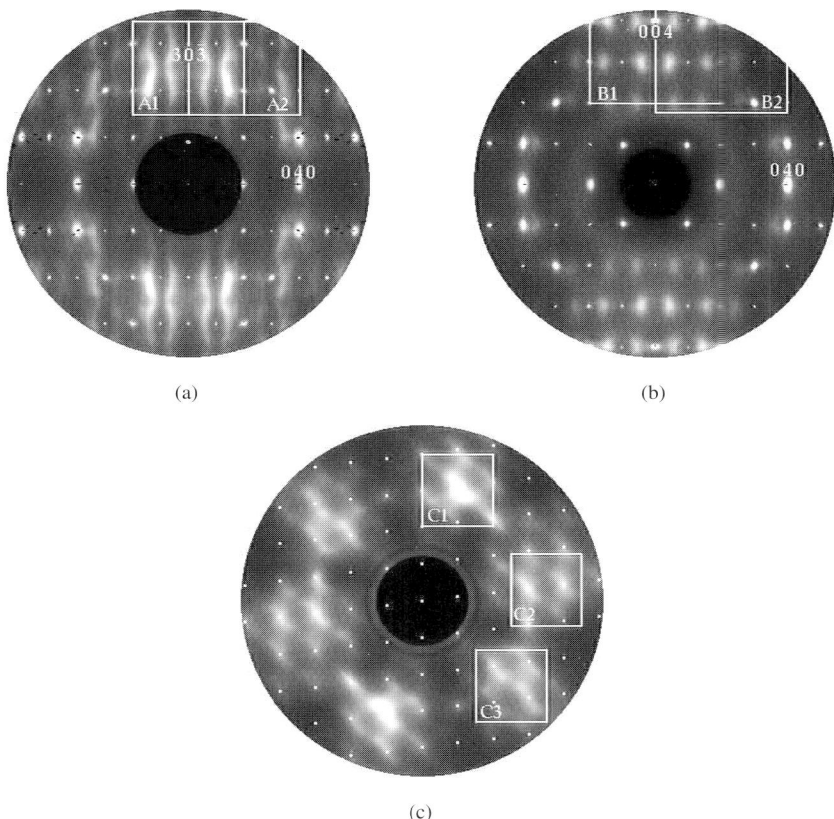

Fig. 14.2 Observed X-ray diffraction patterns of $Fe(CO)_{12}$. (a) $(hk\bar{h})$ section. b) $(0kl)$ section. (c) $(h\frac{1}{2}l)$ section. The white rectangles indicate the regions of the data actually used for fitting (see text for details). In (c) the white dots have been inserted to mark the positions of integral h and l.

is required and this was adopted throughout the study (e.g. see Welberry and Proffen, 1998). Binary $(+1, -1)$ random variables, $\sigma_{i,j,k,l} = \pm 1$, were used to represent the two molecular orientations that are assumed to occur at each site i, j, k, l. Here i, j, k specify the unit cell and l specifies one or other of the two sites within each cell. Hereafter the two different orientations are referred to as '+' and '−'. In the first stages of the study the model used only this set of binary occupancy variables with the molecules confined to the average positions. Subsequently further (continuous) random variables were added to the model. The variables $X_{i,j,k,l}$, $Y_{i,j,k,l}$ and $Z_{i,j,k,l}$ described rigid-body centre of mass translations of the whole molecules, while $\varphi_{i,j,k,l}$ were used to describe molecular librations.

14.3.1   *Ordering of $(+/-)$ orientations*

The MC energy used to specify local ordering of the molecular orientations was of the Ising model form,

$$E_1 = \sum_n a_n \sigma_{i,j,k,l} \sigma_{i_n,j_n,k_n,l_n}. \tag{14.1}$$

Here the summation is over all n-types of neighbour of a particular site. In this case eight neighbouring vectors and their symmetry equivalents were included. Details of these vectors are give in Table 14.2 and Fig. 14.3. Unique values of the parameters a_n are required for these eight vectors which comprise four vectors between nearest-neighbours along each type of body diagonal, three vectors between next-nearest-neighbours in each of the axial directions and one additional vector between third-nearest-neighbours in the $[1\,0\,1]$ direction.

Table 14.2   Definition of the eight intermolecular vectors used in the analysis. See also Fig. 14.3.

Contact Type	Vectors to centre of neighbouring molecules from central molecule at (000)	
1	$\frac{1}{2}[\bar{1}\,\bar{1}\,\bar{1}]$	$\frac{1}{2}[\bar{1}\,1\,\bar{1}]$
2	$\frac{1}{2}[1\,\bar{1}\,1]$	$\frac{1}{2}[1\,1\,1]$
3	$[1\,0\,0]$	$[\bar{1}\,0\,0]$
4	$\frac{1}{2}[1\,\bar{1}\,\bar{1}]$	$\frac{1}{2}[1\,1\,\bar{1}]$
5	$\frac{1}{2}[\bar{1}\,\bar{1}\,1]$	$\frac{1}{2}[\bar{1}\,1\,1]$
6	$[0\,0\,1]$	$[0\,0\,\bar{1}]$
7	$[0\,1\,0]$	$[0\,\bar{1}\,0]$
8	$[1\,0\,1]$	$[\bar{1}\,0\,\bar{1}]$

The starting configuration was a random distribution in which each of the $\sigma_{i,j,k,l}$ variables was either $+1$ or -1 with equal probability. Hence $\langle \sigma_{i,j,k,l} \rangle$ was zero.

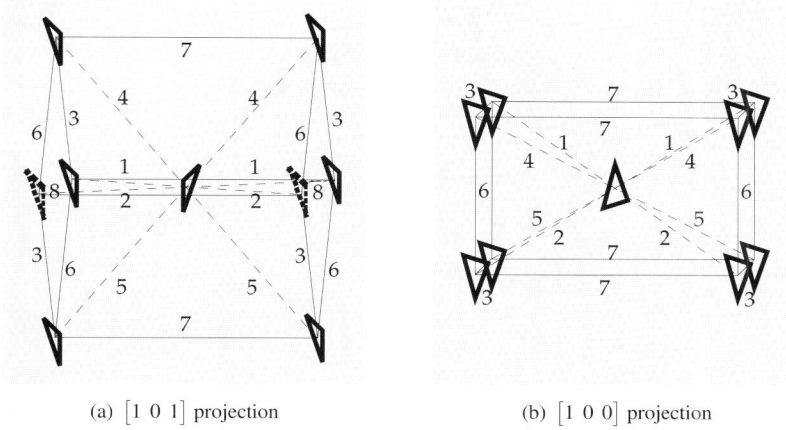

(a) $[1\ 0\ 1]$ projection   (b) $[1\ 0\ 0]$ projection

Fig. 14.3   Definition of vectors linking near-neighbour Fe_3 sites. See also Table 14.2.

The MC iteration was carried out by choosing a pair of sites at random and interchanging the variables in the two sites. In this way the overall equal concentration of the two molecular orientations was maintained.

For use in the iterative fitting process described later it would have been possible to use the interaction energies, a_n, as the system parameters. However, it has been a frequent practice in previous studies to use a feed-back mechanism in the MC runs in order to obtain specified values of correlations along given interaction vectors (see Welberry *et al.*, 1993a). Thus, rather than the primary quantities a_n, actual correlation values $c_n = \left\langle \sigma_{i,j,k,l} \sigma_{i_n,j_n,k_n,l_n} \right\rangle$ were used as system parameters. A given MC simulation therefore has as input the set of target correlations c_n and the corresponding set of energies, a_n, is determined during the course of the iteration. For all MC occupancy simulations 50 cycles of iteration were used. A cycle is defined as usual as that number of individual MC steps required to visit each site once on average (i.e. in the present case $32 \times 32 \times 32 \times 2$).

14.3.2   *Centre of mass displacements*

Any local ordering of site occupancies such as that described above is invariably accompanied by local 'size-effect' relaxation displacements which depend on the particular type of intermolecular contact, that is, whether neighbouring sites are occupied by molecules in orientations $(++)$, $(+-)$, $(-+)$ or $(--)$. Though such 'size-effect' displacements are generally small they can contribute substantially to the diffuse scattering signal. In the present case the system is not simply comprised of single atoms but of a complete molecular shape. Even if the molecule is assumed to be rigid, local relaxation may involve not only a shift of the centre of mass, but also a rotation away from its average orientation.

In its most general form such a relaxation would require the use of many more parameters than is feasible with the computational resources available. In the present context some quite drastic simplifications needed to be made. Consequently, the motion of each molecule was restricted to consist simply of a rigid-body translation of the centre of mass together with a single librational degree of freedom. In this way it was possible to limit the number of parameters required to define the relaxation model in the MC simulation to manageable proportions.

For the centre of mass shift an energy of the form,

$$E_2 = \sum_{n,m} \left(d_{n,m} - d_{\text{ave}} (1 + \varepsilon_{n,m}) \right)^2 \qquad (14.2)$$

was used. Here, as before, the summation is over intermolecular vectors from a molecule at site n to its neighbour at site m. In this case only the four shortest vectors along which molecules are physically in contact were considered (vectors 1, 2, 4 and 5 in Table 14.2). $d_{n,m}$ is the instantaneous value of the distance between the centres of mass and d_{ave} is the corresponding equilibrium distance in the average structure. $\varepsilon_{n,m}$ is a 'size-effect' parameter which takes different values according to whether the two sites joined by the vector are $(++)$, $(-+)$, $(+-)$ or $(--)$. For each vector, $\varepsilon_{n,m}$ is specified by two variable parameters g_m and u_m,

$$\varepsilon_{n,m}(++) = g_m + u_m$$
$$\varepsilon_{n,m}(+-) = \varepsilon_{n,m}(-+) = -g_m \qquad (14.3)$$
$$\varepsilon_{n,m}(--) = g_m - u_m.$$

Though four different combinations of neighbour exist only two variables are needed since it was assumed that $(+-)$ and $(-+)$ are equivalent and the overall sum of shifts must be zero. The form of eqn (14.3) allows for both symmetric (g_m) and antisymmetric (u_m) distortions.

14.3.3 *Orientational relaxation*

The analysis of the thermal ellipsoid tensors carried out by Braga *et al.* (1994) indicated that the Fe_3 triangles appeared to be undergoing librations about an axis parallel to the pseudo two-fold axis of the triangle passing through $Fe(1)$. Consequently it was assumed that a rotation about the pseudo two-fold axis of the triangle was a reasonable one. Such a model could be tested against alternative models.

If the random variable $\varphi_{i,j,k,l}$ is used to represent the angular variation about the pseudo two-fold axis an expression for the way in which the rotations of neighbouring molecules are linked by a size-effect like expression analogous to that for the centre of mass shifts given by eqn (14.2) above, may be formulated as,

$$E_3 = \sum_{n,m} \left(\Delta \varphi_{n,m} - \xi_{n,m} \right)^2. \qquad (14.4)$$

Here $\Delta \varphi_{n,m}$ is the difference in the value of the orientational variable at site n and that at a neighbouring site m. $\xi_{n,m}$ is a 'size-effect'-like parameter which takes different values according to whether the two sites joined by the vector are $(++)$, $(+-)$, $(-+)$ or $(--)$. As for the centre of mass 'size-effect' (see eqn (14.3)) this is allowed to have symmetric and antisymmetric components:-

$$\xi_{n,m}(++) = \gamma_m + \nu_m$$
$$\xi_{n,m}(+-) = \xi_{n,m}(-+) = -\gamma_m \qquad (14.5)$$
$$\xi_{n,m}(--) = \gamma_m - \nu_m.$$

This formulation thus requires two independent parameters, γ_m and ν_n, in addition to the centre of mass size-effect parameters, g_m and u_m, for each intermolecular vector. In the final model used in the study such orientational parameters were used on only the intermolecular vectors 1 and 2 given in Table 14.2. Although it was considered that further improvement of the fit of the model to the observed data could most likely be obtained by including similar parameters for the vectors 4 and 5, this would have involved a further 50% increase in computation time for each cycle of refinement and has not been pursued.

14.3.4 *Refinement procedure*

The relaxation of the molecular centres of mass and orientations, as described above, was applied subsequent to the occupational ordering and used a second stage of MC simulation. For this part of the simulation a further 50 cycles of iteration were carried out in which the values of the continuous random variables $X_{i,j,k,l}$, $Y_{i,j,k,l}$, $Z_{i,j,k,l}$ and $\varphi_{i,j,k,l}$ were adjusted using the system energy $E = E_2 + E_3$ while the occupancy variables, $\sigma_{i,j,k,l}$, remained constant.

14.4   Calculation of diffraction patterns

The calculation time for obtaining diffraction patterns from the MC models is proportional both to the number of atoms in real-space as well as the number of points in reciprocal-space at which calculations are performed. In the present case it was not feasible to use either the full complement of atoms nor to utilise the full set of recorded data, and some quite drastic approximations had to be made.

For the real-space part of the problem, to use all 27 atoms $(3Fe + 12C + 12O)$ in each molecule represents a nine times more expensive calculation than to use just the 3 Fe atoms, and throughout the present work only the Fe atoms were used. At first sight this may appear a rather drastic approximation. It may, however, be justified by the following considerations. First, since the atomic scattering factors for Fe are much larger than those for C and O, the scattering from the Fe atoms will tend to dominate. Second, since the lighter atoms are distributed throughout the unit cell rather than being concentrated in specific positions, their contribution would tend not to produce the rather simple and distinctive diffuse scattering patterns that are observed.

Two distinct situations were considered possible:

1. The whole molecule (Fe's and ligands) behaved as a rigid molecule. In this case leaving out the C's and O's would affect only the overall molecular structure

factor contribution to the scattering and not the detail of intensity variations within a reciprocal unit cell.

2. The behaviour of the CO ligands was not related in any simple way to the behaviour of the Fe's. If this were the case the motion of any one ligand might similarly not be related in any simple way to any other. It then might be expected that, with this multiplicity of inter- and intra- CO to CO vectors, both in terms of distance and mutual orientation, these would contribute rather broad featureless diffuse intensity quite unrelated to the Fe scattering.

In either case it seemed likely that using only the Fe's to calculate the diffraction patterns to be fitted to the observed data should provide at least a good first approximation to the problem and at least give some indication as to whether 1 or 2 above pertained.

14.5 Least-squares

The basic least-squares method seeks to minimise the sum of squares of the differences between a set of observed and calculated quantities. In the present case the set of observed quantities consists of diffuse scattering intensities measured at individual pixels in the diffuse sections shown in Fig. 14.2. The corresponding calculated quantities are the suitably scaled values of the intensity obtained at corresponding points in reciprocal-space from an MC simulation of a model system having system parameters p_i. That is the goodness-of-fit χ^2 is minimised, where

$$\chi^2 = \sum_{h,k,l,m} \omega_{hklm}(\Delta I)^2 \tag{14.6}$$

and,

$$\Delta I = I_{\text{obs}} - \left(b_m + f_m I_{\text{obs}}\right). \tag{14.7}$$

Here the summation is over all are non-integral reciprocal points h, k, l corresponding to individual pixels in the m measured sections of data. f_m is a scale and b_m a background correction applied to section m. (Note, b_m and f_m are determined directly, as described by Proffen and Welberry, 1997, and are not included as parameters in the least-squares matrix). ω_{hklm} is the weight for the corresponding data point h, k, l of data plane m. The weights used in the work described here were taken as $\omega_{hklm} = 1/I_{\text{obs}}$.

Increments Δp_i to be applied to the model parameters are calculated using the following expression

$$\Delta p_i = \sum_{l=1}^{\text{npar}} \mathbf{A}_{il}^{-1} \mathbf{B}_l. \tag{14.8}$$

The matrix $\mathbf{A}$ and the vector $\mathbf{B}$ involve the differentials of ΔI with respect to each of the variables p_i. $\mathbf{A}$ is a symmetric matrix and $\mathbf{B}$ is a vector,

$$\mathbf{A}_{ij} = \sum_{hklm} \omega_{hklm} \frac{\partial \Delta I}{\partial p_i} \frac{\partial \Delta I}{\partial p_j} \tag{14.9}$$

$$\mathbf{B}_i = \sum_{hklm} \omega_{hklm} \Delta I_{\text{trial}} \frac{\partial \Delta I}{\partial p_i}. \tag{14.10}$$

It is also convenient to define a correlation matrix as a measure of how dependent the parameters used in the least-squares process are with each other. The correlation matrix $\mathbf{C}$ is defined as

$$\mathbf{C}_{ij} = \frac{\mathbf{A}_{ij}^{-1}}{\sqrt{\mathbf{A}_{ii}^{-1} \mathbf{A}_{jj}^{-1}}}.$$
(14.11)

Note that these correlations are not to be confused with the occupational correlations present in the disordered structure.

14.6 Estimation of the differentials, $\partial \Delta I / \partial p_i$

The differentials are computed as follows. If $\mathbf{p} = (p_0, p_1, p_2, p_3, \dots p_i, \dots, p_n)$ is the current set of system parameters, the differential is estimated by performing two complete MC simulation and diffraction pattern calculations using parameter sets, $\mathbf{p}^+ = (p_0, p_1, p_2, p_3, \dots p_i + \delta_i, \dots p_n)$ and $\mathbf{p}^- = (p_0, p_1, p_2, p_3, \dots p_i - \delta_i, \dots p_n)$, where δ_i is a suitably chosen small increment. The differential is then taken as

$$\frac{\partial \Delta I}{\partial p_i} = \sum_{hklm} \frac{(\Delta I_{p+} - \Delta I_{p-})}{2\delta_i}.$$
(14.12)

If the calculated diffraction patterns were infinitely accurate then it would be best to choose δ_i as small as possible. However, each calculation of ΔI is only an approximation to the true value corresponding to a given model parameter set. Both the MC simulation itself and the ensuing diffraction pattern calculation result in inaccuracies. The size of the model crystal is one limiting factor. For a chosen system size of $32 \times 32 \times 32 \times 2$ molecular sites the normal statistical variations in the MC simulation lead to lattice averages such as the correlation coefficients c_n having an accuracy of $\sim 1/\sqrt{32 \times 32 \times 32 \times 2} \approx 0.004$. Though using a larger system size would improve this accuracy a corresponding increase in computer time (and memory) would be incurred.

Calculation of the diffraction pattern from the results of the MC simulation is performed using the program *DIFFUSE* (Butler and Welberry, 1992). The diffracted intensity from a large simulation array is computed as the average of the intensity from a large number of small sub-regions (or lots) of the model crystal, chosen at random. The purpose of performing the calculation in this way is to smooth out high-frequency variations in the pattern. The size of an individual lot needs to be sufficient to include all significant non-zero correlations. Optimal results are obtained when every point in the array is sampled ~ 1–2 times on average. Further details of this sampling and the effect on the quality of the diffraction pattern can be found in Welberry and Proffen (1998). In the present case an ellipsoidal lot-size extending over $5 \times 5 \times 5$ unit cells was used throughout. That is, each lot included $\sim \frac{4}{3}\pi \times 2.5^3 \simeq 65$ cells. In order to completely cover the whole array once on average ~ 500 lots are required. Most of the work presented here was carried out with this degree of coverage. For final calculations the number of lots was increased to 900, providing some improvement of quality at the expense of proportionately more computer time. In order to obtain still higher quality patterns it would be necessary to increase the MC simulation array size, or average over a number of separate runs.

Initial experiments involving only occupancy correlation variables were carried out in which the value of δ_i was chosen to be 0.025 (this was considered the lower limit of what would provide a significant difference between MC simulations with $\mathbf{p}^+$ and $\mathbf{p}^-$). However, it was found that this value of δ_i lead to rather poor estimates of the differentials and the matrix $\mathbf{A}$ was clearly badly determined, leading to poor convergence of the least-squares process. For later runs a value of $\delta_i = 0.05$ was adopted and these gave much better convergence.

For the later computer runs in which centre of mass displacements were used the value of δ_i was chosen to be 0.03, corresponding to a 3% change in the intermolecular distances. This value is comparable to the magnitude of size-effect shifts found in other studies. For these later stages of the study, the length of time taken for the MC calculation of each cycle was such as to preclude a great deal of further experimentation, and since this value appeared to yield satisfactory results, it was retained for the rest of the study. For the runs in which the angular variables, $\varphi_{i,j,k,l}$, were used δ_i for these variables was chosen to be 0.01 (in radians). This value gives rise to atomic shifts comparable to those resulting from the centre of mass displacements, and for similar reasons this value also was retained for the rest of the study. Figure 14.4(a) shows an example of the difference between the diffraction patterns of two MC simulations for $\mathbf{p}^+$ and $\mathbf{p}^-$ where the variable changed is one of the size-effect parameters u_m. Figure 14.4(b) shows the progress of refinement for one section of data.

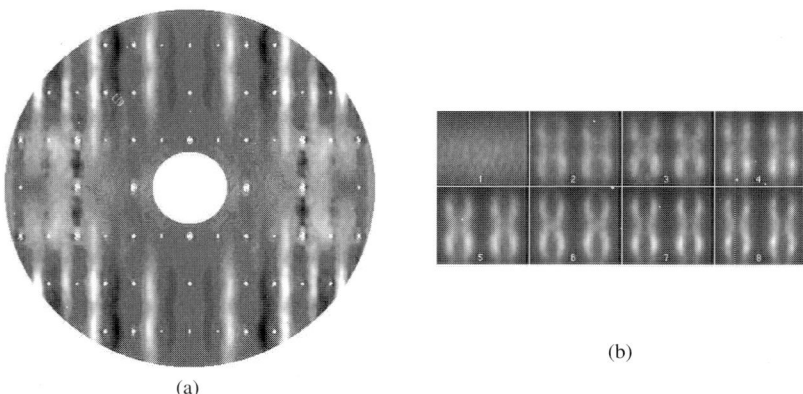

(b)

(a)

Fig. 14.4 (a) An example of the difference $I_{p+} - I_{p-}$ (b) Showing the progress of refinement for one section.

14.7   Progress of refinement

Initial refinements were carried out using only A1 and B1 regions from the $(h\,k\,\bar{h})$ and $(0\,k\,l)$ data and only occupancy correlation parameters along the vectors 1 to 7, since it was thought that this would be sufficient to obtain a solution. It became apparent that, although good fits to the restricted data sets could be obtained, the solutions for the correlation parameters varied considerably depending on the starting configuration. In addition it was found that these solutions gave calculated values for the $(h \frac{1}{2} l)$ section pattern quite unlike the observations (see Fig. 14.5). At this point the correlation matrix (eqn (14.11)) clearly indicated that some of the parameters were highly correlated. Consequently it became necessary to include in the refinement regions C1, C2 and C3 from the $(h \frac{1}{2} l)$ data in order to break these dependencies. Further attempts at refinement then indicated that the initial 7 parameters were insufficient to provide a fit to the observations and the interaction vector 8 was added. Refinement then proceeded satisfactorily to give a solution for the occupancy correlations. Calculated patterns for the complete reciprocal sections for this solution are shown in the left-hand column of Fig. 14.6.

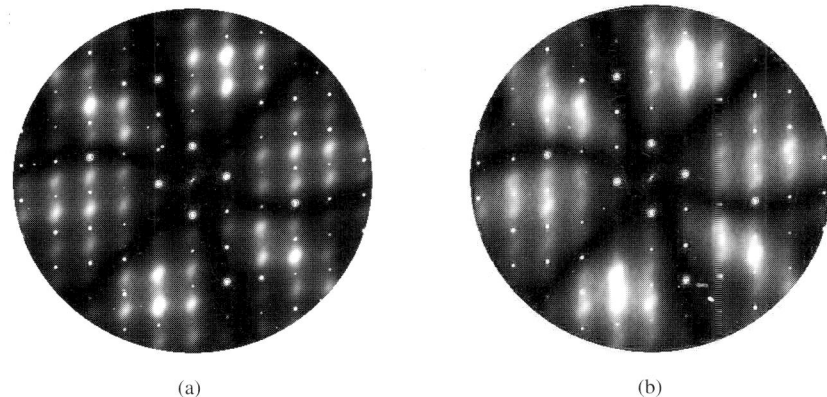

(a)   (b)

Fig. 14.5 $(h \frac{1}{2} l)$ patterns calculated from two different solutions which give good agreement for the $(h k \bar{h})$ and $(0 k l)$ sections. [Note the $(h 0 l)$ Bragg peaks have been superimposed on the plots in order to show the reciprocal lattice positions].

The regions of data A1 and B1 were replaced by the regions A2 and B2 and then refinement of the centre-of-mass and orientational relaxation parameters was carried out while keeping the occupational correlation parameters constant. Note that these new regions include part of the pattern that shows strong asymmetry in the diffuse bands of intensity seen in Fig. 14.2. Four relaxational parameters $(g_m, u_m, \gamma_m, v_m)$ on each of the nearest-neighbour vectors 1 and 2 were included. Computational resources precluded the possibility of including any more parameters. This lead to a final solution in which 16 parameters had been used (eight occupancy correlations and eight relaxational parameters). Calculated patterns for the complete reciprocal sections for this final solution are shown in the right-hand column of Fig. 14.6. Note the improvement over the occupancy-only fit, particularly in terms of the relative intensity of the different fringes labelled 1...6 in Figs. 14.6(a) and (b). The final R factor for $\sim 40,000$ data points was $\sim 21\%$, where $R = (\chi^2 / \sum \omega I_{obs}^2)^{1/2}$.

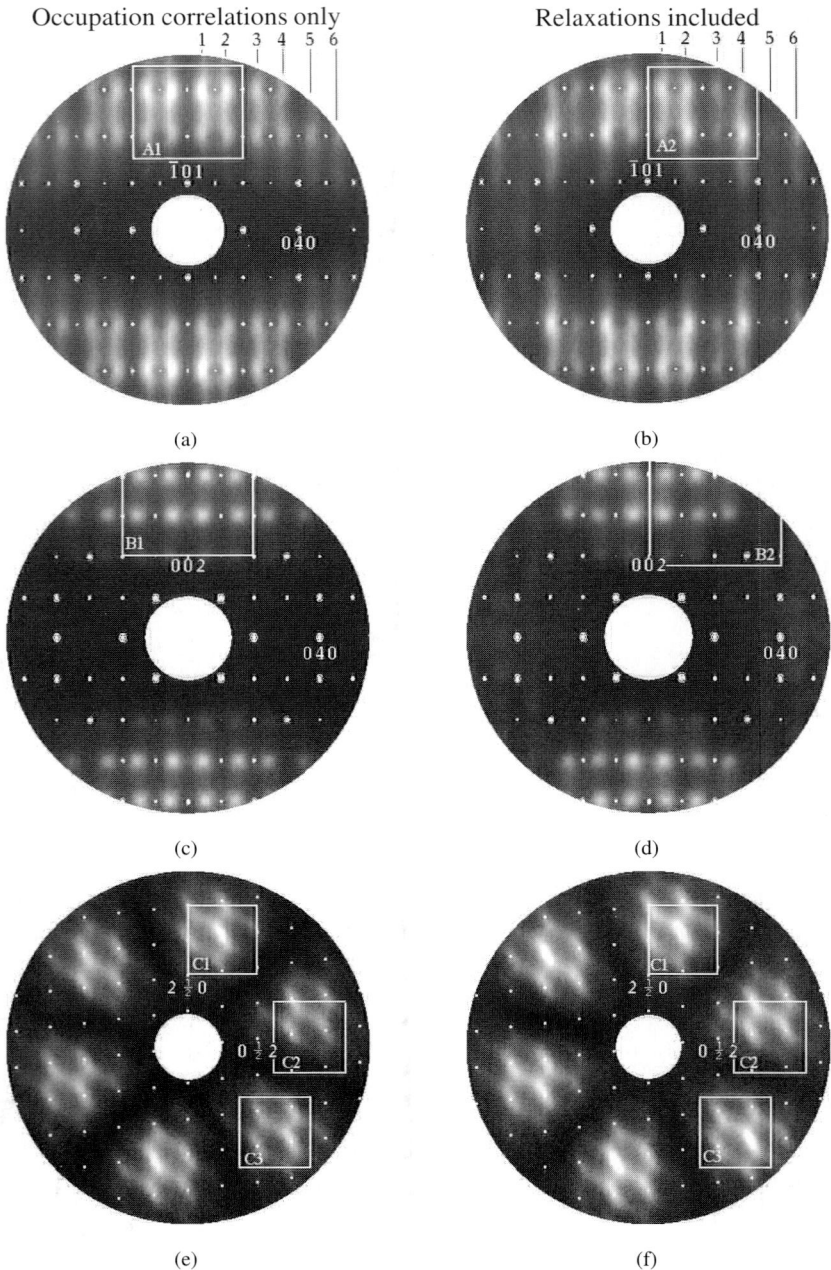

Fig. 14.6 Calculated patterns from final model. The left-hand column, (a), (c), (e), contain occupancy correlations only, whereas the right-hand column, (b), (d), (f) includes displacive relaxations.

14.8   Discussion of solution

In order to be able to visualise exactly what this local relaxation means it is convenient to make some statistical plots of the positions and orientations of the molecules averaged over the whole simulation. Figure 14.7 shows such a probability plot for all of the molecules (both $(+)$ and $(-)$) in the structure. The solid triangles represent the positions of the average coordinates obtained from the crystal structure determination. This plot does not show any correlation information but simply shows the 1-body distribution function for each atomic site. This may be compared with the plots of the anisotropic displacement parameters given by Braga *et al.* (1994). The refinement clearly indicates that the atomic site Fe(1) is reasonably localised and isotropic while sites Fe(2) and Fe(3) show a very large anisotropy. As a quantitative measure of this anisotropy the variance $\left\langle \varphi_{i,j,k,l}^2 \right\rangle$ can be computed. The value obtained for the final MC solution is $\sim (15^\circ)^2$. This is a little larger than the value of $130 = (11.4^\circ)^2$ estimated by Braga *et al.* (1994) from a rigid-body analysis of their anisotropic displacement parameters.

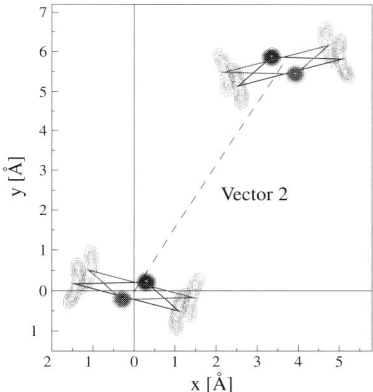

Fig. 14.7   The average of all $(+)$ and $(-)$ molecules in each molecular site.

To see the effect of the local relaxation that has resulted from the analysis, similar plots to Fig. 14.7 may be made by averaging over only those cells which contain $(++)$, $(+-)$, $(-+)$ or $(--)$. Such plots are shown in Fig. 14.8. The pair of sites within one unit cell corresponds to the inter-site vector 2. Similar plots for pairs of sites joined by vector 1 are shown in Fig. 14.9. Here it is clearly seen that the value of $\varphi_{i,j,k,l}$ for a particular molecule is strongly influenced by the occupancy of the neighbouring site. For some combinations of neighbouring occupancies, for example, Fig. 14.8(a), Fig. 14.9(d) the effect is quite dramatic while for other combinations, for example, Fig. 14.8(d), Fig. 14.9(a), it is rather less so.

14.9   Conclusion

In this chapter the feasibility has been demonstrated of automatic least-squares refinement of a Monte Carlo model of a disordered structure by quantitative comparison of its

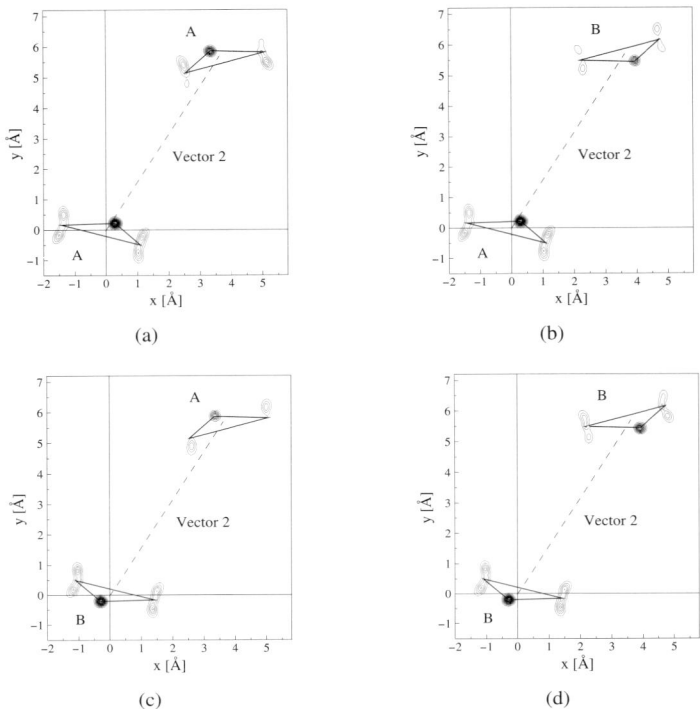

Fig. 14.8   Average of all cells with (a) $(++)$; (b) $(+-)$; (c) $(-+)$; (d) $(--)$ pairs of molecules along vector 2.

calculated diffraction pattern with observed diffuse scattering data. A model involving 16 independent parameters was refined using a subset of the observed data consisting of in excess of ~ 40000 data points. Of the 16 parameters used, eight described the correlation of the $(+)$ and $(-)$ occupancy variables along different neighbouring inter-site vectors, four described the dependence of centre of mass relaxation on the occupancy of neighbouring sites and the final four parameters described, similarly, the dependence of the molecular orientational relaxation on the occupancy of neighbouring sites.

The particular course that was followed during this investigation was influenced by certain decisions made at a quite early stage of the study and the fact that limited (albeit generous) computational resources precluded the possibility of simultaneously following alternative routes. The decision to use the limited regions of data A1, B1, etc. indicated in Fig. 14.2 was influenced by the desire to maintain the full resolution of the observed data and by prior experience in making visual comparisons between patterns, where the emphasis tended naturally to concentrate on the most intense features. In retrospect, it seems that it would perhaps have been more appropriate to use a data set of lower resolution but covering more of reciprocal-space. A similar choice to break the problem into two parts by first refining the occupancy correlations and then devel-

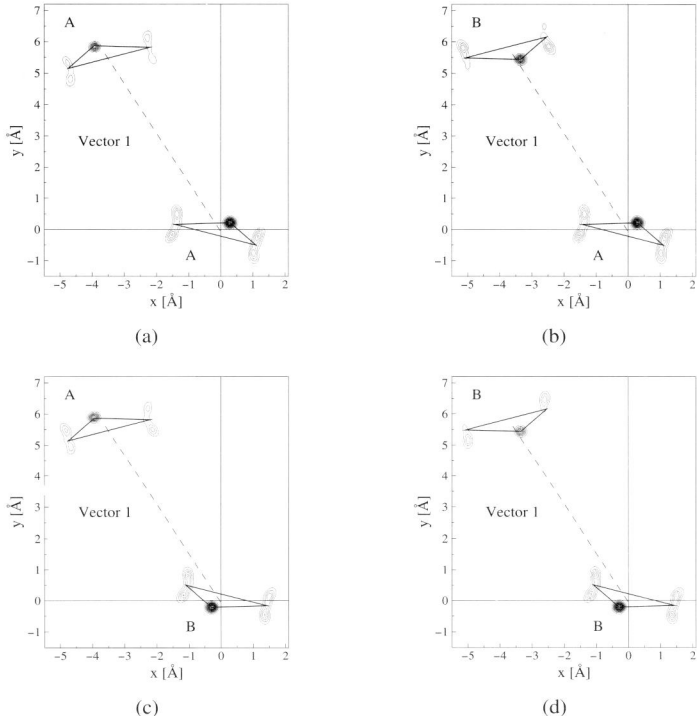

Fig. 14.9 Average of all cells with (a) $(++)$; (b) $(+-)$; (c) $(-+)$; (d) $(--)$ pairs of molecules along vector 1.

oping a relaxation model, was again influenced by prior experiences with qualitative modelling. This in turn dictated that the regions of data A1, B1 were used in the first stage of analysis since they were seemingly not affected by relaxation, and that these were later replaced by the regions of data A2, B2 when relaxational displacements were involved. This raises the question of whether in future work the possibility of refining simultaneously both occupational and relaxational parameters should be considered.

The methodology described here is a natural extension of the previous more qualitative approach to the interpretation of diffuse scattering patterns, in which visual comparisons of patterns were made and the model parameters adjusted using more subjective criteria. Apart from the more obvious benefits of a quantitative treatment one significant advantage of the new approach is that it allows more objective assessment of how the analysis is progressing. For example it was readily apparent from the behaviour of the early refinements that the two sections of data chosen for the initial study were inadequate to determine the model correlation parameters uniquely, and that further data was required. Similarly when additional data was added it was immediately apparent that the initial set of seven model parameters was unable to describe this additional data and a further parameter was required. Such judgements would have been difficult to make using the previous qualitative approach.

Even for the relatively simple model described here a single refinement represents a formidable computational task and consumed many hours of CPU time on the fastest computers available. However, as computational resources become ever faster, more and more complex problems should be amenable to solution by this technique.

FURTHER APPLICATIONS OF THE AUTOMATIC MONTE CARLO METHOD

15.1 Introduction

The iron carbonyl example described in Chapter 14 was chosen as the first example to be studied by the automatic Monte Carlo (MC) refinement method because of its extreme simplicity. This work was carried out at a time when the computational task involved was only just becoming a viable possibility. Since that first study, computational power has continued to increase according to Moore's law (e.g. see Voller and Porte-Agel, 2002) and the complexity of problems to which the method may be applied has increased accordingly. It should be stressed that even now the computational task is still a formidable one and simplifications and approximations still have to be made. The extra computational power can be used not only to tackle more complex problems but also to increase the quality of the computation, both in terms of the sample size and number of MC cycles of the simulation, the number and resolution of reciprocal lattice sections included in the analysis, but also the quality of the calculated diffraction pattern that can be obtained from the simulation coordinates.

In this chapter, two contrasting examples are described. The first of these is benzil, $C_{14}H_{10}O_2$. Although benzil has been known for a long time to exhibit strong diffuse scattering the structure is basically a perfectly ordinary ordered structure. Each molecular site contains only one basic molecular orientation and the observed diffuse scattering is purely thermal in origin. Large amplitude atomic motions occur because of the internal flexibility of the molecules. The second example, p-methyl-N-(p-chloro-benzylidene)aniline (ClMe), $C_{14}H_{12}ClN$, is closely related to the MeCl example that was discussed in Chapter 9. As for MeCl, ClMe is highly disordered with each molecular site containing the molecule in one of four different basic orientations. A plot of the average structure, as determined by Bragg analysis, shows a very confused superposition of peaks. The strong diffuse scattering in ClMe, though having its origins in the

Benzil   ClMe   MeCl

occupational disorder, also has a strong component of displacement disorder caused by the molecules being displaced from their average position as they relax to best accommodate the occupational disorder. The observed scattering does not change substantially as the temperature is lowered, indicating that these displacements are basically static. ClMe, therefore represents a much more complex system to challenge the automatic refinement method.

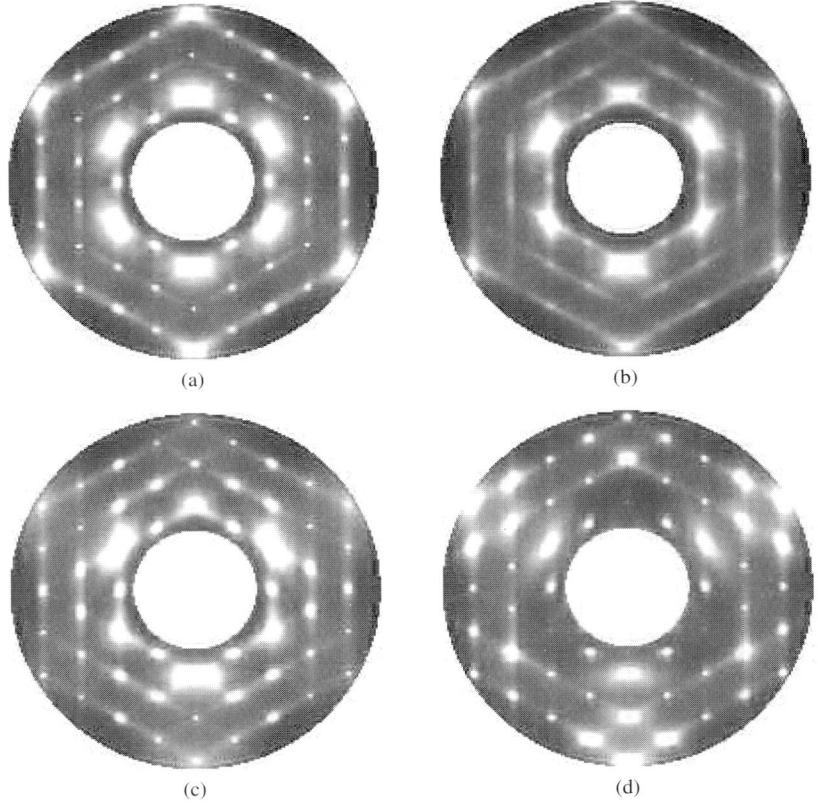

Fig. 15.1   Observed X-ray diffraction patterns of benzil. (a) $(h\,0\,l)$ section. (b) $(h\,\frac{1}{2}\,l)$ section. (c) $(h\,1\,l)$ section. (d) $(h\,2\,l)$ section.

15.2   Benzil, $C_{14}H_{10}O_2$

15.2.1   X-ray diffuse scattering data

Diffuse scattering data for benzil were recorded using the position sensitive detector system described in Section 1.2.1. Data were collected on reciprocal planes $(h\,k\,0)$, $(h\,k\,\frac{1}{2})$, $(h\,k\,1)$, $(h\,k\,\frac{3}{2})$, $(h\,k\,2)$, $(h\,k\,\frac{5}{2})$, $(h\,k\,3)$, $(h\,k\,4)$. The PSD was used in a single setting which was chosen to span the range of approximately $26.0°$–$77.2°$ in 2θ, using $CoK\alpha$ radiation. Initially, sections of data consisted of 400×400 pixels, at which resolution a single pixel corresponds approximately to the experimental resolution. For use in the analysis, however, the data were rebinned into smaller arrays of 200×200 pixels. Because of the limitations of available computer time only four of these sections were used in the data analysis. These sections of data, displayed as grey scale images, are shown in Fig. 15.1.

15.2.2 Structure specification

Figure 15.2(a) shows a drawing of the average structure of benzil viewed down **c**. The structure can be considered as consisting of three molecular layers related by the 3_1-screw axis in the space group $P3_121$. These are shown in different shades of grey for clarity. Figure 15.2(b) shows a plot of a single molecule referred to local molecular axes, u, v, w, and also shows the labelling of atoms. The small black circle marks the position of two dummy atoms a and b which were used to define the molecular orientation. b is located at the mid-point of the C1—C2 bond and a is located directly above it along the two-fold axis of the molecule. The vector between a and b defines the w-axis of the molecular coordinate system. The v-axis is along the C1—C2 bond and the u-axis is normal to it.

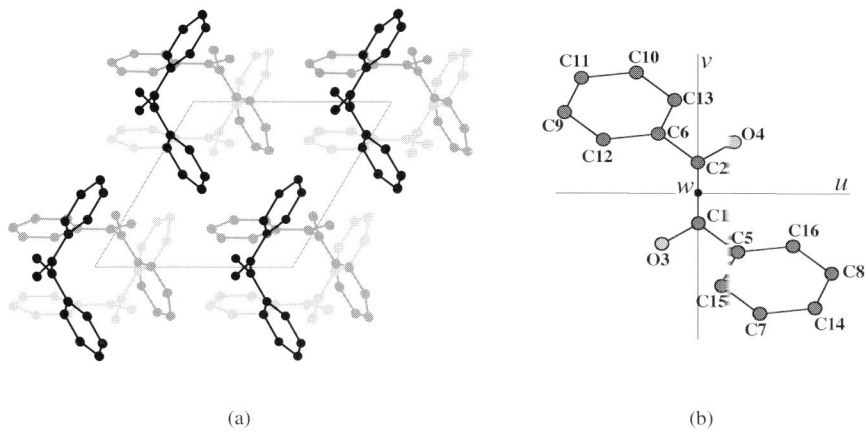

(a)   (b)

Fig. 15.2 (a) The structure of benzil viewed down **c**. The structure consists of three symmetry-related layers of molecules normal to **c** which are draw in three different shades of grey. The molecules are also seen to occur in spiral columns. (b) Single benzil molecule in local coordinate frame. Note, H atoms have been omitted in (a) and (b).

For the MC simulation the molecule is treated as a number of rigid fragments with the only degrees of freedom being the rotations around the C1—C2, C1—C5 and C1—C6 bonds. This is achieved by using a z-matrix description of the molecular coordinates. The z-matrix used is shown in Table 15.1. Starting from the dummy atoms a and b the coordinates for each successive atom k are generated using the distance k–l from a previously defined atom, the angle k–l–m made with two previously defined atoms and the dihedral angle k–l–m–n made with three previously defined atoms. In Table

Table 15.1   The z-matrix representation of the molecule used in the MC simulation. See text for details.

Atom k	Bond length $k\text{–}l$ (Å)	Bond angle $k\text{–}l\text{–}m$ (°)	Dihedral angle $k\text{–}l\text{–}m\text{–}n$ (°)	l	m	n
a	–	–	–	–	–	–
b	**0.7615**	–	–	a	–	–
C1	*0.7615*	90.0	–	b	a	–
C2	*1.0769*	45.0	180.0	a	b	C1
O3	**1.215**	**115.8**	54.4*	C1	b	a
O4	*1.215*	*115.8*	*54.4**	C2	b	a
C5	**1.449**	**123.6**	**177.9**	C1	C3	b
C6	*1.449*	*123.6*	*177.9*	C2	C4	b
C7	**2.410**	150.0	**172.0***	C5	C1	b
C8	*2.410*	60.0	180.0	C7	C5	C1
C9	*2.410*	150.0	*172.0**	C6	C2	b
C10	*2.410*	60.0	180.0	C9	C6	C2
C11	*1.391*	30.0	180.0	C9	C10	C6
C12	*1.391*	30.0	180.0	C6	C9	C10
C13	*1.391*	30.0	180.0	C6	C10	C9
C14	*1.391*	30.0	180.0	C7	C8	C5
C15	*1.391*	30.0	180.0	C5	C7	C8
C16	*1.391*	30.0	180.0	C5	C8	C7

15.1 the nine numbers given in bold are quantities whose initial values were adjusted to provide that the atomic coordinates generated by the z-matrix gave the best fit to those obtained from the average structure coordinates. Numbers given in italics are quantities directly related to these nine and all other numbers in plain text were fixed constants. After this initial fit to the average structure coordinates, all the quantities in the z-matrix were fixed as constant for the remainder of the study, with the exception of the dihedral angles indicated by * which allow rotation around the C1—C2 bond (ϕ_1), and rotations of the phenyl groups about the C1—C5 and C1—C6 bonds, respectively (ϕ_2 and ϕ_3).

The molecules are placed in the crystal framework by rotating the coordinates generated from the z-matrix by the appropriate angle and translating them to the appropriate origin in the crystal. Four quaternion variables (one of which is redundant) are used to specify the rotation and three cartesian coordinates to specify the origin. Together with the four variables indicated by * in the z-matrix, this means that a total of 11 variables per molecule are needed to specify the positions of all of the atoms in the crystal. Initially all these variables can be given their average values so that the model crystal is simply a perfectly ordered system with the average structure, or alternatively each variable can be given some random perturbation from its average value. A crystal size of $32 \times 32 \times 16$ unit cells was used making a total of $49,152$ molecules or $786,432$ atoms.

15.2.3 Intermolecular interactions

Each molecule in the average structure is in contact with 15 neighbouring molecules and these contacts comprise a very large number of individual atom–atom contacts. Even if only those distances less than 4.5 Å are used to describe the intermolecular interactions the number of these would be far too large to be contemplated for the MC simulations. Since the molecule itself consists of a small number of rigid fragments all that is required to adequately model the interactions is a sufficient number of atom–atom contacts to define the equilibrium spacing and mutual orientation of each pair of neighbouring molecular fragments. That is, a small subset of all those contact vectors less than 4.5 Å are used as 'effective' interactions. It was found sufficient to use only those atom-atom contacts involving the atoms O3, O4 and C7, C8, C9 and C10. Each inter-fragment contact was then represented by 2 to 3 atom–atom contact vectors. The total number of these vectors connected to any individual molecule was 34. These 34 vectors, comprising 10 symmetry-inequivalent vector types, are given in Table 15.2. At a later stage, four additional contact vectors (one additional vector type) involving atoms C11 and C14 were added. Of the 11 final different types of contact vector some were between molecules in the same molecular layer, some between molecules in different layers and others were between molecules within the spiral columns of molecules shown in Fig. 15.2(a). Some of these different types of interaction are shown in Fig. 15.3.

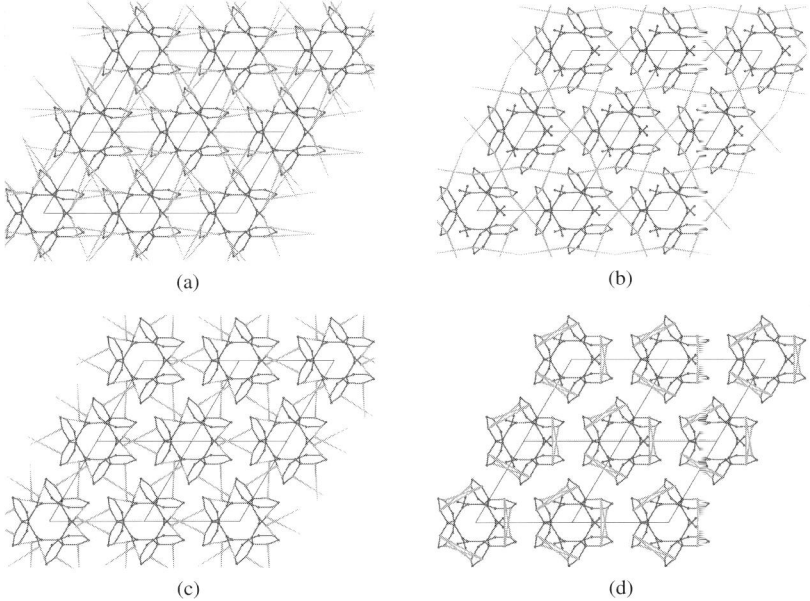

Fig. 15.3 Plots showing the different kinds of intermolecular vectors used to represent the effective interactions. (a) within molecular layers (vectors 1,2), (b) and (c) between molecular layers (vectors 3,4,5) and (d) within spiral columns (vectors 8,9).

15.2.4 *MC simulation*

Interactions between the molecules were included along each of the vectors given in Table 15.2. These were in the form of Hooke's law (harmonic) springs with force constants F_i corresponding to the 10 (subsequently 11) different vector types. The inter-molecular contribution to the energy of the system was thus,

$$E_{inter} = \sum_{\substack{\text{all contact} \\ \text{vectors}}} F_i(d_i - d_{ave})^2. \qquad (15.1)$$

Here d_i is the instantaneous length of a particular vector of type i and d_{ave} is the equilibrium length of that vector. Values for d_{0i} are also given in Table 15.2. The internal flexibility of the molecule was initially assumed to involve no energy penalty but such intra-molecular forces were subsequently included. In the final model a simple torsional force was included for each of the torsion angles ϕ_1, ϕ_2, ϕ_3. This was of the form,

$$E_{intra} = \sum_{\substack{\text{all} \\ \text{molecules}}} F_{12}[(\Delta\phi_1)^2 + (\Delta\phi_2)^2 + (\Delta\phi_3)^2], \qquad (15.2)$$

where the same torsional force constant was used for all three angles. Here $\Delta\phi_i$ are the deviations of the angles ϕ_i from their mean values.

The total energy of the crystal is then expressed in the form,

$$E_{total} = E_{inter} + E_{intra} \qquad (15.3)$$

and this was used in the application of the normal MC algorithm as follows. A site in the crystal was chosen at random and the set of 11 variables defining the conformation, orientation and position of the molecule, was subjected to a small random increment. The contribution to the energy of all terms that are dependent on the altered variables is computed before and after the change is made. The energy difference, $\Delta E = E_{new} - E_{old}$ was used to decide whether the new configuration should be kept, or the system returned to the original configuration. A pseudo-random number, η, chosen uniformly in the range $(0, 1)$, was compared with the transition probability, P, where

$$P = \exp(-\Delta E/kT). \qquad (15.4)$$

T is the temperature and k Boltzmann's constant. If $\eta < P$ the new configuration was accepted, while if $\eta > P$ it was rejected and the system returned to the original state. An MC cycle is defined as that number of such individual steps needed to visit each site once on average. In the present work it was assumed throughout that $kT = 1$ and consequently the force-constants determined were relative to this energy scale.

The procedure was also adopted that at each step only one of the 11 variables (chosen at random) was incremented. In this way it was possible to monitor the acceptance/rejection ratio for each variable and the random step size for each was adjusted accordingly to maintain an acceptance/rejection ratio close to unity. Some initial runs were carried out using 200 cycles of iteration but final analysis utilised 400 cycles.

Table 15.2 The different interaction vectors in benzil. The vectors link the atom given in column 1 of a molecule in symmetry position 1 with the atom given in column 2 of a molecule whose symmetry position is given in column 6

From atom	To atom	x increment	y increment	z increment	Symmetry position	Distance (Å)	Vector type
O3	C14	0	1	0	1	3.42	11
	C7	0	1	0	1	4.50	1
	C8	0	1	0	1	4.01	2
	C8	0	0	0	2	3.64	5
	C9	1	1	0	2	3.35	4
	O4	0	0	0	2	4.04	6
O4	C7	1	1	−1	3	3.35	4
	C10	1	0	0	1	4.01	2
	C10	0	0	−1	3	3.64	5
	O3	0	0	−1	3	4.04	6
	C9	1	0	0	1	4.50	1
	C11	1	0	0	1	3.42	11
C7	C8	0	−1	0	2	3.99	7
	C9	0	0	0	2	4.23	8
	C9	0	0	0	3	4.46	10
	C10	0	0	0	2	3.73	9
	O4	1	0	0	2	3.35	4
	O3	0	−1	0	1	4.50	1
C8	C7	1	0	−1	3	3.99	7
	C9	0	0	0	2	3.73	9
	C10	0	−1	0	1	4.34	3
	C10	1	0	0	1	4.34	3
	O3	0	−1	0	1	4.01	2
	O3	0	0	−1	3	3.64	5
C9	C7	0	0	−1	2	4.46	10
	C7	0	0	−1	3	4.23	8
	C8	0	0	−1	3	3.73	9
	C10	−1	0	−1	3	3.99	7
	O3	0	1	−1	3	3.35	4
	O4	−1	0	0	1	4.50	1
C10	C7	0	0	−1	3	3.73	9
	C8	−1	0	0	1	4.34	3
	C8	0	1	0	1	4.34	3
	C9	0	1	0	2	3.99	7
	O4	−1	0	0	1	4.01	2
	O4	0	0	0	2	3.64	5
C14	O3	0	−1	0	1	3.42	11
C11	O4	−1	0	0	1	3.42	11

Broad features of a pattern are reproduced with relatively few cycles, while features corresponding to relatively long-range effects take many cycles to settle down. For benzil it was apparent that to reproduce the strong diffuse peaks that surround the Bragg peaks as well as the rather narrow diffuse lines that are apparent in the sections of data in Fig. 15.1, rather long computation times are necessary.

15.2.5   Automatic fitting of MC model

Once the MC model described above had been established the automatic MC fitting methodology described in Chapter 14 was used to adjust the model to give the best fit to the observed data. One difference that was made from that earlier work was that the differentials, $\partial \Delta I / \partial p_i$, were calculated using a modified version of eqn (14.12), namely

$$\frac{\partial \Delta I}{\partial p_i} = \sum_{hklm} \frac{\left(\Delta I_{p+} - \Delta I_p\right)}{\delta_i}. \tag{15.5}$$

Here ΔI_p is the value of ΔI determined with the current set of parameters. By using ΔI_p instead of ΔI_{p-} the number of individual MC simulations is reduced. This advantage is off-set by the fact that the value of δ_i that needs to be used to get the same magnitude of differences is correspondingly larger. Figure 15.4 shows examples of the difference in calculated intensity when three different force constants were incremented. The intensity distributions themselves were virtually indistinguishable by eye but these difference patterns show that changing the different force constants affects the diffuse pattern in clearly distinguishable ways. It should be stressed that for this to work the quality of the calculated patterns needs to be sufficiently good. This requires both a sufficiently large simulation crystal size and a high quality diffraction pattern calculation.

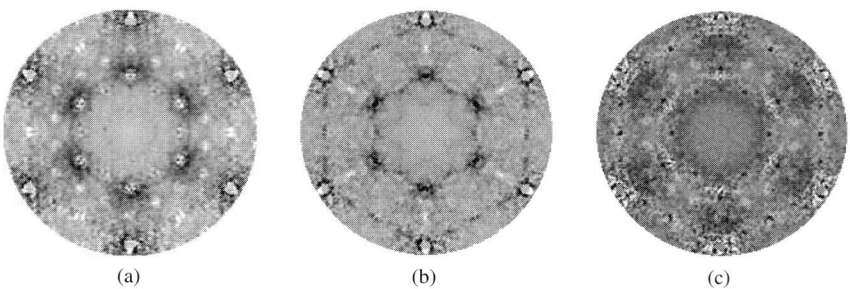

| (a) | (b) | (c) |

Fig. 15.4 Illustration of the calculation of the differentials, $\partial \Delta I / \partial p_i$. (a), (b) and (c) are plots of the difference in intensity between a $(h\,0\,l)$ diffraction pattern calculated from an MC simulation using a particular set of parameters, p_i, and patterns calculated from simulations calculated when three different force constant parameters have been incremented. Light and dark show increases or decreases in the calculated intensity from the original.

At the start of the refinement the system energy was considered to be dependent on only the first 10 force constants, F_i, acting along the corresponding vectors defined in the final column of Table 15.2. Torsional force constants for the three angles ϕ_1, ϕ_2, ϕ_3

were initially assumed to be zero. It was apparent from these early cycles of refinement that the narrow diffuse lines that are seen in all the sections of diffuse data originate from a coupled motion of neighbouring molecular fragments along the $[100]$, $[010]$ and $[110]$ directions. The coupling occurs by transmission between the O3 or O4 atoms of one molecule and the phenyl group of the neighbour. This corresponds to the contact vectors types 1 and 2 in Table 15.2 (see also Fig. 15.3(a)), and the value of F_1 in particular refined to a large value. However, it was clear from the geometry calculations that, while the F_1 and F_2 interactions were able to model this effect reasonably well, a direct interaction to the terminal carbon atom of the phenyl group might be more appropriate since it involved a much shorter distance (3.41 Å compared to 4.50 Å and 4.01 Å for vectors 1 and 2). Consequently an additional force constant F_{11} was added along this vector. Although F_{11} refined to a value comparable to the value for F_2, any tendency for it to take over the rôle previously played by F_1 and F_2 was not very apparent during the number of cycles of refinement that it was feasible to carry out. Moreover it did not result in significant improvement of the fit.

Two changes to the initial model did, however, result in significant improvement of the fit. The first was the inclusion of an overall average Debye–Waller factor, $\mathbf{B}=8\pi^2\langle u^2\rangle$. This corrected a discrepancy between the relative intensities of high and low-angle peaks in the calculated pattern and resulted in a reduction of the agreement factor, $R = (\sum \omega (\Delta I)^2 / \sum \omega I_{obs}^2)^{1/2}$, of $\sim 2\%$. The second was including a torsional force constant for the three angles, ϕ_1, ϕ_2, ϕ_3. This latter inclusion resulted in a reduction in R of $\sim 3\%$. A final agreement factor of 14% was obtained for a model in which 11 spring constants, 1 torsional force constant and 1 Debye–Waller constant were refined.

The final values of the MC parameters are given in Table 15.3 and Fig. 15.5 shows the diffraction patterns calculated from the MC simulation using these parameters. The agreement with the observed patterns of Fig. 15.1 is generally very good, as would be expected with an overall R value of $\sim 14\%$. The agreement obtained in this study is considerably better than any other study to date in which the automatic MC refinement method has been used. This can perhaps be attributed to a number of factors. First, the disorder in benzil is describeable purely in terms of thermal disorder without the need to consider occupational disorder and the consequent cross-interactions between occupancy and displacement. Second, the force model used, though involving the idea of 'effective' interactions, used a sufficient number of individual atom–atom interactions that the separation and mutual orientation of individual molecular fragments were well defined, thereby providing a most effective model of the real intermolecular forces. Finally, the calculated diffraction patterns that were used in the least-squares process were of a better quality than those used in earlier studies. This was partly due to the longer computation times that it was possible to use but also because of the high symmetry of benzil which allowed smoother calculated patterns to be obtained by averaging.

The analysis of benzil has shown quite clearly that the diffuse lines that feature so prominently in the observed diffraction patterns are due to strong longitudinal displacement correlations transmitted from molecule to molecule via the network of strong interactions, F_1, as depicted in Fig. 15.6(a) (but also including F_2 and F_{11}). *Longitudinal correlation* means that the direction of displacement and the direction of correlation are

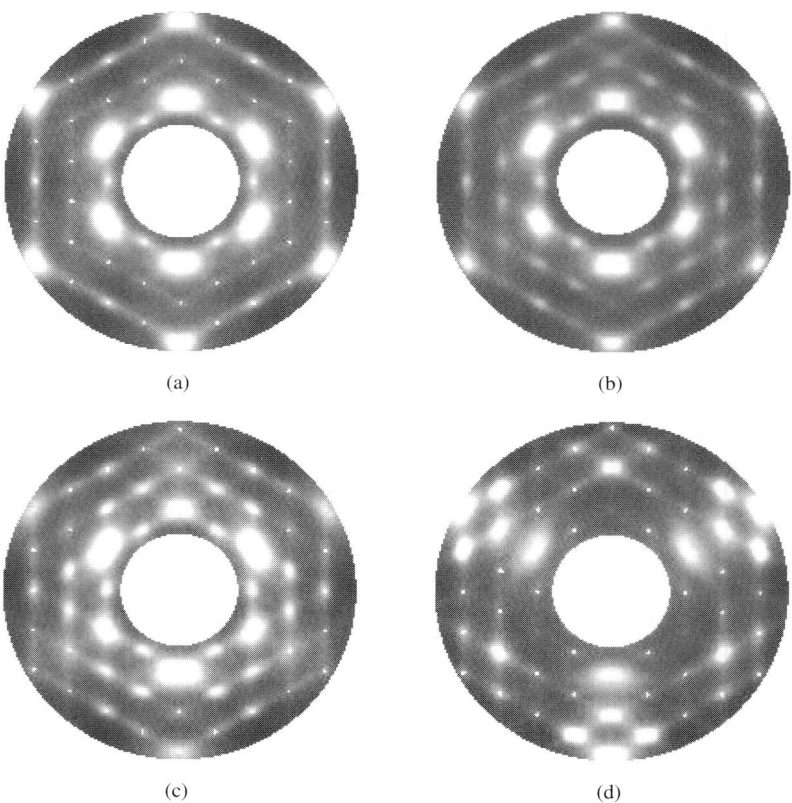

Fig. 15.5 Diffraction patterns of benzil calculated from the final MC simulation. (a) $(h\,0\,l)$ section. (b) $(h\,\frac{1}{2}\,l)$ section. (c) $(h\,1\,l)$ section. (d) $(h\,2\,l)$ section. Cf. corresponding observed patterns in Fig. 15.1 .

Table 15.3   Final values of the parameters used in the MC simulation of benzil

Parameter	Value	Esd		Parameter	Value	Esd
F_1	120.0	3.3		F_8	76.8	9.7
F_2	48.6	4.7		F_9	120.4	6.3
F_3	15.7	3.6		F_{10}	80.3	3.1
F_4	85.7	2.8		F_{11}	73.8	7.4
F_5	34.8	4.1		F_{12}	0.159	0.014
F_6	16.5	2.0		B_{overall}	6.17	0.09
F_7	145.6	8.8				

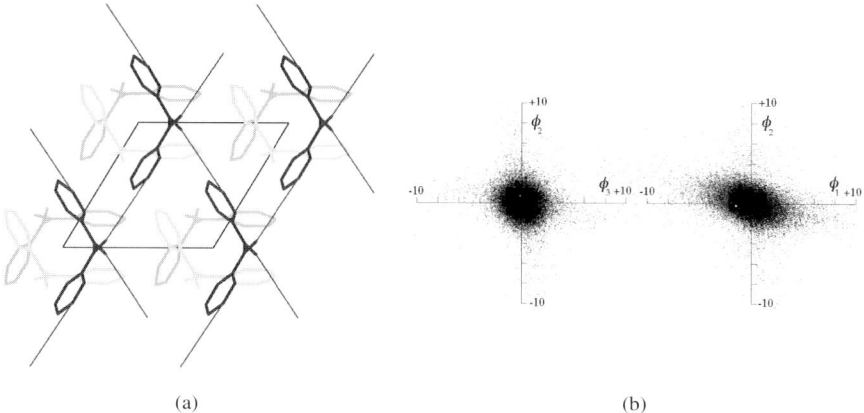

(a)   (b)

Fig. 15.6 (a) Plot showing the 2D network formed in the benzil simulation by intermolecular springs having the strongest in-plane force constant, F_1. (b) Plots of the distribution of the torsion angles ϕ_2 vs. ϕ_3 and ϕ_1 vs. ϕ_2. Angles are in degrees away from their mean values.

the same. It seems likely that the molecular basis for this is in fact hydrogen bonding between the O3 or O4 atom on one molecule and the para H atom (i.e. that attached to C11 or C14) of the phenyl ring of a neighbouring molecule. The diffuse lines in the diffraction patterns result from the fact that the rhomb-shaped cells of this network undergo a low-energy shearing mode. It should be noted that each vertex of the network involves the intra-molecular ϕ_1 torsion angle.

Figure 15.6(b) shows plots of the distribution of ϕ_2 vs. ϕ_3 and ϕ_1 vs. ϕ_2 for all of the molecules present in the final simulation. Since the model contains no direct forces correlating the angles, any departure from a circularly symmetric distribution is indicative of the fact that the external inter-molecular forces act on the molecules to produce correlation between the angles. Figure 15.6(b) shows that there certainly is correlation between ϕ_1 and ϕ_2 (and also ϕ_1 and ϕ_3) but very little between ϕ_2 and ϕ_3.

15.3 *p*-methyl-*N*-(*p*-chloro-benzylidene)aniline (ClMe)

15.3.1 *X-ray diffuse scattering data*

The second example, *p*-methyl-*N*-(*p*-chloro-benzylidene)aniline (ClMe), $C_{14}H_{12}ClN$, is much more complex than the benzil example described above and it would seem necessary that to unravel such a complex problem an extensive set of diffuse scattering data would be required. However, the study of ClMe pre-dated that of benzil so that computational resources were more limited and with the necessity to also use a substantially more complex MC model it was necessary that the number of data points to be utilised was kept to a minimum. For this reason the study was carried out using only a single section of data, namely the $(h\,0\,l)$ section. It is also worth noting that the structure of ClMe is very similar to that of MeCl (see Chapter 9) for which much useful information was obtained from a single section of data, albeit using a simpler method of analysis.

The $(h\,0\,l)$ diffuse scattering data for ClMe is shown in Fig. 15.7 alongside that of MeCl. The two patterns are clearly very similar, though there are some points of difference. The strong regions of scattering close to Bragg peak positions at B, C, and F are clearly very similar in the two patterns but the shapes of the peaks at A and E are quite different. Note also that whereas for MeCl both sets of peaks near A and E are reasonably symmetric, for ClMe those around A are quite asymmetric. Of particular note are the features labelled g which are relatively weak for MeCl but much more prominent in ClMe. These g features were not modelled at all in the analysis of MeCl described in Chapter 9.

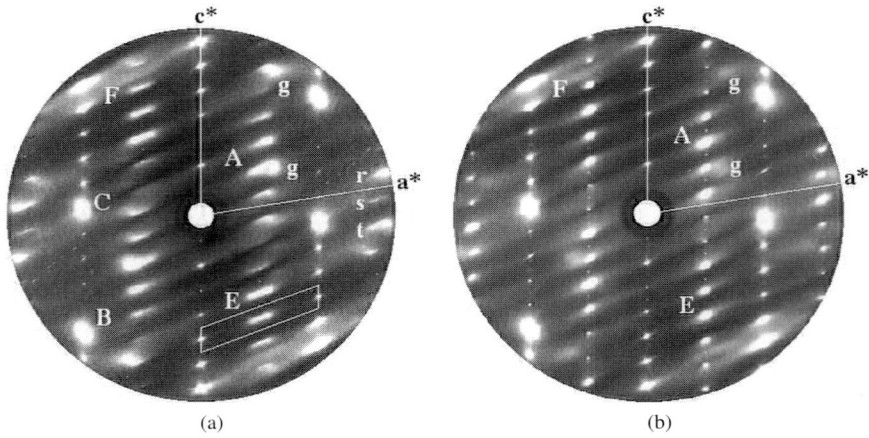

(a)   (b)

Fig. 15.7   A comparison of the $(h\,0\,l)$ sections of the diffraction patterns of (a) ClMe and (b) MeCl. The various markings on the two figures are referred to in the text.

The images shown in Fig. 15.7 which contain 400×400 pixels were recorded, as for the benzil data, using the position sensitive detector system described in Section 1.2.1. These are excessively large for the direct-fitting methodology described in Chapter 14 and for the study of ClMe the number of data points was reduced by rebinning the data to form a more manageable but lower resolution array of (200×200 pixels). The PSD was used in a single setting which was chosen to span the range of approximately $3.5°$–$54.6°$ in 2θ, using Cu Kα radiation.

15.3.2   *Structure specification*

Haller *et al.* (1995) determined the average structure of ClMe to be space group $P2_1/n$ with a unit cell of $a = 5.971, b = 7.411, c = 27.420$ Å, $\beta = 99.13°$. However the structure is very close to being $P2_1/a$ with a cell of half the c-spacing (i.e. $c = 13.710$ Å as reported earlier by Bar and Bernstein (1983)), and for modelling the disorder this simpler cell, in which there are two molecules only, is assumed.

Figure 15.8(a) shows a drawing of the structure of ClMe viewed down **b**. In this figure the molecules are shown in only a single orientation. The average structure determination shows a confused superposition of four different orientations. The four different orientations are shown in Fig. 15.9 referred to a local molecular coordinate system with orthogonal axes X', Y', Z'. Figure 15.9(a) shows the superposition of the same molecular shape in four different orientations. It is seen that the departures of the molecule conformations from planarity are quite small. Figure 15.9(b) shows the four orientations separately, together with the values of two binary $(+1, -1)$ random numbers that are used to represent these basic orientations. σ is used to represent an end-to-end flipping of the molecule while ρ is used to represent a side-to-side flipping.

The central bond in ClMe is a chemical double bond and so unlike in benzil there is no rotation about this bond. The two phenyl rings are in principle able to rotate about the single bonds connecting them to the central $-C=N-$ moiety. However, the fact that the molecule is observed to be close to planar in the average structure determination, together with the fact that in projection down **b** any out-of-plane motions would be difficult to observe, suggested that treating the whole molecule as a single rigid body would be a reasonable approximation to be made in this study. In that case the whole molecule can then be replaced in the MC simulation by a simple 4-atom *motif* which defines the orientation and position of the molecule. Figure 15.9(a) shows this 4-atom *motif* in black superposed on the actual molecules. Since each of the four different molecular orientations is defined relative to the axes defined by the *motif* the actual atomic coordinates at any crystal site can be retrieved from a knowledge of the position and orientation of the *motif* and the values of the two random variables ρ and σ. Thus to specify the coordinates of all the atoms in a given crystal site, 9 variables are required: 3 centre of mass coordinates, 4 quaternion parameters and the two random variables ρ and σ.

15.3.3 *Intermolecular interactions*

Each molecule in ClMe is surrounded by 14 neighbouring molecules and by using direct interactions between the four dummy atoms which comprise the *motif* the large number of real atom–atom interactions can be replaced by a small number inter-*motif* interactions. These are shown in Fig. 15.9(b). From the central *motif* there are 26 such vectors but some of these are related by symmetry so that only eight different kinds of spring force constants, F_i, are required. The vectors along which these different force constants act are given in Table 15.4.

For the case of benzil, in which only one molecular orientation occurs in each average site position, eqn (15.1) is all that is required to perform the MC simulation. For ClMe account has also to be taken of the fact that the equilibrium length of the springs will be different depending on which of the four molecular orientations occupy the two sites (n and m) linked by the vector. Consequently it is necessary to use an equation similar to eqn (14.2). Here eqn (15.1) is modified to give

$$E_{\text{inter}} = \sum_{\substack{\text{all contact} \\ \text{vectors}}} F_i \big(d_i - d_{\text{ave}}(1 + \varepsilon_{n,m}) \big)^2. \tag{15.6}$$

where $\varepsilon_{n,m}$ are size-effect parameters that depend on the particular pair of molecular orientations at sites n and m. The $\varepsilon_{n,m}$ provide that some interatomic distances are longer

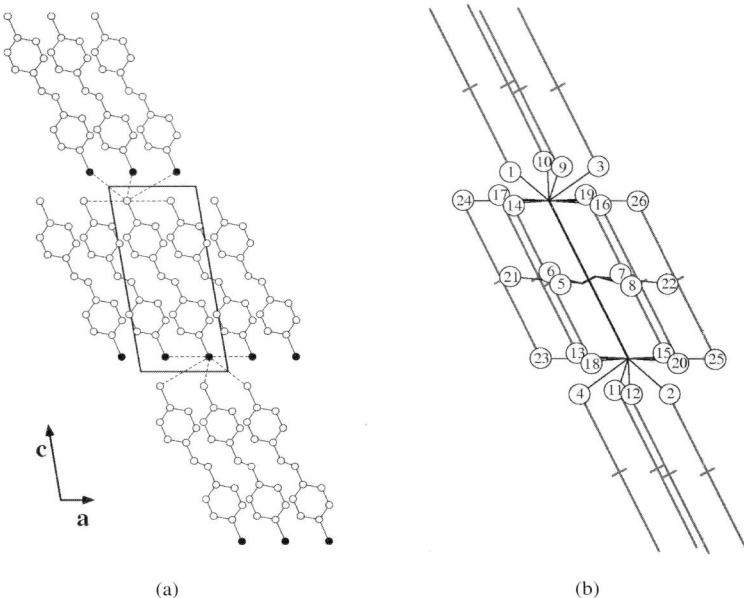

(a)   (b)

Fig. 15.8 (a) Plot of the structure of ClMe viewed down b. (b) Schematic diagram of the environment of a central molecule of ClMe showing the 26 different vectors along which springs linking the central molecule to its 14 neighbours are placed to represent the effective molecular interactions. In (b) the view is slightly rotated to reveal the neighbouring molecular sites which occur above and below the height of the central molecule.

Table 15.4 Summary of the force constants, types of size-effect parameters and occupancy correlation constants used in the MC simulation of ClMe

Spring constant	Interaction vectors	Determined force constant	Size-effect dependent on	Occupancy correlation constant	Determined occupancy correlation
F_1	1–2	131.8	End-to-end		
F_2	3–4	437.1	End-to-end		
F_3	5–8	70.0	Side-to-side	a_1, b_1	−0.5
F_4	9–12	72.3	End-to-end	a_2, b_2	+0.225
F_5	13–16	0.0	End-to-end		
F_6	17–20	272.2	End-to-end		
F_7	21–22	77.8	Side-to-side	a_3, b_3	+0.17
F_8	23–26	304.1	End-to-end		
			Cross-correlation	c_0	+0.65

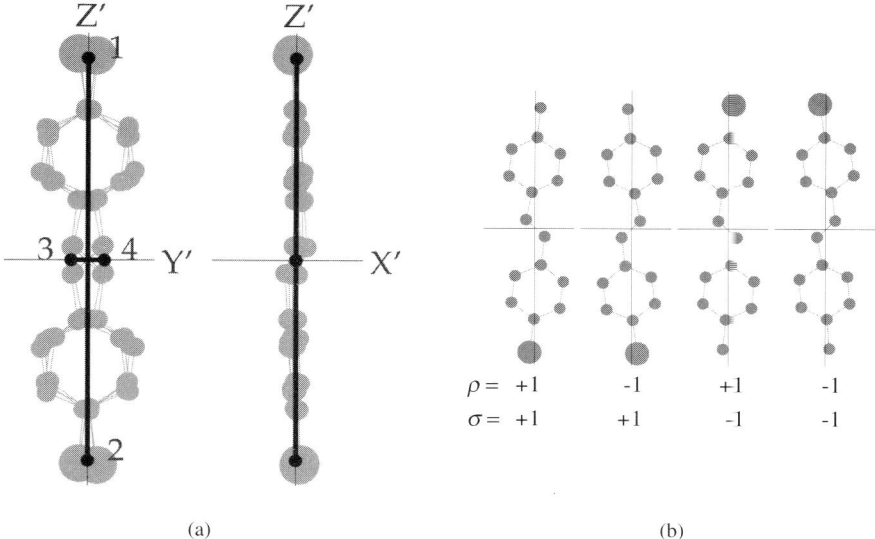

(a)   (b)

Fig. 15.9   (a) The four-atom motif used to represent the ClMe molecules in the MC simulation. (b) The four different orientations of the ClMe molecule and the values of the random variables used to represent them.

than the average while others are shorter. Since each site contains one of four different molecular orientations this means that there are 16 different possible values of $\varepsilon_{n,m}$ required for each different interaction vector (though there will be some constraints required to ensure the average is maintained). It is not feasible to attempt to determine such a large number of parameters particularly with a limited set of observed data. Even if full 3D data were available and it was computationally feasible to use it, it is by no means certain that sufficient information would be present in the data to unambiguously distinguish between so many of these parameters. In the present situation with both limited data and computational resources it is necessary to make further substantial simplifications in order to proceed.

The simplification that was adopted was as follows. For those vectors involving contacts between the ends of the molecular *motif*, that is, involving pseudo-atoms 1 and 2, it was assumed that $\varepsilon_{n,m}$ only depended on the value of the variable σ which defines the end-to-end disorder. Similarly, for those vectors involving contacts between the sides of the molecular *motif*, that is, involving pseudo-atoms 3 and 4, it was assumed that $\varepsilon_{n,m}$ only depended on the value of the variable ρ which defines the side-to-side disorder (see also Table 15.4). This simplification means that $\varepsilon_{n,m}$ has only four possible values for each vector instead of the original 16. Moreover, in order that the average length of the vector is maintained, one constraint must be applied to these four values so that only three independent parameters are required per vector making 24 parameters in total.

15.3.4   *Occupancy distributions*

Equation (15.6) provides the MC simulation with the means to allow the molecular positions and orientations in ClMe to relax locally given the distribution of basic molecular orientations that are present. The MC model also needs to provide the means by which the basic occupancy distributions (i.e. the distributions of ρ and σ) can be determined. The MC energies used to specify local ordering of these variables can be written in the form of Ising models. If (i, j, k, l) are indices defining a site l in unit cell i, j, k and (i_n, j_n, k_n, l_n) similarly define a neighbouring site then these models can be formulated as a sum over the neighbouring sites n,

$$E_{ss} = \sum_n a_n \rho_{i,j,k,l} \rho_{i_n,j_n,k_n,l_n} \tag{15.7}$$

$$E_{ee} = \sum_n b_n \sigma_{i,j,k,l} \sigma_{i_n,j_n,k_n,l_n} \tag{15.8}$$

$$E_{se} = \sum_n c_n \sigma_{i,j,k,l} \rho_{i_n,j_n,k_n,l_n}. \tag{15.9}$$

The a_n, b_n, c_n, are interaction parameters which govern whether particular neighbouring molecular orientations are correlated. One such set of three parameters is required for each symmetry-independent intermolecular vector along which significant correlation might be expected to occur. In the present case, after some preliminary testing it was found sufficient to include occupancy correlation parameters along only three types of vector. These are indicated in column 5 of Table 15.4. It should be noted that for the case of E_{se} a zeroth term c_0 is possible. This corresponds to the situation where, in a given site, the side-to-side orientation depends on its end-to-end orientation. A c_0 parameter was accordingly included in the MC simulation.

In principle the MC simulation should be performed using eqn (15.6) and eqns (15.7)–(15.9), simultaneously. This is not viable in practice, however, since any changes to the occupancy distributions occurring as a result of eqns (15.7)–(15.9) will require many iterations using eqn (15.6) to allow the new configuration to attain an energy minimum. The MC simulation is therefore carried out in two stages. The first stage uses eqns (15.7)–(15.9) to determine the occupancy distribution and the second stage uses eqn (15.6) to allow local relaxation of positions and orientations.

15.3.5   *MC simulation*

15.3.5.1   *Occupancy ordering*   For diffuse scattering data such as that observed for ClMe there is clearly a large component which is due to displacement disorder. This presents difficulties when attempting to determine occupancy correlations. If a fit to the whole data is attempted with a model that contains no displacement disorder then the occupancy variables will adjust to try to fit peaks which are predominantly due to the displacement disorder. For example, the strong TDS-like peaks that surround some of the strong Bragg peaks might well induce spurious positive correlations between the molecular orientations. In order to avoid this kind of problem a better strategy is to try to fit the occupancy distribution parameters using only those parts of the data that can be reasonably assumed to be least affected by displacement disorder. Since the intensity distribution due solely to occupancy correlations is periodic (apart from the variation due to the molecular structure factors) it is only necessary, for one type of occupancy variable, to find one reciprocal cell where the displacement disorder is negligible. In the present case the region chosen was that outlined in white in the vicinity of the label E in Fig. 15.7(a). The elongated diffuse peak in this reciprocal cell is quite symmetric (unlike those near label A) and is distant from the strong TDS-like peaks (labels B, C and F) that surround the Bragg peaks.

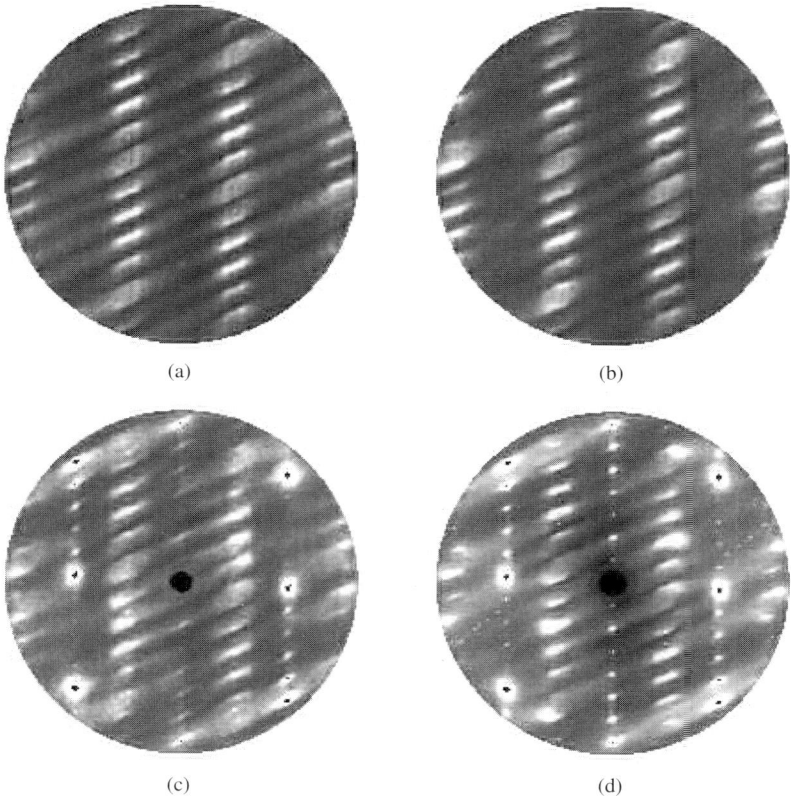

(a)   (b)

(c)   (d)

Fig. 15.10 Diffraction patterns calculated from the MC Simulation. (a) Using occupancy correlations for the end-to-end disorder, zero correlations for the side-to-side disorder and no displacements. (b) Using occupancy correlations for both end-to-end and side-to-side disorder and no displacements. (c) Final calculated diffraction pattern including side-to-side correlations, end-to-end correlations, a cross correlation between these two, and size-effect relaxation displacements. (d) Observed X-ray data for comparison.

Peaks such as those near label E are easily demonstrated to be due to correlations between the end-to-end flipping of the molecules so that use of this limited region of data can allow the determination of the b_i parameters of eqn (15.8). Values for the correlations determined in this way are given in the final column of Table 15.4. In order to determine the a_i parameters of eqn (15.7) in a similar way a comparable region in the pattern needs to be found where the diffuse scattering can be attributed to the side-to-side disorder and which is similarly unaffected by displacement disorder. This is simply not available in the single section of data being used. Consequently two different possible ways of treating the side-to-side disorder were considered. In the first case it was assumed that the side-to-side flipping was completely random. In the second it was assumed that the same correlations applied to both the side-to-side and the end-to-end disorder. Figures 15.10(a) and (b), respectively, show calculated patterns for these two cases using the same correlations given in the final column of Table 15.4. Though these two calculated patterns look very similar there are some differences. In particular it can be seen that the relative intensity of the peaks that are labelled r, s and t in Fig. 15.7(a) are different in the two patterns. In Fig. 15.10(b) the middle of the three peaks, s, is weak, while in Fig. 15.10(a) it is comparable in intensity to r. From this point of view it seems that the second model, in which the ρ correlations are set to be the same as the σ correlations, agrees better with the observed pattern. Consequently this model was adopted and kept constant throughout the rest of the study.

15.3.5.2 *Relaxation displacements*   Use of the automatic fitting procedure for determining the displacement distributions was carried out using MC simulations which included the 8 force constants, F_i, and the 24 size-effect parameters $\varepsilon_{n,m}$ as variable parameters. The a_i and b_i parameters were kept fixed at their determined values but the cross-correlation parameter c_0 was allowed to refine. As the refinement proceeded it was found that F_5 tended to go towards zero and so in the final model this, together with its three corresponding size-effect parameters, was removed from the model altogether. It is possible that this effect is caused by the fact that all of the vectors of the type F_5 are superimposed on vectors of the type F_6 in this projection and consequently the single $(h0l)$ section of data does not contain information to resolve them.

The final model refined to an agreement factor $R = (\sum \omega (\Delta I)^2 / \sum \omega I_{obs}^2)^{1/2}$ of 32.8%. The diffraction pattern calculated from the final model is shown in Fig. 15.10(c) with the observed data shown in Fig. 15.10(d) for comparison. Values of the refined size-effect parameters for each spring type are given in Table 15.5. Although the R value achieved is not nearly so good as that achieved for benzil this is undoubtedly a reflection of the numerous approximations that had to be made for this much more complex problem. Nevertheless it is clear from a visual comparison of the observed and calculated patterns (Fig. 15.10) that most of the different diffuse features in the pattern have been modelled quite well and that the analysis has satisfactorily determined the key features of the problem. As available computational resources improve and more extensive data sets can be incorporated it is likely that some of the approximations that were made could be lifted and this should allow a better fit to be obtained.

Table 15.5 Force constants F_i and size-effect parameters $\varepsilon_{n,m}$ for the eight different types of spring used in the final MC simulation of ClMe

Spring, i	F_i	ε_{--}	ε_{-+}	ε_{+-}	ε_{-+}
1	131.8	0.1017	−0.0012	0.0054	−0.1063
2	437.1	−0.0426	−0.0149	0.0487	−0.0091
3	70.0	−0.3133	0.3023	0.4036	−0.5264
4	72.3	0.0434	0.1021	0.0421	−0.1330
5	0.0	0.0000	0.0000	0.0000	0.0000
6	272.2	0.4237	−0.0189	0.0063	−0.3880
7	77.8	0.0301	0.0706	−0.2134	0.0415
8	304.1	−0.0432	0.1584	−0.0594	−0.0257

15.3.6 Results for ClMe

Although the fitted model for ClMe is by no means perfect some interesting results emerged from the analysis. The first of these was that the cross-correlation parameter c_0 refined to quite a large value. What this means is in fact that one of the four molecular orientations shown in Fig. 15.9(b) appears not to occur at all in the structure. This is the orientation corresponding to $\rho = +1$ and $\sigma = -1$. Figure 15.11(a) shows a plot of the positions of the atoms in all unit cells superimposed on a single unit cell. Figure 15.11(b) shows a similar plot but in this only atoms corresponding to when $\sigma = +1$ and $\rho = +1$. This occurs for (50%) of the molecules. Figure 15.11(c) shows a similar plot for $\sigma = -1$ and $\rho = +1$ (20% of the molecules) and Fig. 15.11(d) shows a similar plot for $\sigma = -1$ and $\rho = -1$ (30% of the molecules). Note that these %'s are consistent with the occupancies reported for the side-to-side and end-to-end disorder. That is, $P(\rho = +1) = (20\% + 50\%) = 70\%$. $P(\rho = +1) = (50\% + 0\%) = 50\%$.

The second interesting result relates to the size-effect parameters. Figure 15.12 shows plots of the distribution of vector lengths in the final simulation for two particular vectors, 3 and 19. For vector 3 it is seen that the length of the vector is longer than average when the two linked molecules (n, m) are in orientations corresponding to $\sigma_n = +1$ and $\sigma_m = -1$ but are smaller than average for other combinations. The particular combination $\sigma_n = +1$, $\sigma_m = -1$ corresponds to the orientations when the Cl ends of the two molecules are in contact. This clearly shows that the size-effect occurs because the Cl's try to avoid each other. A similar effect is observed for the vector 19, except that here the Cl—Cl combination is longer than average but the Me—Me combination is shorter than average.

It is also interesting to see just how the size-effect on vector 3 influences the diffraction pattern. Figure 15.13 shows a plot of the observed diffuse scattering placed alongside and in the same orientation as a diagram showing the arrangement of vectors. A line is drawn in the direction of vector 3 that extends to overlap the diffraction pattern. A line normal to vector 3 and passing through the centre of the diffraction pattern is also drawn. It is seen that close to this second line the diffuse peaks, such as those indicated

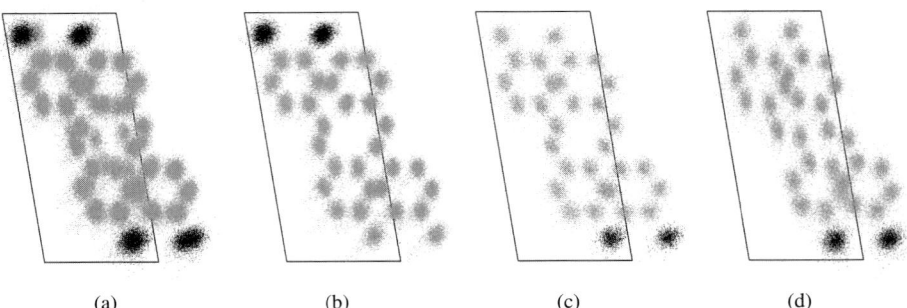

Fig. 15.11 Plot of the average unit cell obtained by superposing on a single unit cell the positions of atoms from all unit cells in the MC simulation. Black corresponds to the Cl atom, grey to C atoms. (a) Includes all molecular orientations. (b) Molecular orientations corresponding to $\sigma = +1$, $\rho = +1$ (50%). (c) Molecular orientations corresponding to $\sigma = -1$, $\rho = +1$ (20%). (d) Molecular orientations corresponding to $\sigma = -1$, $\rho = -1$ (30%). The refined cross-correlation parameter between ρ and σ resulted in there being 0% of the molecular orientation corresponding to $\sigma = +1$, $\rho = -1$.

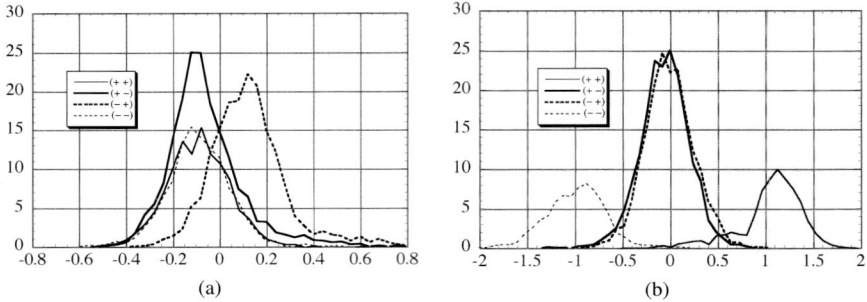

Fig. 15.12 Plots of the distribution of vector lengths for two vectors, 3 and 19 (see Fig. 15.8), in the final MC simulation, as a function of the type of intermolecular vector, $\sigma_n - \sigma_m = ++, -+, +-, --$. In each case the curve corresponding to the largest distance occurs when the contact is essentially between a Cl atom in one molecule with a Cl atom in the other.

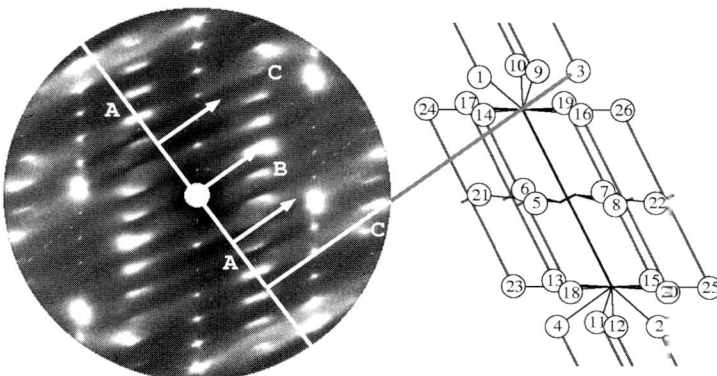

Fig. 15.13 Diagram showing how size-effect distortions along vector 3 cause asymmetry in the diffuse peaks of the diffraction pattern. The effect is a minimum along a line through the origin normal to vector 3 and increases away from this line. Thus at A the peaks are symmetric while at B and C they are increasingly asymmetric.

by the letter *A*, are symmetric, indicating that they are not appreciably affected by the size-effect distortion. As one moves away from this line in the direction of the arrows the amount of asymmetry increases progressively, for example, at B and C.

15.4 Conclusion

The two examples described in this chapter represent opposite extremes of what currently can be achieved by using the automatic refinement method to study diffuse scattering problems. The benzil example is essentially a study of thermal diffuse scattering (TDS) of a perfectly ordinary molecular crystal. The outcome was a model which gave a truly quantitative description of the observed diffuse scattering not far short of the levels of agreement routinely available for Bragg scattering. What was revealed by this study was some insight into which intermolecular interactions are important and how they lead to the particular molecular motions which cause the diffuse scattering. Of particular note in benzil is the fact that the motion involves the coupling of one of the internal torsional modes of the molecule with a low energy lattice mode involving the shearing of a hydrogen bonded network of molecules. Also revealed was the fact that as the central torsion angle varies the phenyl rings also adjust their orientation in a correlated way. The scattering in benzil is by no means unusual and the automatic refinement method has reached a point in its development where this kind of information is obtainable for any molecular crystal for which single crystals are available. This is information over and above that revealed by the analysis of Bragg peaks using conventional crystallography. As the methodology improves the level of detail that can be revealed can only get better.

With four molecular orientations possible in each molecular site and with large local relaxational displacements accompanying the disorder the second example of ClMe represents a much more complex problem, and probably represents as complex a system as it is currently feasible to tackle. At the time when the work described was carried out computational resources were strictly limited and some quite drastic approximations had to be made. These included using a much simplified model and a much restricted set of observed data. The outcome in this case was a model giving a much poorer quantitative agreement with the observed data. Nevertheless the much simplified model did reproduce the variety of quite complex diffraction features and the results were convincing enough to suggest that use of more extensive data and removing some of the simplifying assumptions could lead to a much more quantitative result. With the much enhanced computational resources available now and in the future, a convincing quantitative solution to this kind of problem should be possible.

Despite the above limitations of the ClMe study described, some interesting and useful results emerged. A key feature of the model was a size-effect parameter applied to one of the strongest inter-molecular spring constants. This was along a vector linking sites at the ends of a pair of molecules and the determined parameters revealed that when the contact involved the Cl substituent in both of the molecules then the ends were forced apart relative to the average site separation. It was this size-effect that resulted in the highly asymmetric observed diffuse peaks. The refinement also indicated that the end-to-end disorder and the side-to-side disorder were correlated so that in effect only three of the four possible orientations of the molecule were present. There is no doubt that with more extensive data and an improved model even more detail of this complex system can be revealed.

One final point regarding the automatic MC refinement method should perhaps be made. In all the published examples that have utilised the method to date a solution has been obtained by starting from a disordered configuration with random occupancies and, if required, random displacements from the average positions. So far this seems to have enabled smooth refinement towards a final solution. In the main, though with the possible exception of the ClMe system described above, the systems have been relatively simple and it is possible that for more complex systems the model parameter space will be sufficiently complex that the refinement will only find a local minimum that may be distant from the true minimum. In this case it is necessary that, in addition to the *structure refinement* afforded by the automatic MC methodology, a *structure solution* stage of the analysis is required. Some recent work by Weber and Bürgi (2002) in which a genetic algorithm has been used to explore the parameter space is a possible way in which such a *structure solution* might be achieved. In any case the future of the use of MC simulation to test and refine model systems seems assured.

DISORDER INVOLVING MULTI-SITE INTERACTIONS

16.1   Introduction

In much of the work described in Chapters 9–13, and in particular Chapter 14 on the automatic refinement of Monte Carlo (MC) models it is apparent that the models used most often have been based on 'pair-interactions' of various kinds. For binary variables representing site occupancy pair-interaction Ising models have frequently been used while for continuous variables representing atomic displacements either similar direct Ising-like pair-interactions have been used or pairs of atoms or molecules have been linked together by 'springs'.

The fact that such models have been successful in producing model calculated diffraction patterns in generally good agreement with observation could perhaps be supposed to be testimony to the fact that for the kinds of examples discussed a description in terms of pair-interactions is valid. However, it should always be remembered that the diffracted intensity only contains information about pair correlations and so a model built of pair-interactions is always likely to be able to reproduce the observed intensity distribution, provided sufficient numbers of interactions are used. The question really is whether the model derived in this way is physically meaningful or would, perhaps, a simpler model involving multi-site interactions be physically or chemically more appropriate? It will be recalled that in an earlier chapter (Section 3.4) it was shown that a 1D model involving second-nearest-neighbour interactions, including a 3-body interaction, could give an identical diffraction pattern to a third-nearest neighbour model involving only pair-interactions.

In this chapter studies of systems are described in which a treatment solely in terms of pair-interactions is clearly not appropriate.

16.2   Oxygen/fluorine ordering in $K_3MoO_3F_3$

$K_3MoO_3F_3$ is one member of a rather large group of oxyfluoride phases which are reported to crystallise either in the ideal cubic ($Fm\bar{3}m$) A_2BMX_6 elpasolite (ordered perovskite) structure type or in closely related modulated variants thereof (see Fig. 16.1). The materials are of interest because of their interesting ferroelectric and ferroelastic phase transition behaviour, but the focus of the discussion here relates to the question of O/F ordering in them. Although the optical and dielectric properties of these materials have been extensively studied the corresponding crystallographic properties have been far less well understood.

Like many other oxyfluoride systems random anion site disorder has invariably been reported for the anion, or X site, positions of these phases. This is despite the fact that

O and F are quite distinct species that ought to provide a strong driving force for local crystal chemical ordering and associated structural distortion. However, the fact that O and F are similar in scattering power for any of the common diffraction techniques has considerably complicated the search for direct evidence of O/F ordering. (See Withers *et al.*, 2003, for more details).

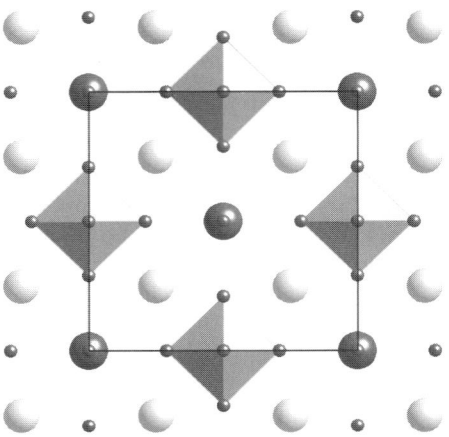

Fig. 16.1   Schematic diagram of the ideal cubic ($Fm\bar{3}m$) A_2BMX_6 elpasolite structure (section from $z = 0.2$ to 0.8). The small dark balls are X anions, the larger light balls are the A cations, the larger dark balls are the B cations and the MoX_6 octahedra are shown as grey shaded octahedra.

16.2.1   *Observed diffuse scattering patterns*

In the case of $K_3MoO_3F_3$ a recent electron diffraction study (Brink *et al.*, 2002) has reported the existence of a complex three-dimensional continuous diffuse intensity distribution presumably arising from the O/F ordering and/or the associated structural distortions which act to improve the local crystal chemistry. Figure 16.2 shows examples of 2D sections of this 3D diffuse distribution. The 'sharp', highly structured, essentially continuous nature of the diffuse scattering is immediately apparent. These remarkable patterns are not uncommon. For example, similar diffuse distributions have been observed in non-stoichiometric vanadium carbide, VC_{1-x}, and other substoichiometric transition metal carbides and nitrides (see Sauvage and Parthé, 1972). These authors showed that the shape of the diffuse curves in reciprocal-space were approximated quite closely by the equation:

$$\cos \pi h + \cos \pi k + \cos \pi l = 0. \tag{16.1}$$

Sauvage and Parthé (1972) used the shape of the observed diffuse distribution in VC_{1-x} to extract a conventional series of two body Warren–Cowley short-range order (SRO) parameters. Subsequently de Ridder *et al.* (1976, 1977) interpreted the significance of eqn (16.1) in terms of a six-body octahedral cluster relationship—namely that each transition metal ion should always be surrounded by the average number of C or

N atoms. That is, there should, as far as possible, be only one vacancy in the nearest-neighbour octahedron of available sites.

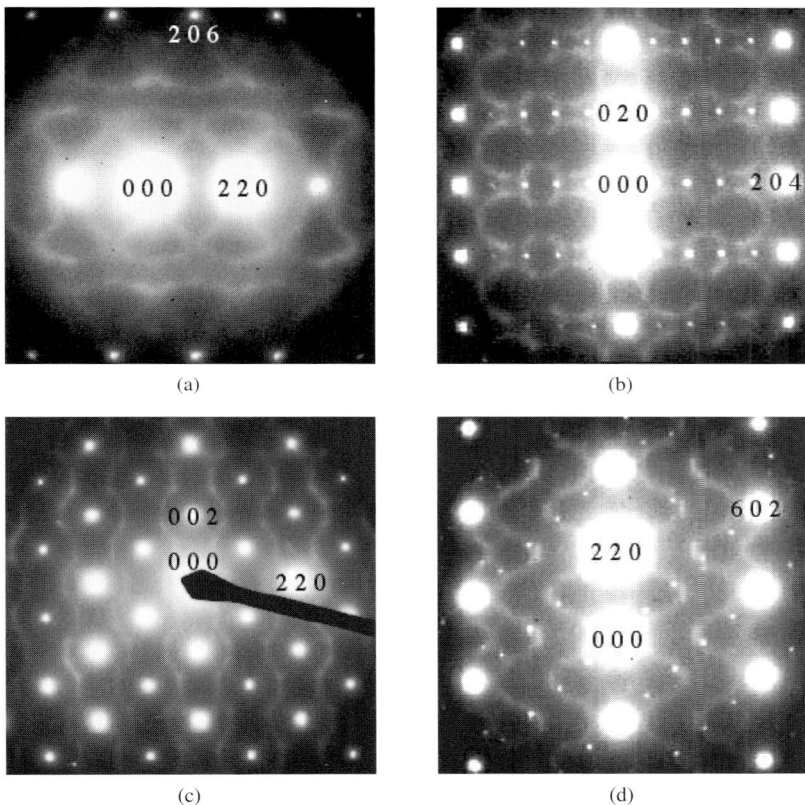

(a)

(b)

(c)

(d)

Fig. 16.2 Electron diffraction photographs characteristic of $K_3MoO_3F_3$. (a) $\langle 3\,\bar{3}\,1\rangle$ zone axis. (b) $\langle 2\,0\,\bar{1}\rangle$ zone axis. (c) $\langle 1\,\bar{1}\,0\rangle$ zone axis. (d) $\langle \bar{1}\,1\,3\rangle$ zone axis.

16.2.2 Chemical constraint for $K_3MoO_3F_3$

For the present case of $K_3MoO_3F_3$ the analogous requirement would be that each Mo ion be always surrounded by the average number of anions, that is, 3O's and 3F's. Such a local O/F ordering constraint is chemically sensible but would not by itself, however, be sufficient to generate the observed diffuse scattering. Reference to Fig. 16.1 shows that each MoX_6 is not connected directly to any other (i.e. a given X is not shared by neighbouring Mo's) and so any SRO within one octahedron cannot be propagated unless some further criterion is used. In order for SRO to propagate the octahedral cluster relationship giving rise to the diffuse intensity must be associated with the larger octahedron of 6 MoX_6 octahedra which surround each $K(2)X_6$ octahedron.

In addition it is clear that the observed diffuse scattering cannot be due primarily to O/F ordering because as stated earlier their scattering factors are too similar. Rather it must be assumed that the scattering is primarily associated with cation shifts, most likely Mo cation shifts, associated with the O/F ordering. In two previously studied compounds (FeOF and NbO_2F) it was found that opposite corners of each local octahedron tended to be occupied, one by an O and the other by an F. In turn, this local O/F ordering induced cation shifts away from the F's and towards the O's. Adopting the same scheme here it is seen that, viewed as a whole, any octahedron then has one triangular face that has 3O's, one that has 3F's and all of the other faces have either 2O's and an F or 2F's and an O.

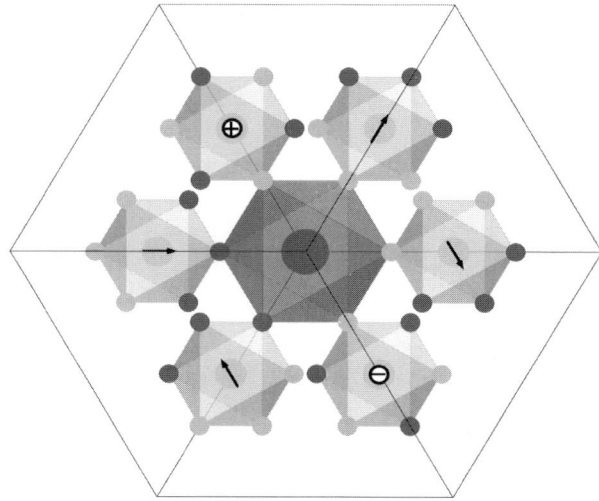

Fig. 16.3   Schematic diagram showing part of the structure of $K_3MoO_3F_3$ viewed down $[1\,1\,1]$. The central $K(2)X_6$ octahedron is surrounded by six MoO_3F_3 octahedra. These have been drawn with different orientations. Small dark circles represent O atoms and small light circles F atoms. Arrows indicate the expected displacement directions for the Mo atoms.

The resultant cation shift is thus directed, in one of the four different $\langle 1\,1\,1 \rangle$ directions, away from the face containing the 3F's and towards the face containing the 3O's. Figure 16.3 shows a view of the structure (down $[1\,1\,1]$) showing six such MoX_6 octahedra surrounding a single K(2) ion. Each of the six octahedra has been drawn with a different configuration of the 3O's and 3F's. For four of the six a black arrow indicates the direction that the enclosed Mo ion would be expected to be displaced. In the other two the displacement would be expected to be either into $(-)$ or out of $(+)$ the plane of the paper. These different octahedral configurations are just different orientations of the $(MoO_3F_3)^{3-}$ ion. Each of the $(MoO_3F_3)^{3-}$ ions has a dipole moment of 3m symmetry directed in either the $+$ or $-$ sense along one of the possible $\langle 1\,1\,1 \rangle$ real-space directions. The direction of the local Mo shift defines both the orientation and sense of the local dipole moment. If all these $(MoO_3F_3)^{3-}$ dipole moments were aligned it would obviously lead to a macroscopic spontaneous polarisation or dipole moment per unit volume. If this is to be avoided it is necessary that the sum of all the individual dipole moments is zero. This is easily achieved on a macroscopic scale, but can it also be achieved on the microscopic (unit cell) scale? Can it be achieved, for example, for the octahedral cluster of six MoX_6 octahedra which surround each $K(2)X_6$ octahedron?

16.2.3 *MC simulation using a simple constraint*

Monte Carlo simulation was used to produce a real-space O/F and associated Mo shift distribution that satisfied as far as possible the constraint that, for the octahedral cluster of six MoX_6 octahedra which surround each $K(2)X_6$ octahedron, the sum of the $(MoO_3F_3)^{3-}$ dipole moments is zero. That such a constraint does indeed give rise to a diffuse distribution closely related to that observed experimentally is apparent from simulated diffraction patterns obtained from the resultant real space distribution as shown in Fig. 16.4. The extraordinary similarity of Fig. 16.4 to Fig. 16.2 is immediately apparent.

The MC simulations were carried out as follows. A simulation volume was used which comprised $32 \times 32 \times 32$ parent elpasolite type unit cells, each of which included 4 $Mo(O/F)_6$ octahedra. The local conformation of these octahedra was defined in terms of three sets of binary $(-1,+1)$ random variables $u_{i,j,k,l}$, $v_{i,j,k,l}$, $w_{i,j,k,l}$ where i,j,k are indices defining the parent unit cell and l the particular $Mo(O/F)_6$ octahedron within that unit cell. These binary variables served two purposes. Firstly they were used to define which type of anion (O or F) was the nearest-neighbour to the central Mo ion in each of the three mutually orthogonal directions. Secondly, they were used to define a displacement shift from its ideal average position of the Mo cation itself. Thus, for example, a value of $u_{i,j,k,l} = +1$ for a Mo at $(0,0,0)$ would mean that the anion site at $(x,0,0)$ was occupied by O and that at $(-x,0,0)$ by F. At the same time the Mo itself would be displaced away from the F and towards the O. Conversely for a value of $u_{i,j,k,l} = -1$ the O and F would be interchanged and the Mo displacement would be reversed. Similarly the variables $v_{i,j,k,l}$ and $w_{i,j,k,l}$ corresponded to analogous arrangements of O/F neighbours and the associated Mo displacements in the $y-$ and $z-$ directions respectively. (In all the calculations the value of x was assumed to be 0.20 and the magnitude of the Mo shift was assumed to be 0.011 (fractional coordinate shifts) which corresponds to ± 0.095 Å.

The above formulation automatically constrains every $Mo(O/F)_6$ octahedron to contain 3O and 3F. This means that each octahedron is in one of eight different orientations where the triangular face containing 3Os points towards one of the eight $\langle 1\,1\,1 \rangle$ directions, as does the associated Mo displacement vector. (Since $K(2)(O/F)_6$ octahedra occur between these $Mo(O/F)_6$ octahedra and share the same anions, it is interesting to consider how the distribution of O and F in these octahedra is affected by the anion ordering in the Mo octahedra. Table 16.1 contains some statistics of this distribution before and after the simulation. The quantity minimised in the MC simulation was an energy defined as

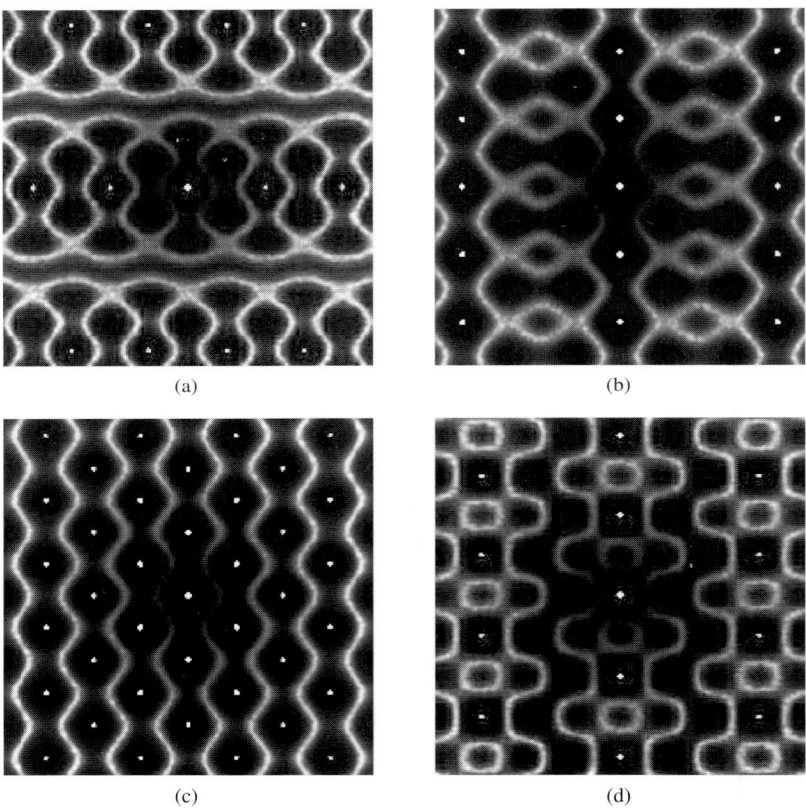

(a)   (b)

(c)   (d)

Fig. 16.4 Diffraction patterns of $K_3MoO_3F_3$ calculated from the MC simulation using the simple constraint eqn (16.2). (a) $\langle 3\bar{3}1 \rangle$ zone axis. (b) $\langle 20\bar{1} \rangle$ zone axis. (c) $\langle 1\bar{1}0 \rangle$ zone axis. (d) $\langle \bar{1}13 \rangle$ zone axis.

$$E_{\text{local}} = \sum K[(\bar{u})^2 + (\bar{v})^2 + (\bar{w})^2], \tag{16.2}$$

where,

$$\bar{u} = \sum_1^6 u_{i,j,k,l}, \qquad \bar{v} = \sum_1^6 v_{i,j,k,l}, \qquad \bar{w} = \sum_1^6 w_{i,j,k,l}. \tag{16.3}$$

Here $\bar{u}$, $\bar{v}$ and $\bar{w}$ are the respective sums of the six $u_{i,j,k,l}$, $v_{i,j,k,l}$ and $w_{i,j,k,l}$ random variables corresponding to the six nearest-neighbouring Mo sites that surround a given K(2) site. That is, the energy E_{local} attempts to provide that the net displacement of Mo cations is constrained to zero over the first octahedral shell surrounding any given K(2) ion.

Simulation was carried out for 200 MC cycles (that number of individual MC steps required to visit each Mo site once on average). Each random variable was initially set to be arbitrarily $+1$ or -1. At each step in the iteration the random variables at two different i, j, k, l sites were interchanged and the effect on the total energy of the system computed. In this way the total numbers of $+1$ and -1 values was maintained throughout the simulation. As the iteration proceeded various lattice averages were monitored and these are given in Table 16.1. Diffraction patterns were calculated from the final distributions using only the Mo positions. These calculations, made with the program DIFFUSE (Butler and Welberry, 1992), obtained the diffraction pattern by taking the average of a large number of sub-regions (lots) of the main simulation array. In the present case each computed pattern was obtained as an average of 400 individual lots, each of size $10 \times 10 \times 10$ unit cells.

Table 16.1 Fractions of the different anion configurations for the K(2) octahedra before and after the ordering of the MoO$_3$F$_3$ octahedral orientations

Configuration	6O	5O+F	4O+2F	3O+3F	2O+4F	O+5F	6F
Initial random	0.016	0.093	0.234	0.315	0.233	0.093	0.016
Final	0.022	0.104	0.228	0.292	0.229	0.104	0.022

The figures given in Tables 16.1 and 16.2 demonstrate a number of points. Firstly it is apparent that the ordering of the MoO$_3$F$_3$ orientations has only a relatively small effect on the frequencies with which different anion arrangements occur in the K(2) octahedra, although the changes that do occur tend to reduce the number of those configurations which have equal numbers of F and O and increase those that are predominantly O or predominantly F. Attempts were made to incorporate an additional term in the MC energy which would constrain the distributions of anions in the K(2) octahedra to comprise only (2O+4F), (3O+3F) and (4O+2F) combinations. It proved quite feasible to do this but the resulting effects on the diffuse scattering patterns were quite deleterious.

The second point demonstrated by the figures in Tables 16.1 and 16.2 is that it has not been entirely possible to satisfy for all K(2) sites the rule that the displacements of

Table 16.2 Mean values of the displacement variables $u_{i,j,k,l}$ for the six Mo atoms around a given K(2) site. Values obtained for the $v_{i,j,k,l}$ and $w_{i,j,k,l}$ displacements are the same within expected statistical error

Lattice Average	$\langle \lvert \sum_{1}^{6} u_{i,j,k,l} \rvert \rangle$	$\langle \lvert \sum_{1}^{6} u_{i,j,k,l} \rvert^{2} \rangle^{1/2}$
Initial random	1.876	2.449
Final	0.447	0.947

the six Mo atoms that surround any given K(2) ion should all sum to zero (i.e. the energy $E_{\text{local}} = 0$ see eqn (16.2)). Nevertheless the lattice averages in real-space (along with the simulated diffraction patterns in reciprocal-space) show that substantial progress towards this ideal has been achieved. Increasing the number of cycles would undoubtedly enable even further progress to be made although ever more slowly and at the expense of increased computational time.

16.3   Short-range order in $(Bi_{1.5}Zn_{0.5})(Zn_{0.5}Nb_{1.5})O_7$

In this section studies are described on one of the phases that is found to occur in the $Bi_2O_3 - ZnO - Nb_2O_5$ (BZN) ternary system (see also Withers *et al.*, 2004). The nominal composition of the phase of interest is $(Bi_{1.5}Zn_{0.5})(Zn_{0.5}Nb_{1.5})O_7$ and this is of the conventional cubic $A_2B_2O_6O'$ pyrochlore structure type.

16.3.1   *Observed diffuse scattering patterns*

Figure 16.5 shows electron diffraction patterns from BZN taken close to various different zone axes. So far BZN has only been available in the form of powder samples so that observation of X-ray diffuse scattering has not been feasible. Even obtaining the electron diffraction patterns is difficult because the scattering is weak and must be observed in the presence of very strong Bragg peaks and the examples shown are by no means perfect. Nevertheless the patterns are clearly seen to show highly structured diffuse scattering. Figure 16.5(d) shows most clearly that this consists of diffuse blobs at points in reciprocal-space defined by $G \pm \langle 001 \rangle^*$, where G is a reciprocal lattice vector of the $Fd\bar{3}m$ pyrochlore cell. In other sections it is apparent that there are in addition more extended diffuse streaks connecting the blobs. Also note that in Fig. 16.5(b) the diffuse scattering appears to be in the form of a *motif* of scattering around each of the allowed Bragg peak positions (e.g. (400), (044) and (004)). This is somewhat reminiscent of the *motifs* discussed in wüstite (see Section 12.2). Also similar to the wüstite example is the fact that the *motifs* show an asymmetry which is characteristic of 'size-effect' distortions.

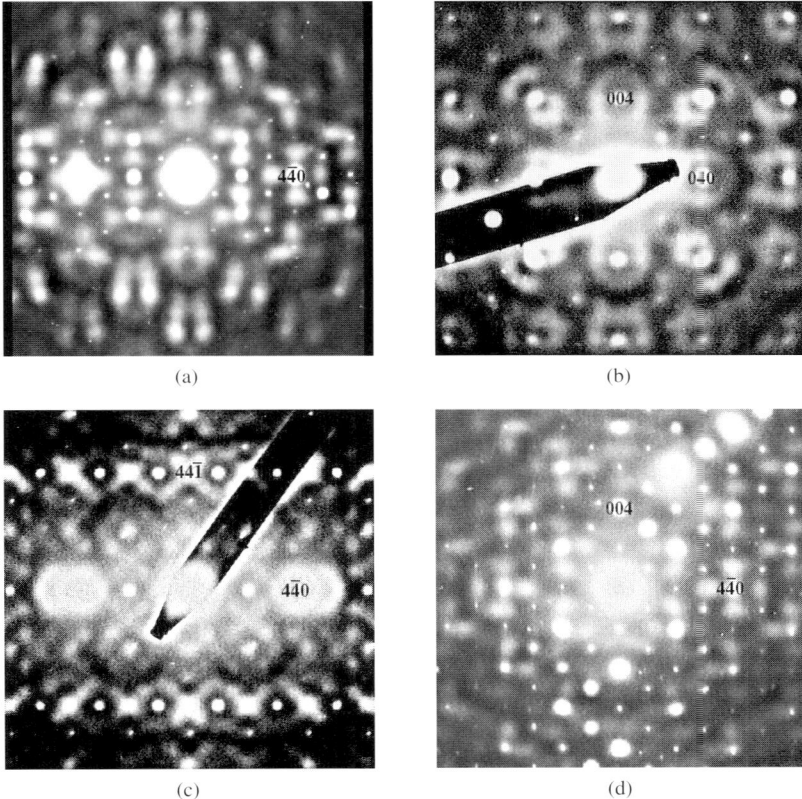

Fig. 16.5 Electron diffraction photographs characteristic of BZN. (a) $\langle 551 \rangle$ zone axis. (b) $\langle 001 \rangle$ zone axis. (c) $\langle 118 \rangle$ zone axis. (d) $\langle 110 \rangle$ zone axis.

16.3.2 *Chemical considerations*

In order to understand the details of the work described in this section it is important to realise that this rather complex structure is comprised of two interpenetrating substructures, each of which is much simpler. These are a B_2O_6 octahedral substructure and an $O'A_2$ tetrahedral substructure. This latter has the anti-cristobalite structure. The two substructures are shown in Fig. 16.6.

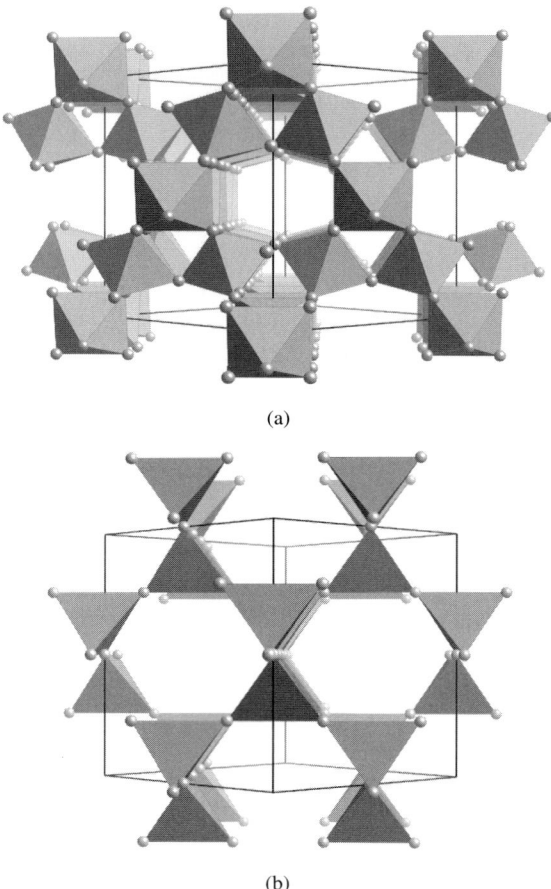

(a)

(b)

Fig. 16.6 The two different sublattices in BZN viewed approximately down $[1\,1\,0]$. (a) The B_2O_6 corner-connected octahedral substructure; (b) The $O'A_2$ corner-connected tetrahedral anti-cristobalite substructure. Note how the tetrahedra in (b) fit in the voids in (a).

16.3.2.1 *The B_2O_6 octahedral substructure* The B site contains $\frac{1}{4}Zn^{2+}$ and $\frac{3}{4}Nb^{5+}$. Despite their quite different charges, apparent valence (AV) calculations (see Brese and

O'Keeffe (1991)) suggest that the preferred bond length to the surrounding oxygens in the octahedral environment is 1.98 Å for Nb^{5+} and 2.11 Å for Zn^{2+}. It would be expected that the B—O distance of an *average* B site would be $(\frac{1}{4} \times 2.11) + (\frac{3}{4} \times 1.98) = 2.01$ Å. This is in fact very close to the B—O distance observed in the average structure determination. The fact that these two preferred B—O distances do not differ a great deal from the average implies that if these two ions are disordered on the B sites the surrounding oxygens would not need to relax very much. This is reflected in the relatively small atomic displacement parameters observed for the B and O sites in the average crystal structure determination. For these reasons it seems unlikely that this substructure is the major contributor to the observed diffuse scattering.

16.3.2.2 *The $O'A_2$ tetrahedral anti-cristobalite substructure* In this substructure it is the O' that occupies the site at the centre of the tetrahedron and the A cations that are at the corners. On average there are 3 Bi^{3+} and 1 Zn^{2+} per tetrahedron. The O' anion in this substructure is tetrahedrally coordinated by four A cations at a distance of 2.287 Å. This distance is determined solely by the cubic lattice parameter a. For Bi^{3+} in the A site, the ideal O'—Bi distance is 2.35 Å whereas for Zn^{2+} in the A site, this ideal O'—Zn distance is 1.96 Å. The weighted average (2.25 Å) of these two ideally preferred distances is again close to the observed average. However, unlike in the B_2O_6 octahedral substructure, the two ideal bond lengths in this substructure differ markedly from each other and from the mean. This is reflected in the much larger displacement parameters obtained for the A and O' sites in the average structure determination. There thus appears to be strong grounds for expecting that there will be significant cation ordering and a much greater need for structural relaxation in this substructure. To see what form this might take consideration is first given to how well the valence of the O' ion is satisfied for different configurations of the cations in a given tetrahedron.

16.3.2.3 *Valence considerations for the O' site* If there is disorder in the A cation sites it is interesting to note the relative apparent valence of the O' at the centre of the tetrahedron for different combinations of the Bi^{3+} and Zn^{2+} ions at the corners. Table 16.3 shows the apparent valence values computed on the assumption that all the ions are fixed on the ideal average sites of the anti-cristobalite substructure. These values clearly show that there should be a strong preference for each tetrahedron in the structure to have the 3Bi+Zn stoichiometry.

Table 16.3 Apparent valence of the O' ion for different A_4 stoichiometries

Configuration	4Bi	3Bi+Zn	2Bi+2Zn	Bi+3Zn	4Zn
Apparent valence	2.38	1.99	1.60	1.22	0.83

16.3.2.4   *Valence considerations for the* A_4 *ions*   Although Table 16.3 shows that the O' valence is well satisfied in the average regular tetrahedron if the stoichiometry is $3Bi+Zn$, the same is not true for the cations. Ideally the $O'-Bi$ distance would like to be 2.35 Å and the $O'-Zn$ distance would like to be 1.96 Å instead of the average 2.29 Å. This can only be achieved by a combination of ion shifts involving all five ions. First the O' must be displaced toward the Zn and away from the Bi but at the same time the lengths of the edges of the tetrahedron will change depending on whether the edge is $Bi-Bi$ or $Bi-Zn$. Although calculation of the exact magnitude of these shifts is complex, as it necessarily involves the other O ions to which the A ions are bonded and the detailed geometry of the B_2O_6 substructure, it is clear that the shifts from the average positions would be substantial and quite capable of explaining the 'size-effect' distortions evident in the observed diffraction patterns. (See Withers *et al.*, 2003, for more details).

16.3.3   *MC simulaton of occupancy disorder*

Given the fact the average structure refinement showed that the atomic displacement factors were an order of magnitude larger for the A and O' ions than for the B and O ions, in carrying out computer simulations only the the corner-connected $O'A_2$ tetrahedral substructure was considered, and it was assumed that all other contributions to the diffuse scattering would be negligible in comparison. The aim was to investigate the possible ordering of the Bi and Zn ions on the A site and the subsequent structural relaxation. A model crystal was set-up comprising $32 \times 32 \times 32$ unit cells each of which contained 16 A cation sites and 8 O' sites (i.e. 786,432 atomic sites in total). During the MC simulation only the A cation sites were in fact used with the O' ions being subsequently inserted within each A4 tetrahedron prior to calculation of the diffraction pattern. That is, the positions of the O' ions were assumed to depend on the positions of the surrounding cations (see later).

16.3.3.1   *Ordering of Bi and Zn*   Random variables x_{ijkl} were used to represent the occupancy of each A cation site in the model lattice. The indices i, j, k refer to a particular unit cell and l to the site within the cell. A value of $x_{ijkl} = 0$ corresponds to that site being occupied by a Zn and $x_{ijkl} = 1$ to a Bi. Initially all of the x_{ijkl} were chosen randomly to be 1 with a probability of 0.75 and 0 with a probability of 0.25. In all subsequent simulations these concentrations were maintained since at each step in the MC iteration the variables from two randomly chosen A-sites were interchanged. In order to try to order the Bi and Zn cations on the A-sites an MC energy of the form

$$E_1 = \sum_{\substack{\text{all} \\ \text{tetrahedra}}} (N_{Bi} - 3)^2 \tag{16.4}$$

was used. Here N_{Bi} is the number of A-sites within a given tetrahedron that are occupied by Bi. The MC iteration was then carried out as follows. Two different A-sites i, j, k, l were chosen at random and the energy E_1 was calculated by summing over those tetrahedra which contained the targeted A-sites. After interchanging the two selected sites the energy was recalculated and the difference, $\Delta E = E_{\text{new}} - E_{\text{old}}$, obtained. If $\Delta E < 0$

then the new configuration was accepted, while for $\Delta E > 0$ the new configuration was accepted with probability $P = \exp(-\Delta E/kT)$, but otherwise the system was returned to its original configuration. Iteration was carried out for 200 cycles where a cycle consists of that number of individual MC steps required to visit each site once on average. A temperature $kT = 0.125$ was used throughout all simulations described here. This value was chosen after some initial trials to assess the efficiency with which completeness of ordering could be achieved within a 200 cycle simulation. Table 16.4 gives the fraction of individual tetrahedra containing 0, 1, 2, 3 or 4 Bi ions in the initial distribution and after ordering with energy E_1. Figure 16.7 shows two example diffraction patterns calculated from the ordered distribution.

Table 16.4 The fraction of individual tetrahedra containing 0, 1, 2, 3 or 4 Bi ions in the initial distribution and after ordering in various different ways.

Number of Bi, n	0	1	2	3	4
Random	0.004	0.050	0.211	0.424	0.312
E_1 ordering	0.0	0.0	0.016	0.967	0.017
$\frac{1}{2}E_1 + \frac{1}{2}E_2$ ordering	0.0	0.0	0.042	0.916	0.042
$\frac{2}{3}E_1 + \frac{1}{3}E_2$ ordering	0.0	0.0	0.024	0.952	0.024

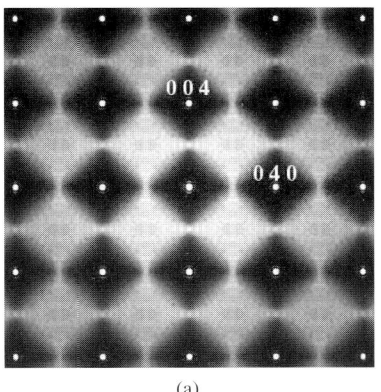
(a)

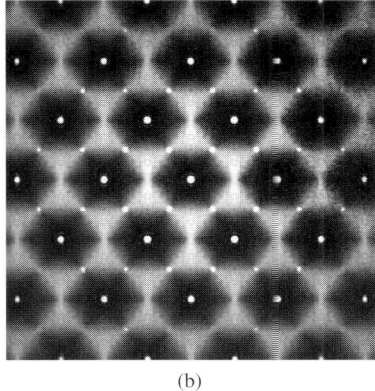
(b)

Fig. 16.7 Calculated diffraction patterns for the MC simulation in which the energy E_1 was used to order the tetrahedral configurations. (a) $\langle 1\,0\,0 \rangle$ zone axis. (b) $\langle 1\,1\,0 \rangle$ zone axis.

It is clear from Fig. 16.7 that although local ordering of the Bi/Zn ions within the tetrahedra produces structured diffuse scattering this is insufficient to produce the sort of longer-range ordering observed in the electron diffraction pictures (see Fig. 16.5. To achieve this a longer-range interaction acting within each of the four different sublattices of the fcc anticristobalite structure is required. Consequently a second energy term was added to the MC energy. This was of the form,

$$E_2 = \sum_{\substack{\text{all } \frac{1}{2}\langle 1\,1\,0\rangle \\ \text{neighbours}}} \delta_{m,n}. \tag{16.5}$$

Here $\delta_{m,n}$ is unity if both the target site and the neighbouring site are occupied by Zn and is zero otherwise. This term therefore has the effect of ensuring that there are very few Zn ions in close proximity to a particular Zn ion along any of the $\langle 1\,1\,0\rangle$ rows of cations. Figure 16.8 shows a local configuration comprising several Bi_3Zn tetrahedra. It is seen that starting from a given central Zn ion next-nearest-neighbouring tetrahedra can each be in one of three different configurations. Of those shown the Zn atoms indicated by arrows are ones for which $\delta_{m,n} = 1$ and the energy E_2 will give an energy penalty. Equation (16.5) was only one of a number of different (but similar) ordering schemes that were tried and was adopted because it gave results which were qualitatively most similar to the observed patterns.

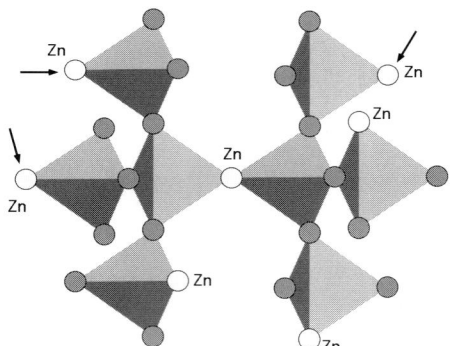

Fig. 16.8   Schematic diagram showing the propagation of disorder in the corner-connected $O'A_2$ tetrahedral anti-cristobalite substructure. The Zn atoms indicated by arrows are ones occurring at a vector distance $\frac{1}{2}\langle 1\,1\,0\rangle$ from the central Zn atom.

Two simulations were carried out using E_2 in addition to E_1. In the first case the total MC energy used was $\frac{1}{2}E_1 + \frac{1}{2}E_2$, while in the second the energy was $\frac{2}{3}E_1 + \frac{1}{3}E_2$. The effect on the $[1\,0\,0]$ diffraction pattern of adding this second energy term is shown in Fig. 16.9. In Fig. 16.9(a), which corresponds to the $\frac{1}{2}E_1 + \frac{1}{2}E_2$ case, it is seen that there is a motif of eight diffuse diffraction spots forming a square centred around the $\langle 2\,2\,0\rangle^*$, $\langle 6\,2\,0\rangle^*$ and $\langle 4\,4\,0\rangle^*$ positions. In Fig. 16.9(b), which corresponds to the $\frac{2}{3}E_1 + \frac{1}{3}E_2$

case, this *motif* is more blurred with the peaks at the corners of the square having been absorbed into a more continuous streak.

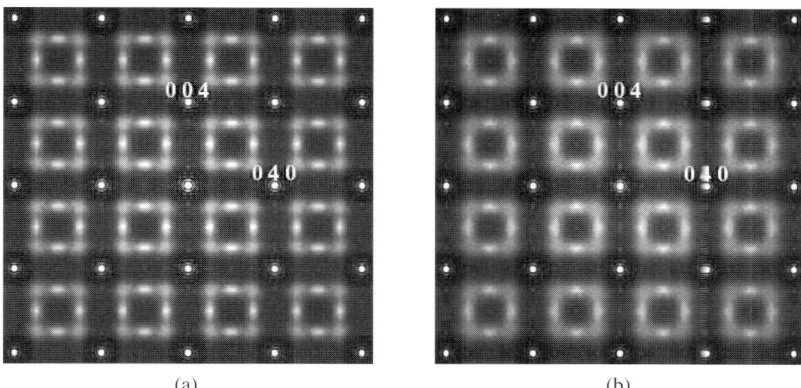

(a) (b)

Fig. 16.9 Calculated $\langle 1\,0\,0\rangle$ zone axis diffraction patterns for the MC simulation in which the energy E_2 was added in differing proportions to the original E_1 energy. (a) Energy $= \frac{1}{2}E_1 + \frac{1}{2}E_2$. (b) Energy $= \frac{2}{3}E_1 + \frac{1}{3}E_2$.

16.3.4 *MC simulaton of size-effect distortions*

The diffraction patterns in Figs. 16.7 and 16.9 were obtained from simulations in which ordering of cations had been carried out but with no local relaxation of the cation positions. That is, all cations were on their perfect anti-cristobalite lattice sites. In order to assess the effects of local relaxation a further stage of MC simulation was carried out in which neighbouring cation–cation distances (forming the edges of the tetrahedra) were subject to an MC energy of the form

$$E_3 = \sum_{\substack{\text{n.n. cation} \\ \text{vectors}}} [d - d_0(1 + \varepsilon_{m,n})]^2. \qquad (16.6)$$

Here d is the instantaneous length of a given cation–cation distance, d_0 is the average cation–cation distance and $\varepsilon_{m,n}$ is a 'size-effect' parameter. For the simulations described here $\varepsilon_{m,n}$ was taken as $+0.06$ for Bi—Bi vectors, -0.06 for Bi—Zn vectors and 0.0 for Zn—Zn vectors (of which there were very few). The value of ± 0.06 corresponds approximately to the magnitude of the distortion required to satisfy the apparent valence calculations for Bi and Zn discussed earlier.

With the occupancy distribution kept fixed, 200 MC cycles of iteration were performed using the energy E_3 to allow the positions of the cations to relax. This was carried out as follows. A cation site was selected at random and the energy E_3 computed by summing over all cation–cation vectors involving that site. The position of the site was then subjected to a small random shift and the energy computed again. If

the new energy was lower then the shifted position was accepted, otherwise the previous position was resumed. [This essentially corresponds to performing the relaxation at
$kT = 0$]. Two hundred cycles of iteration were found sufficient to achieve a configuration that was considered close to fully relaxed. Diffraction patterns were then computed
for comparison with the unrelaxed patterns described above. For each of the tetrahedra
an oxygen atom was inserted near the centre with a position that was calculated as the
weighted mean of the four surrounding cations. In this calculation a 'weight' of 0.47
was used for a Bi and 0.53 for a Zn so that in a tetrahedron with 3 Bi and 1 Zn the
O atom would be shifted towards the Zn corner by an amount corresponding to that
required to best satisfy the local apparent valence requirements (see earlier).

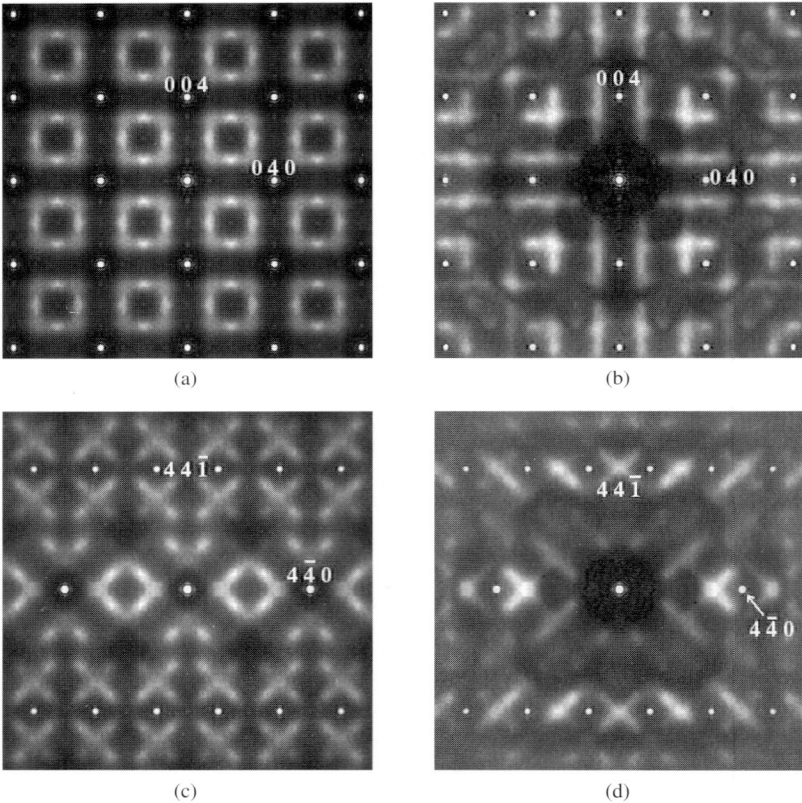

Fig. 16.10   Calculated diffraction patterns of BZN for two different zone axes showing the effect
of adding 'size-effect' distortions. (a) and (b) $\langle 1\,0\,0\rangle$ zone axis; (c) and (d) $\langle 1\,1\,8\rangle$ zone axis; (a)
and (c) without size-effect; (b) and (d) with size-effect.

Figure 16.10 shows the effect on the diffraction patterns of applying the size-effect
relaxations. Examples are shown for the $\langle 1\,0\,0\rangle$ and $\langle 1\,1\,8\rangle$ zone axes. The figures show

that the distortion has had a marked effect on the distribution of intensity in the patterns. All of these examples used the same occupancy distribution obtained using the energy $\frac{2}{3}E_1 + \frac{1}{3}E_2$.

In addition to observing the changes in the diffraction patterns it is also interesting to show plots of the distribution of atomic positions in unit cells of the fully relaxed simulation. Figure 16.11(a) shows a plot of the positions of the Bi, Zn and O' atoms from all of the unit cells in the simulation superposed onto a single unit cell. The distribution is shown in projection down $[1\,1\,0]$ since this most clearly shows some interesting features that occur. First, the cation sites show a very strong anisotropy indicating that the mean-square displacements are in the form of a flat disc. This in good agreement with the anisotropic average structure refinement described earlier. A second feature is that the displacements in the plane of the disc are substantially greater for the Bi ions (dark grey) than for the Zn ions (light grey). This is something that would be difficult to detect in an average structure determination using Bragg reflections. Finally it is seen that the method of placing the O' atoms (black) within each tetrahedron using the weighted mean described above has resulted in a distribution for that site which has a triangular appearance. In Fig. 16.11(b) we show a drawing of the structure for comparison.

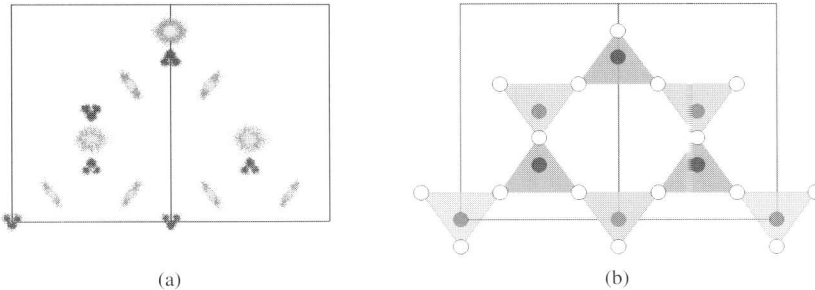

(a)   (b)

Fig. 16.11 (a) Plot of the positions of the Bi, Zn and O' ions from all of the unit cells in the MC simulation superposed onto a single unit cell, viewed down $[1\,1\,0]$. O' ions are plotted as black dots, Bi ions as dark grey dots and Zn ions as light grey dots. (b) The corresponding average structure drawing.

Finally Fig. 16.12 shows calculated diffraction patterns for the same four zone-axes as shown for the observed electron diffraction patterns in Fig. 16.5. Although the agreement is by no means perfect all the main features of the patterns have been qualitatively reproduced by the simple model. It is clear that any more quantitaive description would require a much more complex treatment involving both the $O'A_2$ tetrahedral substructure and the B_2O_6 octahedral substructure as well as the interactions between them.

16.4 Conclusion

In both the examples described in this chapter it has been shown that rather simple models involving a multi-site interaction were able to explain the form of really quite

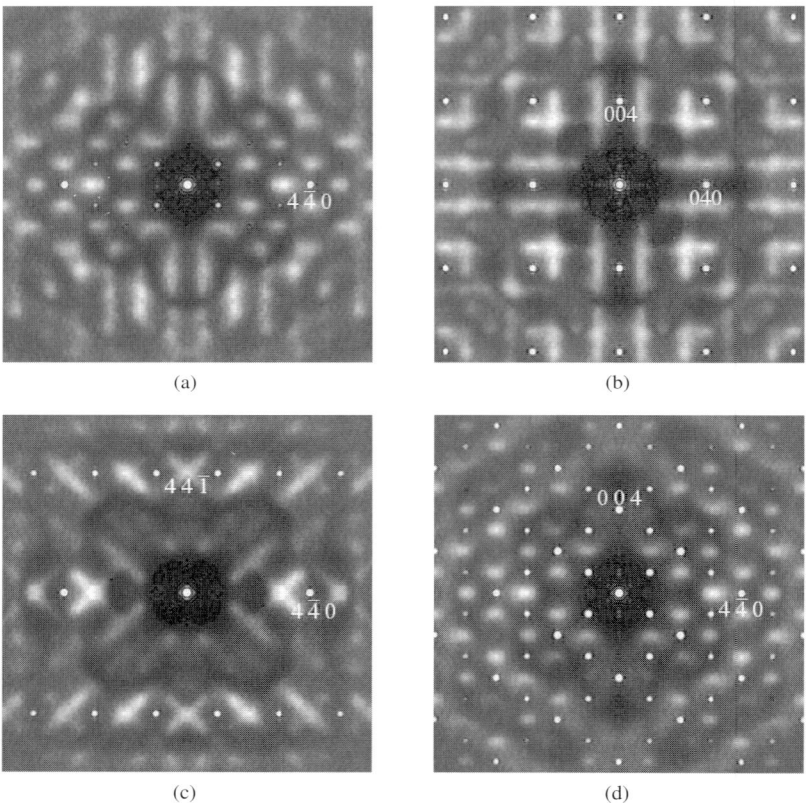

(a)

(b)

(c)

(d)

Fig. 16.12 Calculated diffraction patterns of BZN for the four zone axes corresponding to the observed electron diffraction patterns shown in Fig. 16.5. (a) $\langle 5\,5\,1 \rangle$ zone axis; (b) $\langle 0\,0\,1 \rangle$ zone axis; (c) $\langle 1\,1\,8 \rangle$ zone axis; (d) $\langle 1\,1\,0 \rangle$ zone axis.

complex diffraction patterns. In the first case the multi-site interaction took the form of a constraint applied to the orientations of the six MoO_3F_3 octahedra which surround a particular K(2) site. It isn not sufficient to describe the interaction in terms of *pairs* of such octahedra. All six must must arrange themselves in concert to try to reduce the net dipole moment to zero. In the second case a very similar multi-site (in this case four-site) interaction needed to be invoked to produce the basic cation ordering on the $O'A_2$ tetrahedral sub-lattice in BZN. Again it was not sufficient to invoke interactions that depended simply on *pairs* of neighbouring Bi/Zn cations but required all four A sites in the tetrahedra to be considered simultaneously. In both cases it is the requirement that the local chemistry of the constituent ions be satisfied that is the driving force for the local order.

It is clear from the diffraction patterns that result from these simple multi-site inter-actions that the two-body correlation fields that are generated (the only thing that the diffraction experiment detects) are very complex. It would clearly be missing the whole basis for the disorder and local order if a description solely in terms of pair-interactions were used to try to analyse the observed data.

In an earlier chapter on stabilised zirconia (Chapter 13) it was mentioned that a multi-site interaction might be a more appropriate model to use for ordering the oxygen vacancies instead of the pair-interaction model which used five different types of near-neighbour pair-interactions. To side-step the need for a multi-site interaction model a modulation wave synthesis method was used to generate a correlation field more like the observed one. Although together these methods gave considerable insight into the origins of the diffuse scattering in stabilised zirconias they were less than satisfactory in terms of providing a definitive physico-chemical explanation for the behaviour. These ideas are developed further for stabilised zirconias in the following chapter.

It seems likely also that consideration of multi-site models would further advance the understanding of both the mullite (Chapter 11) and wüstite (Chapter 12) exam-ples. In the wüstite case many features of the diffraction patterns could be satisfactorily described and understood using pair-interaction models. However, these models were unable to account at all for the complex (albeit very weak) scattering in the incommen-surate layers such as that shown in Fig. 12.3(d). It is likely that further developments in the understanding of wüstite can be made by considering the detailed charge-balance and bonding requirements of the Fe^{3+} and Fe^{2+} ions in and around the defect clusters. In the mullite case the emphasis of the work described was toward extracting as quan-titatively as possible the pair correlation information contained in the data. Although attempts were made to build a pair-interaction MC model which reproduced the diffrac-tion patterns these were only partially successful.

17

STRAIN EFFECTS IN DISORDERED CRYSTALS

17.1   Introduction

In previous chapters, (Chapters 8–16) diffuse X-ray scattering patterns from a wide variety of different materials have been discussed. Although the various systems illustrate the diversity of the diffuse scattering patterns that are observed in real materials, to this point no mention has been made of the fact that quite different systems can sometimes display features that are rather similar. One particularly distinctive feature that has been observed in a number of quite different materials is a diffuse 'ring' or 'doughnut' shaped region of scattering, with a dark (or low intensity) centre. Two such patterns have already been presented for the ceramic materials, mullite (Fig. 11.2(c)) and stabilised zirconia (Fig. 13.7(a)). Though these two materials are both hard ceramic materials their structures are very different.   Even more surprising is that a very similar diffraction feature has also been observed in a very different kind of material, the organic inclusion compound didecylbenzene/urea (see Mayo *et al.*, 1998). These examples are shown together for comparison in Figs. 17.1(a)–(c).

Figure 17.1(d) shows a fourth example, which at first sight seems very different. This is part of the diffraction pattern of 1,3-dibromo-2,5-diethyl-4,6-dimethylbenzene (Bemb2), seen earlier in Fig. 8.2, where there is a broad band of diffuse scattering with a dark 'hole' in it. This feature was discussed earlier in Chapter 8 in terms of occupancy modulations, but in this chapter the phenomenon is re-examined by considering it to have the same physical origins as the diffuse 'rings' of the other examples.

What can be causing these diffuse 'rings' in such different materials? In metal alloys diffuse rings of this kind have been attributed to spinodal decomposition (Cahn, 1967). In this it is supposed there is a balance between the ability of atoms to diffuse to produce phase separation and the elastic forces of the lattice. More recently Butler and Hanley (1999) have shown that such features in the diffraction pattern of sol–gel systems can arise from competition between a short-range attractive and a long-range repulsive potential. In this chapter it is shown how this basic concept can be carried over from these sol–gel systems to provide insight into the mechanisms governing disorder in the different crystalline materials mentioned above. The sol–gel systems described by Butler and Hanley (1999) were considered to be continuous media and not confined to a crystal lattice. In the next section the potentials described by these authors are applied to a system of particles on a simple (square) lattice, in order to demonstrate that the diffuse 'ring' effects carry over from continuous media systems to ones where the interacting particles are confined to a crystal lattice. In the subsequent sections the same basic principles are used to describe the effects observed in cubic stabilised zirconias, the inclusion compound didecylbenzene/urea and the pure molecular crystal system, 1,3-dibromo-2,5-diethyl-4,6-dimethylbenzene (Bemb2). Mullite, the other system depicted in Fig. 17.1, has so far not been analysed from this point of view. This is because for the mullite system more complex considerations are involved. In particular in the $0.5c^*$ section, incommensurate modulations with a wave-vector $\mathbf{q} = 0.5c^* + 0.3a^*$ are observed, while in the $0.16c^*$ section there appears to be a tendency to form an incommensurate modulation with a wave-vector in the b^* direction.

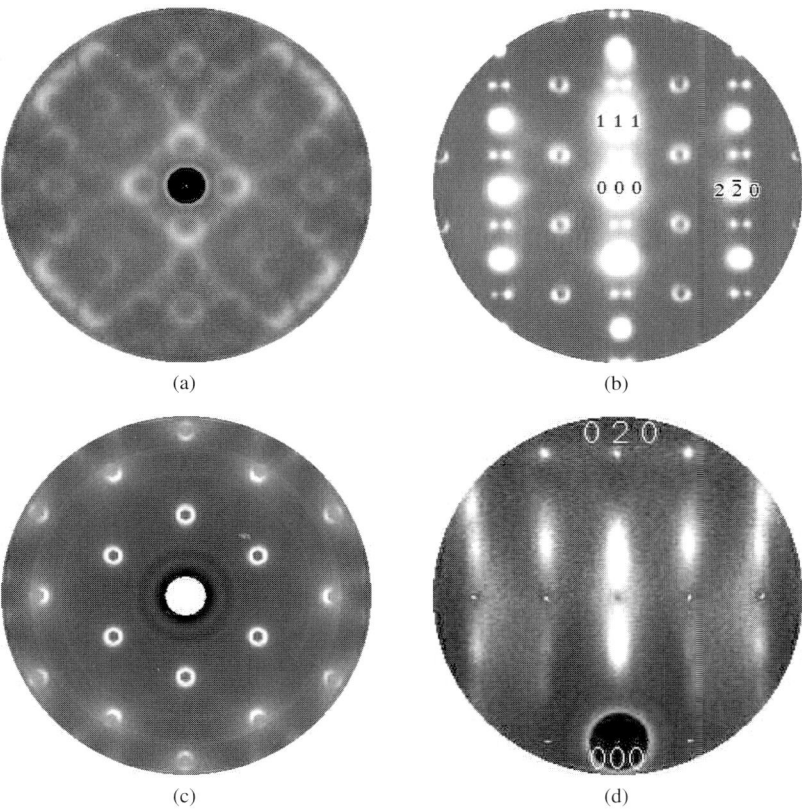

(a)   (b)

(c)   (d)

Fig. 17.1 Sections of the diffraction patterns of various materials exhibiting 'diffuse rings' or 'doughnut' shaped regions of scattering. (a) The $0.7\mathbf{c}^*$ section of the X-ray diffraction pattern of mullite. (b) $\left[1\,1\,\bar{2}\right]$ zone axis electron diffraction pattern of a cubic stabilised zirconia. (c) The $0.35\mathbf{c}^*$ section of the X-ray diffraction pattern of didecylbenzene-urea inclusion compound. (d) Region around $(0\,1\,0)$ of the X-ray diffraction pattern of Bemb2.

17.2 A simple potential used in sol–gel systems

Butler and Hanley (1999) carried out computer simulation experiments on models of sol–gel systems using a potential which was a modified Lennard–Jones potential. This so-called '12-2-6' potential was of the form,

$$U = a\left[1/r^{12} + b/r^2 - (1+b)/r^6\right]. \tag{17.1}$$

This function is plotted in Fig. 17.2. For the normal Lennard–Jones potential $a = 4.0$ and $b = 0.0$. For the modified potential the values of the constants used were $a = 4.8249$ and $b = 0.15$. This value of b provides a small repulsion and the value of a ensures that the well-depth is the same as for the normal Lennard–Jones potential. Note that the modified potential is everywhere greater than the straight Lennard–Jones potential and approaches zero, at high values of r, from above.

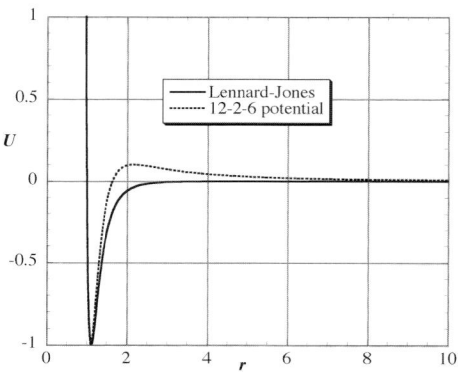

Fig. 17.2 Plot of the potential function, $U = a\left[1/r^{12} + b/r^2 - (1+b)/r^6\right]$, used in the MC simulations of Fig. 17.3. For the normal Lennard–Jones potential $a = 4.0$ and $b = 0.0$. For the 12-2-6 potential used $a = 4.8249$ and $b = 0.15$.

17.2.1 *Simulation on a square lattice*

For the present purposes the potential was used as the interaction between particles confined to a simple square lattice of spacing a_0. In order that the separation of neighbouring points corresponded to the minimum of the potential, a_0 was set equal to $1.1 \times r_0$ where r_0 corresponds to the point where the steep repulsive part of the potential passes through zero. Simulations were carried out using a starting point in which 35% of the lattice sites were randomly chosen to be occupied by particles. The particles were then allowed to interact via the potential given in eqn (17.1). The energy used in the Monte Carlo (MC) simulation was

$$E = \sum_{\substack{\text{vectors } r_{ij} \\ r_{ij} \le 10 a_0}} \frac{a}{kT}\left(\frac{1}{r_{ij}^{12}} + \frac{b}{r_{ij}^2} - \frac{1+b}{r_{ij}^6}\right). \tag{17.2}$$

The summation is over all particles within a radius of $10a_0$ of a given particle. Simulation was carried out for a large number of MC cycles using an array of 256×256 lattice points with cyclic boundary conditions. A value of $kT = 0.32$ was used throughout. The configuration was saved after 50, 1000 and 10,000 cycles and the diffraction pattern of each of these distributions was obtained using a Fast Fourier Transform (FFT) algorithm. Fig. 17.3 shows the results of two comparative simulations using the Lennard–Jones and the '12-2-6' potential at different stages of the iteration. Small representative portions of the real-space distribution are shown for the Lennard Jones case in Figs. 17.3(a)–(c) together with their corresponding diffraction patterns in Figs. 17.3(d)–(f). Similarly, portions of the real-space distribution for the '12-2-6' potential are shown in Fig. 17.3(g)–(i) and their corresponding diffraction patterns in Fig. 17.3(j)–(l).

What is first noticeable about these results is that after a small number of MC cycles the distributions resulting from the two models are fairly similar. The black (35%) phase has begun to separate into 'droplets' and some of these have begun to coagulate into larger domains. Each model has a diffraction pattern showing a broad diffuse ring whose radius is reciprocally related to the spacing between neighbouring 'droplets' in the real-space pattern.

After 1000 MC cycles the two models have begun to show marked differences. The droplets in the Lennard–Jones example have coagulated to form much bigger contorted domains and the diffuse circle has contracted, reflecting the fact that the mean spacing between droplets is now much larger. For the '12-2-6' potential the situation is quite different. There has been some coagulation and the droplets are now elongated and contorted but the mean spacing has not changed enormously. The diffuse ring is now less diffuse but only slightly smaller in radius.

Finally after 10,000 cycles the two models have diverged considerably. In the Lennard–Jones example coagulation has progressed much further, the average spacing between black regions has becomes much bigger and the diffuse ring become even smaller. In contrast for the '12-2-6' potential the black domains have generally become longer and more contorted but their average spacing has not changed much. Similarly the diffuse ring has remained at much the same radius and has become even sharper.

It is easy to see from the progression of pictures in Fig. 17.3 that further iteration will result in complete phase separation for the Lennard–Jones but not for the '12-2-6' potential. In fact in the latter case there appears to be very little further change after about 5000 MC cycles.

17.2.2 *Significance of the modified Lennard–Jones potential result*

Although the physical dimensions involved in sol–gel systems are orders of magnitude larger than the atomic scale effects that are of concern in disordered crystals the physics encapsulated in the simple potential discussed above has clear relevance to these systems too. The essence of the physics is simply that at short distances there is an attractive potential that governs very strongly what happens on a local scale but at slightly larger distances the repulsive hump in the potential curve, even though its height is only very modest compared to the depth of the potential minimum, prevents this local structure being continued. What this means in a crystal is that at a nearest-neighbour level there will

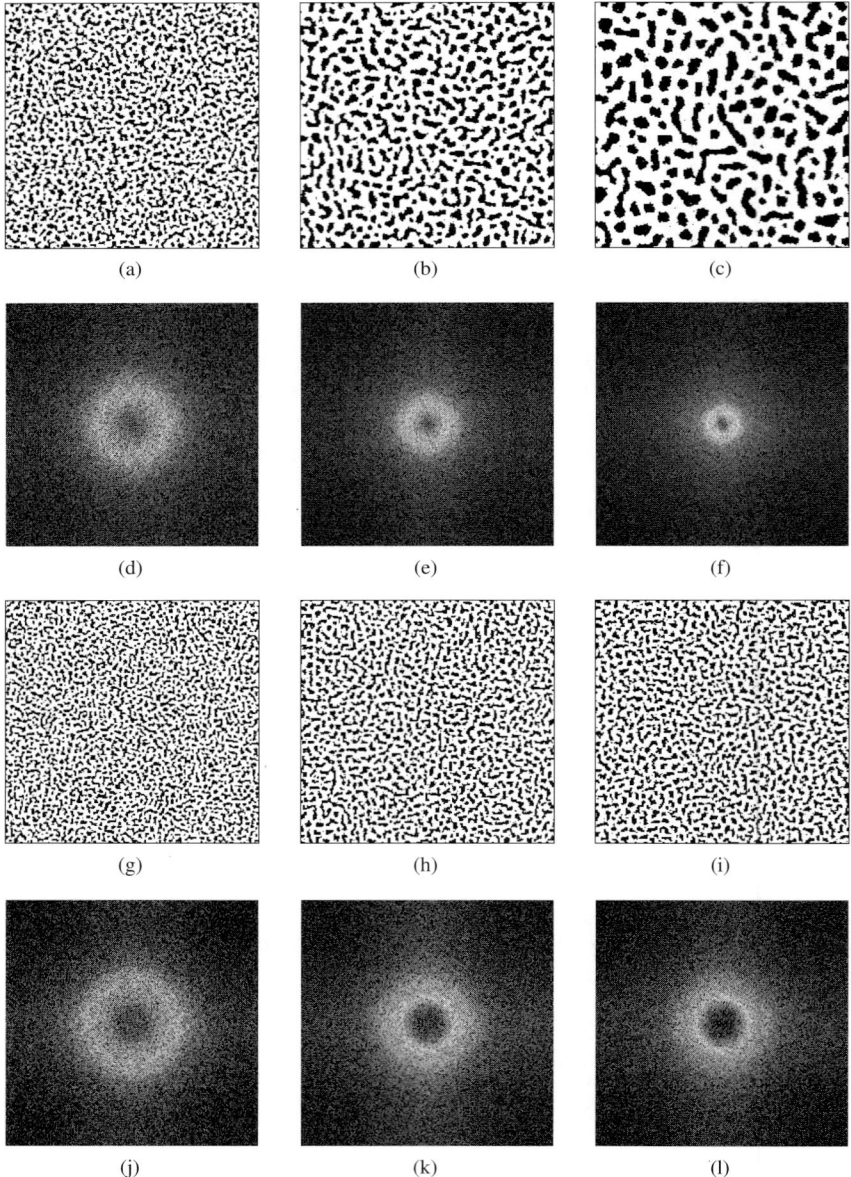

Fig. 17.3 Showing the difference between the Lennard–Jones potential (a)–(f) and the modified potential (g)–(l). (a),(g) correspond to 50 MC cycles. (b),(h) correspond to 1000 MC cycles. (c),(i) correspond to 10,000 MC cycles. Corresponding diffraction patterns appear below each lattice realisation.

be inter-atomic or intermolecular forces dictating what structure the atoms or molecules would like to assume. If this local structure has dimensions that are different from those of the average crystal lattice of which it is part then, as the domain grows, stresses will progressively build up until eventually that particular type of local arrangement cannot be sustained and an alternative has to be found.

In the following sections this basic principle is applied to explain the 'diffuse ring' effects that are observed in CSZs and in the didecylbenzene-urea inclusion compound as well as the 'diffuse hole' in Bemb2. Although in principle interatomic potentials such as the '12-2-6' described above could be utilised to this end, this has not been done because of the magnitude of the calculation that would be entailed for systems more complex than the 2D example given. Instead much simplified potentials are used which nevertheless capture the basic physical idea, without in any way attempting to give a quantitative description of the particular systems.

17.3   Cubic stabilised zirconia

Early work on trying to understand the complex diffuse scattering patterns of cubic stabilised zirconias was described in Chapter 13. A two-stage model was developed which was formulated in terms of the distribution of vacancies on the oxygen sublattice. Since for X-rays the scattering factors of Zr and Y are very similar the distributions of these cations did not feature prominently in the analysis although it was supposed that they could be inferred from the oxygen vacancy distribution. The first stage of the model described the ordering of the vacancies in the oxygen array and the second stage described the way in which the cations relaxed around these vacancy sites.

Two different ways of producing the oxygen vacancy distributions were described. In the first, MC simulation of a pair-interaction Ising-like model was used. This included five different types of interaction terms (ones for nearest-neighbours, next-nearest-neighbours, two types of third-nearest-neighbours and a more distant neighbour). Although qualitatively this reproduced most of the features of the observed diffraction patterns the agreement was by no means quantitative. A second method of producing the distribution of oxygen vacancies was developed in response to the observation in electron diffraction experiments of the 'diffuse rings' that are under discussion here. In this second method the modulation wave synthesis method described in Section 5.8 was used to synthesise directly real-space distributions that gave appropriate diffuse circles in their diffraction patterns. This method, though producing improved agreement with the observed patterns cannot be considered a satisfactory solution to the problem as there is no physical basis for it. For both of these methods the same relaxation model in which cations tended to move away from vacancy sites was used.

In this section the aim is to show that using the basic physical idea established in Section 17.2.2 is sufficient to explain how and why the 'diffuse rings' occur in CSZs. The supposition is that what is occurring is that locally the ions are trying to form a particular structure but the range over which this can occur is limited by the strain that builds up when this structure is constrained to fit on the lattice of the average structure. In order to do this, in contrast to the earlier work, prime consideration is given to the cation distributions.

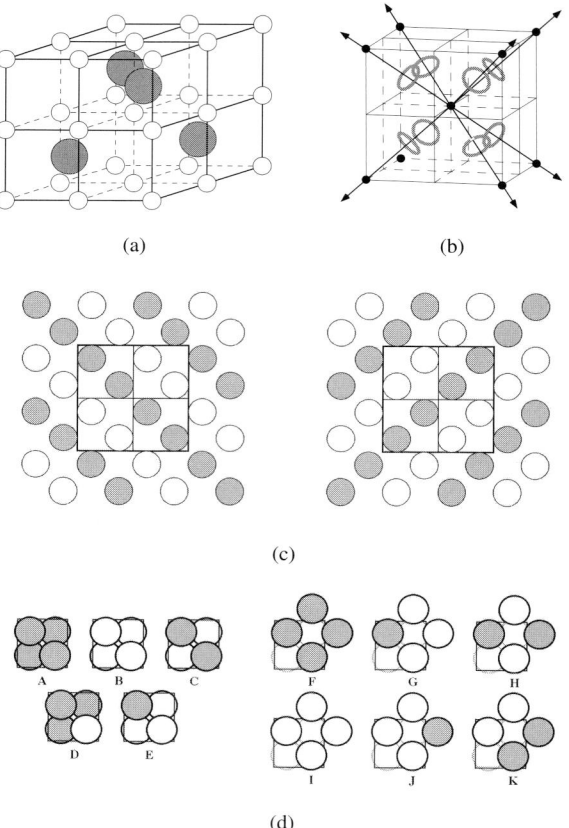

(a)   (b)

(c)

(d)

Fig. 17.4 (a) The average unit cell of CSZs (the fluorite structure). (b) The position of the diffuse circles in the diffraction patterns of CSZs. (c) Successive $[1\,0\,0]$ layers of cations in the pyrochlore structure. The large square is the pyrochlore unit cell and the small square is the fluorite parent cell. (d) Showing the nearest-neighbour tetrahedral clusters, A–E and next-nearest-neighbour square clusters, F–K used in the potential for generating the pyrochlore structure.

17.3.1 Model for local structure

In formulating a model which describes the local structure in disordered CSZs it should be noted that in many systems a stoichiometric fluorite-related phase that commonly occurs is that having the pyrochlore structure of general stoichiometry $A_2B_2O_7$. In the (cubic) pyrochlore structure the A and B cations are ordered to produce a $2 \times 2 \times 2$ superlattice of the basic fluorite parent cell. This superstructure gives rise to extra Bragg peaks at the $\frac{1}{2}\langle 111 \rangle^*$ reciprocal positions—that is, at the points where each of the 'diffuse rings' is centred. It is therefore conjectured that it is the pyrochlore phase that the CSZ is attempting to form.

Figure 17.4(a) shows for reference a perspective drawing of the average CSZ unit cell. The cation sites are shown as large grey circles. Figure 17.4(b) shows the positions of the 'diffuse rings' in the reciprocal-space drawing. Figure 17.4(c) shows the pattern of cation ordering corresponding to the pyrochlore structure. The drawing shows two successive layers of the structure viewed down $[100]$ and it is seen that the two types of cation occur as alternating diagonal rows. In the first layer the rows are along $[110]$ while in the second they are along $[1\bar{1}0]$. Where any two of these rows of cations cross there is a tetrahedron of cations which is comprised of $4A$, $2A + 2B$ or $4B$ cations (other combinations do not occur). It is clear that these nearest-neighbour tetrahedral cation clusters are fundamental to the pyrochlore structure. Similarly characteristic clusters that occur are the $[110]$ squares of next-nearest-neighbour cations. In this case the only combination that occurs is the $2A + 2B$ cluster. Figure 17.4(d) shows all of the possible tetrahedral and square clusters that can occur. Of these clusters only those labelled A, B, C and K occur in the perfect pyrochlore stucture.

Knowledge of these different clusters was used to construct a simple short-range potential, for use in an MC simulation, that would lead to the formation of the pyrochlore structure if no other forces are present. The simple potential that was adopted for this purpose was one based on assigning different relative energies for the different types of cluster. The aim in assigning these energies was to favour the tetrahedral clusters labelled A and B which contain four like-cations and the square clusters which contained two of each type of cation, clusters labelled H and K. The energies used were: $E_A = E_B = -1$; $E_C = E_D = E_E = 0$; $E_F = E_I = +4$; $E_G = E_J = +1$; $E_H = E_K = 0$. These energies are on a scale relative to the simulation temperature of $kT = 2.0$ which was used throughout. Then the local structure contribution to the energy was

$$E_{\text{local}} = \sum_{\substack{\text{all n.n.} \\ \text{tetrahedra}}} (E_i) + \sum_{\substack{\text{all n.n.n.} \\ \text{squares}}} (E_i). \tag{17.3}$$

17.3.2 Origin of strain

In order to understand why a strain term in the energy might give rise to diffuse circles normal to each $\langle 111 \rangle$ direction, Fig. 17.5 shows the cation distribution in pyrochlore for two consecutive close-packed layers normal to $[111]$. It is seen that in the first layer a single A cation is surrounded by six B cations, while in the next layer six A cations surround a single B cation. It is clear from this that if the two ions have substantially different sizes (taking into account the surrounding anions needed to satisfy their valence requirements), then there will be a mismatch in the size of the two layers. It would, therefore, be expected that a strain energy would build-up as the size of pyrochlore domains increases.

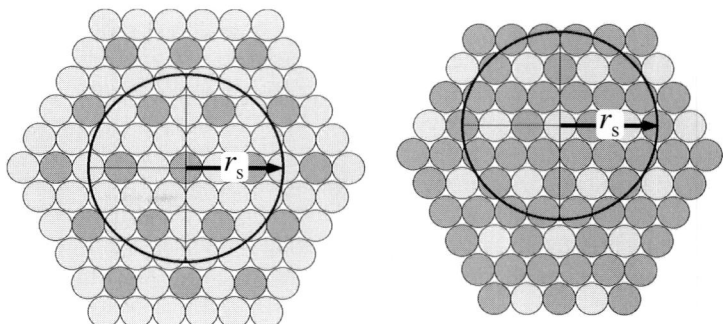

Fig. 17.5 Successive close-packed layers of cations in planes normal to $\langle 111 \rangle$ in pyrochlore. A strain energy is assumed to depend on the ratio of the number of A and B cations within a radius r_s of a given cation.

To model this in a very simple way it is supposed that the strain energy depends on the ratio of A to B cations within a radius r_s of a given cation. r_s was set to $3\times$ the nearest-neighbour cation distance, as indicated by the large circles in Fig. 17.5. If n_A and n_B are the respective numbers of A and B cations within r_s of a given cation the contribution to the MC energy is taken as,

$$E_{\text{strain}} = K_s \left[\frac{(n_A - n_B)}{(n_A + n_B)} \right]^2 . \tag{17.4}$$

The force constant K_s was used to gradually introduce increasing amounts of strain into successive MC simulation runs. Note that strain terms of the type given in eqn (17.4) occur on all four orientations of the $\{111\}$ planes simultaneously.

17.3.3   Results of MC simulation

Monte Carlo simulation was carried out using an array of $32 \times 32 \times 32$ fluorite unit cells. Since computation of the strain energy is computationally intensive, iteration was carried out for only ~ 200 MC cycles. Figure 17.6 shows the resulting real-space distributions of cations on a typical $(1\,1\,1)$ plane from the 3D simulations for three different cases. Alongside each figure is shown the corresponding $\frac{1}{2}\langle 111 \rangle^*$ section of the diffraction pattern calculated from the simulation. For all three examples the strain radius, r_s, was taken as $3\times$ the nearest-neighbour cation distance. For Fig. 17.6(a) the force constant K_s was 5.0, for Fig. 17.6(b) it was 15.0 and for Fig. 17.6(c) it was 25.0. Figure 17.6(a) clearly indicates that large domains of the pyrochlore structure exist and this is borne out in its diffraction pattern, where only slightly broadened Bragg peaks exist at the $\frac{1}{2}\langle 111 \rangle^*$ reciprocal positions. For Fig. 17.6(b) the diffuse peaks have become

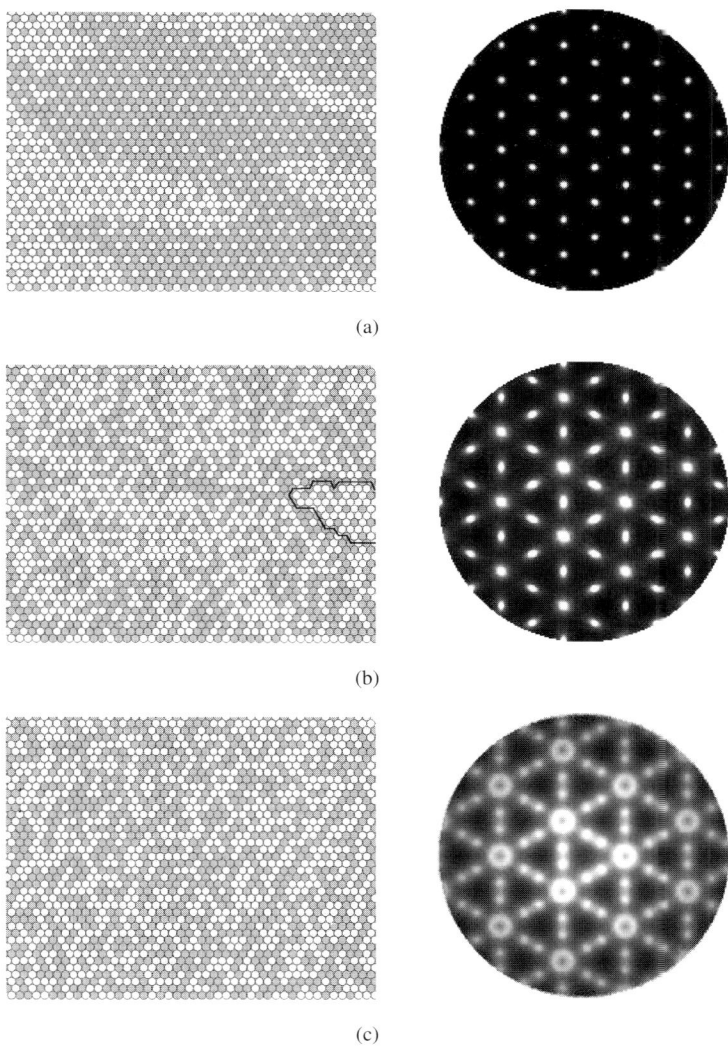

(a)

(b)

(c)

Fig. 17.6 Example regions of the close-packed planes of cations normal to $\left[1\,1\,1\right]$ in CSZ, taken from three MC simulations in which $r_s = 3$, which had different values of the strain term constant $K_s = 5$. For (a) $K_s = 5$; for (b) $K_s = 15$; for (c) $K_s = 25$. On the right is the corresponding diffraction pattern of the reciprocal section calculated from the full 3D simulation.

broader with some definite ellipsoidal shape, while for Fig. 17.6(c) the peaks have re-
solved into definite diffuse circles (and pairs of spots corresponding to the intersection
of the Ewald sphere with diffuse circles oriented out of the plane). It is interesting to note
that, while some small regions of the pyrochlore structure can be seen in Fig. 17.6(b)
(see example outlined) none can be recognised in Fig. 17.6(c).

17.4   The organic inclusion compound didecylbenzene/urea

Urea acts as the host for a large number of different long-chain molecular guests, not
least of which are the *n*-alkanes. One such system was described in Chapter 10. The
particular system of interest in the present context is one in which the guest molecule
was didecylbenzene. This was one of a series of dialkylbenzene molecules that were
studied in order to investigate the ability of the urea to accommodate the bulky benzene
group which is rather too large for the urea channels (see Mayo *et al.*, 1998).

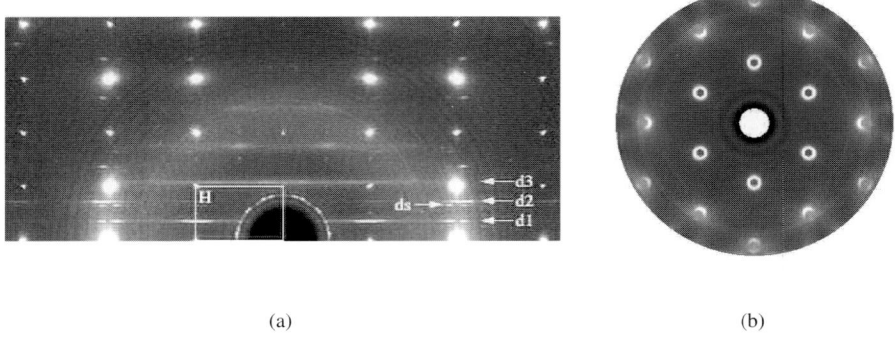

(a)   (b)

Fig. 17.7   Diffraction patterns of didecylbenzene/urea inclusion compound. (a) Pattern normal to
the channel axis showing various diffuse layers. The rectangle marked 'H' indicates the reciprocal
cell of the urea host. (b) Shows the distribution of intensity in the diffuse layer marked 'd1' in (a).

In common with many other urea inclusion compounds the long-chain guests form
pseudo 1D crystals within each urea channel, giving rise to diffuse planes of scattering
normal to the channel axis with the spacing of the planes incommensurate with the urea
c-axis repeat (see Fig. 17.7(a)). Figure 17.7(b) shows the distribution of intensity within
the first (and sharpest) of these diffuse planes which is labelled 'd1' in Fig. 17.7(a).
[Note that there is no zero-level diffuse plane, because in projection down the channel
axis all different orientations of the molecular 'footprint' are superposed, so producing
identical scattering from each channel]. The presence of structure in the diffuse plane
indicates that there is strong interaction between the molecular behaviour in adjacent
channels.

Figure 17.8(a) shows a space-filling drawing of the didecylbenzene molecule together with two possible 'footprints' as viewed down the urea channel. The two arrows
are drawn to indicate the direction of projection used to obtain the two different footprints. In either case the footprint is more elongated than that of a *n*-alkane molecule.

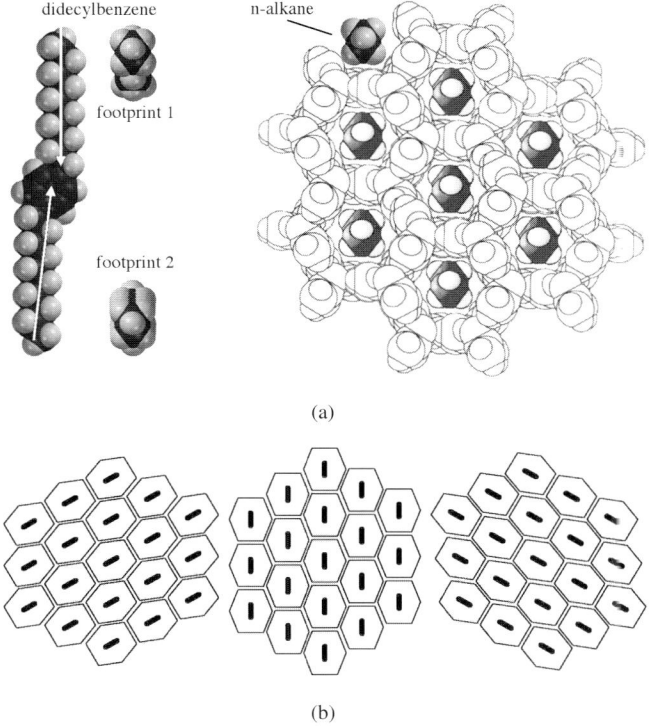

(a)

(b)

Fig. 17.8   Didecylbenzene/urea inclusion compound: (a) shows a space-filling model of a dide-cylbenzene molecule together with its 'footprint' when projected down either of two directions (indicated by the arrows). (a) also shows how the corresponding footprint of the *n*-alkane molecule fits into the urea channels. (b) shows a schematic view of three different orientations of the distorted (exaggerated) urea network when containing the didecylbenzene molecules.

This latter is also shown in the figure and the way it fits rather loosely into the urea channels. Because of this rather loose fit the channels in the n-alkane compounds tend to distort at low temperature into an orthorhombic cell with alkane orientations in neighbouring channels inclined to each other in a herring-bone arrangement (see Welberry and Mayo, 1996). The fact that the diffuse rings of scattering in the d1 layers of didecylbenzene/urea tend to occur around integral h,k positions indicates that the guests in neighbouring channels in this compound do not adopt a herring-bone arrangement but tend to have the same orientation in neighbouring channels. The 'footprint' of didecylbenzene being much more elongated than that of the n-alkanes, will not fit into the urea without a unidirectional elongation of the hexagonal channel cross-section. The only way for such elongated hexagons to pack is if they are all oriented in the same direction. It is imagined, therefore, that locally the structure must try to make domains as depicted in Fig. 17.8(b), of which there are three symmetry-related versions.

As seen from Fig. 17.8(b) the formation of such domains will produce strain in the hexagonal lattice of the average parent structure. As a domain grows larger the strain will gradually increase until a switch to one of the alternative domains becomes necessary. This is a completely analogous situation to that in the previous CSZ example. Locally the elongated footprint of the guest is trying to make a structure in which all the guests are aligned parallel to each other, but the strain that is induced in the hexagonal host lattice tends to suppress this happening over a long range. As before this can be formulated in terms of simple energy functions, E_{local} and E_{strain} to represent these opposing interactions. The local ordering energy is expressed in the form,

$$E_{\text{local}} = \sum_{\substack{\text{nearest} \\ \text{neighbours}}} \delta_{ij}\, \sigma_i \sigma_j \qquad (17.5)$$

Here σ_i, σ_j are 3-state variables defining the three different guest orientations and δ_{ij} is a function which is -1 if $\sigma_i = \sigma_j$ and $+1$ otherwise. For the strain energy, E_{strain}, a radius r_s is again defined. The aim of this term is to maintain, within r_s of a given site, equal proportions of the three different molecular orientations. The form of this strain energy is,

$$E_{\text{strain}} = K_s \left[\left(f_1 - \frac{1}{3} \right)^2 + \left(f_2 - \frac{1}{3} \right)^2 + \left(f_3 - \frac{1}{3} \right)^2 \right] \qquad (17.6)$$

Here f_1, f_2, f_3 are the fractions of sites within the circle of radius r_s of a given site, which have the orientations 1, 2 or 3 respectively.

17.4.1 Results of MC simulation

Simulation was carried out for 200 MC cycles on a 2D hexagonal lattice consisting of 128×128 lattice sites. The temperature used in the simulation was given by $kT = 1.0$ and a suitable value of the strain constant K_s was found by trial and error to be 150. As for the CSZ example the radius r_s was taken as $3\times$ the nearest-neighbour intermolecular distance. [Note, the constant K_s cannot be simply related to that used for the CSZ example because of the rather different form of eqn (17.6) compared to eqn (17.4)]. A small

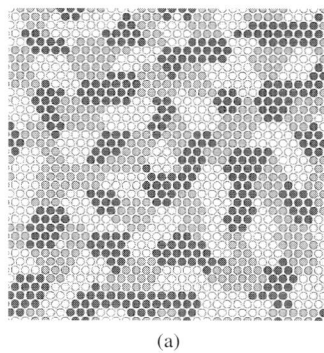 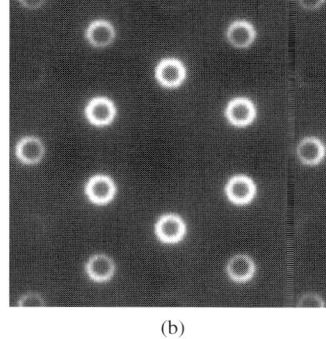

(a)   (b)

Fig. 17.9   (a) Portion of the distribution of the three 'footprint' orientations resulting from the MC simulation. The three shades of grey represent the three orientations shown in Fig. 17.6(a). (b) Diffraction pattern calculated from the distribution shown in (a)

representative portion of the resulting distribution is shown in Fig. 17.9(a). The different coloured circles (white, grey and black) each represents a didecylbenzene molecule in one of the three orientations displayed in Fig. 17.8(b). Figure 17.9(b) shows a diffraction pattern calculated from the distribution. To make this calculation a simplified representation of the molecular scattering factor was used. This consisted of two pseudo-atoms forming a dumbbell shape similar to that used in Welberry and Mayo (1996). A more realistic molecular scattering factor would result in different relative intensities of the diffuse rings seen in the pattern, but would not affect the overall form of the pattern.

17.5   The 'diffuse hole' in Bemb2

17.5.1   *Background*

In the pure molecular crystal compound Bemb2, each molecular site contains the molecule in one of two different possible orientations. The average structure as revealed by Bragg scattering has space group symmetry $P2_1$ but the departures from $P2_1/c$ are very small (see Wood *et al.*, 1984). In this average structure the $1, 3$- and $4, 6$-substituent sites contain almost exactly 50% bromo- and 50% methyl-. If the space group $P2_1/c$ is assumed the average molecular site is centro-symmetric so that the packing of molecular shapes shown schematically in Fig. 17.10(a) satisfies both the 2_1-screw axes and the c-glide planes. In the real crystal, where each site must be occupied by one or other of the two possible molecular orientations, either the 2_1-screw axis or the c-glide plane may be satisfied on a local scale but not both. Figure 17.10(b) shows the arrangement where the 2_1-screw axis is satisfied, while for Fig. 17.10(c) it is the c-glide plane.

On a local level intermolecular interactions are strongly influenced by the molecular dipoles that the molecules possess. In Fig. 17.10(b) these dipole moments are all approximately aligned with each other, giving an unfavourably high contribution to the interaction energy. In Fig. 17.10(c) on the other hand the dipoles in the centre of the cell are approximately aligned antiparallel to those at the corners of the cell, giving a much

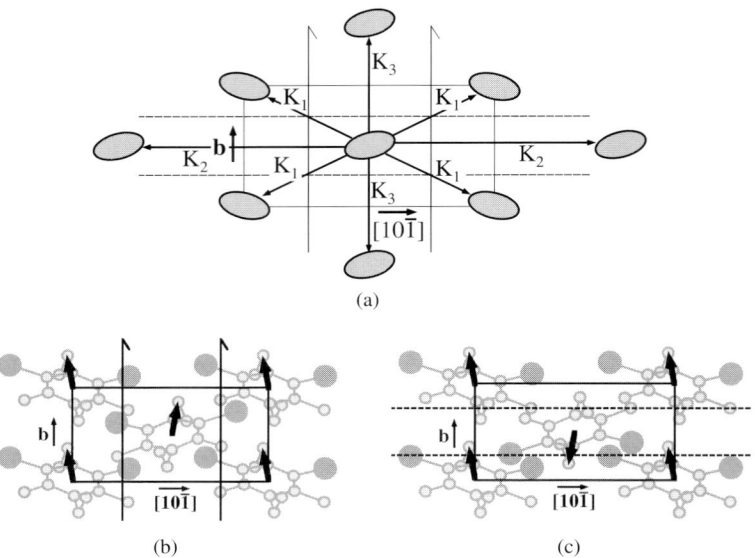

Fig. 17.10 (a) Schematic drawing of the molecular packing of centro-symmetric molecules corresponding to the average structure of Bemb2. (b) Local packing of Bemb2 molecules satisfying the 2_1-screw axis. (c) Local packing of Bemb2 molecules satisfying the c-glide plane. The arrows indicate the directions of the molecular dipole moments.

more stable configuration. Locally, therefore, it appears that the molecules would prefer to form the structure Fig. 17.10(c) where the c-glide symmetry prevails.

Why does this c-glide symmetry structure not take over the whole crystal? It is not easy to see why it should not. However the pioneering work of Kitaigorodsky (1973) showed that molecules are able to pack more efficiently in some particular patterns and symmetries rather than others. For example, they rarely crystallise in space groups that possess a mirror plane since it is difficult to make a dense packing of molecules with a mirror present. This work was extended by Wilson (1990) who presented data on the relative frequencies with which different space groups were formed by molecular crystals. He showed that in the monoclinic system the space group $P2_1$ occurs about 20 times more frequently than the space group Pc and space group $P2_1/c$ occurs a further five times more frequently than that. This clearly implies that there is a substantial stability gain to be had if molecular packing can occur in space group $P2_1$ and even more in space group $P2_1/c$.

What this means for Bemb2 then, is that, though locally the intermolecular forces would like to make the c-glide symmetry structure, the overall packing energy of the average structure would prefer $P2_1/c$. As domains of the Pc structure grow in the average $P2_1/c$ lattice, strain will gradually build up until it is not viable to continue and a change in the packing must occur in order to relieve the strain. The situation is thus entirely analogous to that in the two previous examples. At short distances near neighbour interactions are trying to make one kind of packing, but at slightly longer distances the strain energy builds up sufficiently that this arrangement is no longer viable and the structure cannot propagate further.

17.5.2 MC simulation of the 'diffuse hole' effect

To demonstrate the effect for Bemb2 a simple 2D model corresponding to the $[1\,0\,1]$-axis projection of the structure was constructed. Each molecule in this basal plane is surrounded by eight neighbours involving three symmetry-unrelated vectors (see Fig. 17.10(a)). In order to produce the preferred short-range ordered structure a local interaction energy, E_{local}, was assumed. This had the form:

$$E_{\text{local}} = \sum_{\text{neighbours}} K_i\, \sigma_0 \sigma_i. \qquad (17.7)$$

Here σ_0 is a binary $(+1, -1)$ variable representing the orientation of the target molecule and σ_i is the corresponding variable for one of the eight neighbours. The three independent values of K_i, corresponding to the three symmetry-unrelated vectors shown in Fig. 17.10(a), were chosen by trial and error so that the diffraction pattern calculated from the model gave a qualitatively good agreement with the broad features of the observed diffuse scattering shown in Fig. 17.1(d).

In order to apply a strain term a circular region of radius r_s, around a given molecular site was again used. In this case the fractions of sites, f_i, on each of the two sublattices i that lie within this circle and that are in the orientation corresponding to $\sigma_i = 1$ were used to define the energy. f_1 is the fraction for sites on sublattice 1 (the corners of the unit cell) and f_2 is the fraction for sites on sublattice 2 (the centres of the unit cell). Then the form of the strain term, which attempts to keep the fraction of sites that are in orientation 1 equal on the two sublattices, is given by,

$$E_{\text{strain}} = K_s\,(f_1 - f_2)^2. \qquad (17.8)$$

Note that, defined in this way, the strain term tries to keep the structure, averaged over the circle, to have symmetry $P2_1$. If the fractions f_i are additionally constrained to be 0.5 then the strain term will try to keep the average symmetry to be $P2/c$.

For the Bemb2 case it was necessary for the radius r_s to be much greater than for the CSZ and didecylbenzene/urea examples because the 'diffuse hole' in Fig. 17.1(d) is much smaller than the diffuse rings in Figs. 17.1(b) and (c). Initially a radius of $15\times$ the nearest neighbour intermolecular distance was used. However it was found that this not only gave a diffuse 'hole' as in the observed pattern but also a quite distinct ring of stronger scattering like a miniature version of the rings in Fig. 17.9(b). In order to overcome this the form of eqn (17.8) was changed to include a gradually graded strain by having similar terms which operated over circles of three different radii.

$$E_{\text{strain}} = K_{s1}\,(f_{11} - f_{12})^2 + K_{s2}\,(f_{21} - f_{22})^2 + K_{s3}\,(f_{31} - f_{32})^2. \qquad (17.9)$$

The constants K_{s1}, K_{s2}, K_{s3} operated over circles of radius $10\times$, $12.5\times$ and $15\times$ the nearest-neighbour intermolecular distance and had the values 50, 100 and 200 respectively. With this graded potential the need for the average structure to comply to space group $P2_1$ is strong over distances of $15\times$ the nearest neighbour intermolecular distance but progressively less strong over shorter distances.

MC simulation was carried out on a 2D primitive square lattice of 512×512 sites. The sites of this primitive lattice were then mapped onto the body-centred crystal lattice so that a cell edge of the primitive lattice corresponded to the nearest-neighbour inter-molecular distance in the crystal lattice. 100 cycles of iteration were performed during which fixed values of the strain constants K_{s1}, K_{s2}, K_{s3} were used but the SRO constants of K_i were adjusted using a feedback mechanism in order to achieve predetermined values for near neighbour short-range order parameters (correlation coefficients). These values, which gave good qualitative agreement with the observed pattern, were chosen to be:-

$C_1 = -0.4$ This occurs between molecules at the corners and centres of the unit cell. The negative value indicates a strong tendency for the c-glide local configuration.

$C_2 = +0.4$ This occurs between molecules separated by a whole $[1\,0\,\bar{1}]$-cell repeat. The positive correlation indicates a tendency for neighbouring cells in this direction to be the same. This results in the diffuse scattering being concentrated in bands normal to $[1\,0\,\bar{1}]$.

$C_3 = 0.0$ This occurs between molecules separated by a whole b-cell repeat. Although having a zero value this represents a quite strong restraint to the nearest-neighbour C_1 correlation propagating along b. The zero value gives the strong diffuse band near the centre of the pattern the elongated, rather flat-topped appearance.

The final diffraction pattern calculated from the model including both SRO and strain is shown in Fig. 17.11(b). Note that the 'diffuse hole' appears around all Bragg positions including the systematically absent $(0\,1\,0)$. Figure 17.11(a) shows a small portion of the corresponding real-space distribution. The light grey circles represent molecules in orientation 1 and the dark grey ones in orientation 2. The two regions outlined are small regions of the Pc structure in which the molecule at the centre of the cell is opposed to those at the corner (as in Fig. 17.10(c)). Note that the two outlined regions are in anti-phase with each other. Regions locally satisfying the $P2_1$ symmetry have both the corner and cell-centre sites occupied by the same colour circles. Such regions are hard to find and usually are no bigger than one or two unit cells. Despite this the presence of the 'diffuse hole' indicates that over distances greater than about $15\times$ the nearest-neighbour intermolecular distance the average structure conforms to $P2_1$.

17.6 Conclusion

In this chapter it has been shown that a feature that has been observed in the diffuse scattering patterns of a wide variety of different materials—a diffuse 'ring' or 'dough-nut' shaped region of scattering—can be understood in terms of a simple model that has been borrowed from the field of sol–gel science. In this it is supposed that there is a balance between the local attractive forces that are trying to make a particular structure and a rather longer-range repulsive force. In the disordered crystal context it is believed that

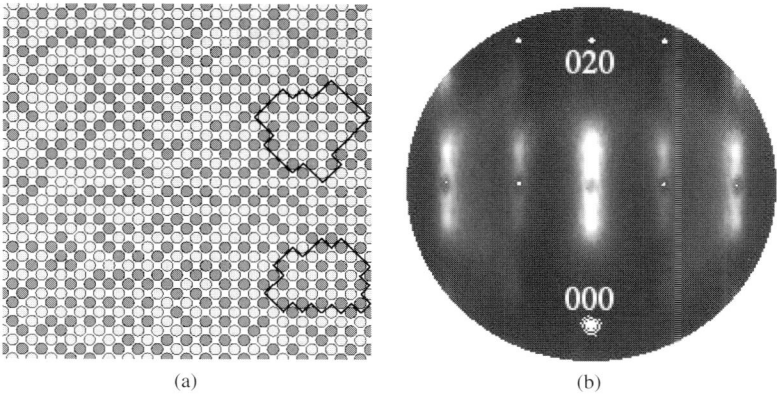

(a)   (b)

Fig. 17.11 (a) Small portion of the real-space distribution representing the disorder in Bemb2. The two shades of grey represent the orientations of the molecule with dipoles pointing up and down respectively (see Fig. 17.10). The outlined regions are both regions having local c-glide symmetry where the dipole at the centre of the unit cell is opposed to those at the corners as in Fig. 17.10(c). However notice that the two regions are in anti-phase to each other. (b) Corresponding diffraction pattern calculated from the MC simulation.

this latter force has its origin in the strain that builds as the preferred local structure tries to fit into the average crystal lattice. Simple MC models using the principle have been used to demonstrate the effect for three quite different examples; cubic stabilised zirconia, the didecylbenzene/urea inclusion compound and the pure molecular compound Bemb2.

There seems little doubt that the similar diffuse feature that is observed in mullite (Fig. 17.1(d)) has similar origins to those of the other examples described here. However, because this system shows additional incommensurate diffraction effects not easily described by the kind of simple model presented here, no attempt has yet been made to model it in this way.

The basic principle established in this chapter is particularly important in the context of the stabilised zirconia problem. The realisation that the very complex structural problem can be explained by such a simple physical idea should enable further progress to be made in understanding the properties of a wide range of these materials. A description in terms of the cation ordering is, from a chemical point of view, a much more natural approach than that used in previous studies where the oxygen vacancies were given prominence. To incorporate the anions into the derived cation distributions should be quite feasible using simple bond-valence criteria, so that a much more complete model should now be accessible.

It is important to point out that the potentials used in the three examples described in this chapter are simple approximations invented for the sole purpose of demonstrating the principle of competing interactions, and they should not be taken out of context. For example it is not suggested that the pyrochlore lattice energy is only dependent on the cation clusters shown in Fig. 17.4(d). The potential involving these clusters merely provided a simple means of generating a pyrochlore lattice using only near-neighbour interactions. Similarly, the device that was used in all three examples, of applying a strain energy via a constraint of a particular lattice-average over a circular domain, was adopted for simplicity and convenience only. A more rigorous treatment would clearly need to involve proper interatomic interactions between all atoms over a large distance range (e.g. using eqn (17.1)), and would necessarily be more computer intensive for real systems than perhaps currently feasible.

MISCELLANEOUS EXAMPLES

18.1 Introduction

In previous chapters in Part III of this book examples have been discussed which illustrate the way in which the study of diffuse scattering has developed in recent years. What has emerged from these studies is that Monte Carlo (MC) simulation has assumed a place of some importance as a tool for aiding both the interpretation and the analysis of diffuse scattering data from a wide variety of different materials that exhibit an equally wide variety of different scattering effects. The course followed through this survey has largely been chronological and examples have been chosen which have had in many cases particularly pedagogical attributes. In this final chapter a number of miscellaneous examples are described which have not readily fitted into the discourse to date but which are interesting in their own right and add to the diversity of the systems for which computer simulation and MC simulation in particular has provided useful insights. In common with the earlier examples these too underline the importance of the underlying theoretical concepts.

18.2 The defect structure of the zeolite mordenite

18.2.1 *Background*

The zeolite mordenite is a widely used industrial catalyst (Sie, 1994; Maxwell *et al.*, 1992) that has long been suspected of framework variability, though the details have been elusive. Its rather complex structure is shown in Fig. 18.1(a) in projection down $[0\,0\,1]$. It can be assembled from a smaller motif (shown in Fig. 18.1(b)), which consists of a four-membered ring (4MR) of tetrahedra and eight legs attached to the outer vertices (12 tetrahedra in all). The 48 tetrahedral sites in the repeating unit cell are divided amongst four identical 4MR motifs (these are labelled A–D for later reference), which stack in columns parallel to the $[0\,0\,1]$ channel axis. Two of the columns in the ideal structure are displaced by half a unit cell length along the channel axis (i.e. by $\frac{1}{2}[0\,0\,1]$) relative to the other two, as distinguished by the light and dark motif shadings in Fig. 18.1(a).

Figure 18.2 shows example diffraction patterns of mordenite (see also Fig. 1.12). The single Laue exposure shown in Fig. 18.2(a) has the $[0\,0\,1]$ axis vertical and it is seen that the $l = 2n + 1$ layer lines contain streaks. These streaks each span roughly the width of the frame from left to right and their position stays relatively constant as the crystal is rotated, indicating that they are in fact the cross sections of large sheets of diffuse intensity in the reciprocal lattice planes normal to $[0\,0\,1]$. Such features were in fact reported in the electron diffraction work of Sanders (1985).

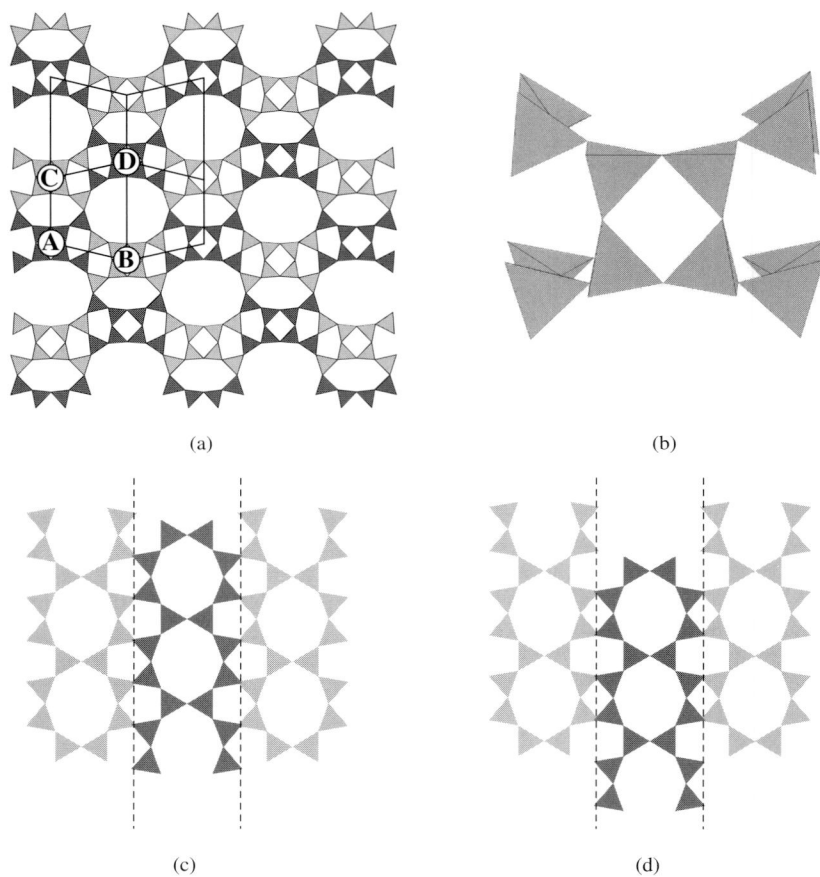

Fig. 18.1 Schematic diagram of the structure of mordenite. (a) The structure viewed down **c** comprised of columns of motifs. The two shadings correspond columns having a relative shift of $\frac{1}{2}[0\,0\,1]$. (b) The structural motif consisting of 12 corner connected tetrahedra. (c) Lateral view of the columns in the normal configuration. (d) Lateral view of the columns containing a fault in which the central dark column has undergone a $\frac{1}{2}[0\,0\,1]$ translation.

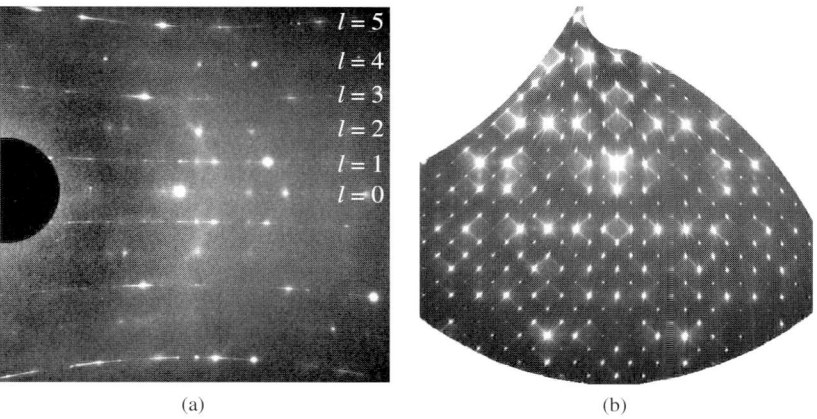

$l = 5$
$l = 4$
$l = 3$
$l = 2$
$l = 1$
$l = 0$

(a)   (b)

Fig. 18.2   X-ray diffuse scattering of mordenite: (a) shows a single frame of the 1000 stationary Laue exposures from which (b) was reconstructed; (b) Part of reciprocal lattice section $(h\,k\,5)$ of mordenite obtained from the $100°$ rotation of the crystal. (Data used in this figure are reproduced with kind permission of Dr Branton Campbell.)

Recently Campbell and Cheetham (2002) showed that the presence of long c-axis chains of material which have been fault-shifted by $\frac{1}{2}[0\,0\,1]$ will give rise to diffuse $l = 2n + 1$ sheets of this kind. The diffuse sheets appear to have a width along $[0\,0\,1]$ no greater than that of the Bragg peaks, which implies that the defect chains span the entire crystal along $[0\,0\,1]$.

Early single-crystal X-ray diffraction studies of natural mordenite samples reported evidence of a dilute concentration ($\sim 2\%$) of 4MRs that appeared to be shifted by $\frac{1}{2}[0\,0\,1]$ relative to their expected locations (Mortier *et al.*, 1975; Schlenker *et al.*, 1979). It is quite difficult to imagine locally displacing a single 4MR by $\frac{1}{2}[0\,0\,1]$ within the framework in a way that maintains both tolerable bond lengths and the chemically essential tetrahedral connectivity but it is perfectly feasible to shift a whole column of the 12-tetrahedra motifs shown in Fig. 18.1(b). How this can be achieved is shown in Fig. 18.1(c) and Fig. 18.1(d). Only very small displacements of the atoms from their ideal positions are required to make the column fit in its new location. This type of columnar defect preserves stoichiometry and tetrahedral coordination, and also allows for reasonable Si—O bond lengths.

Figure 18.2(b) is a reconstruction of the $(h\,k\,5)$ reciprocal layer of mordenite obtained from the set of 1000 Laue exposures, one of which is shown in Fig. 18.2(a). Here it is seen that the diffuse layer has in fact got a great deal of structure. The $(h\,k\,1)$ and $(h\,k\,3)$ sections show similar features. There are at least three distinct types of diffuse features: rather broad features that span the space around several Bragg reflections; star-shaped intensity distributions centred around certain Bragg reflections and narrow streaks of intensity along lines for which $h \pm k =$ integer. The linear streaks appear to join together to form diamonds at some positions, while the broad features give the appearance of filling in the interior of some of the diamonds. The star-like features have arms normal to $[110]$ and $[1\bar{1}0]$. The vertical streaks from some reflections and streaks which appear curved are artefacts due to the blooming effect described in Section 1.3.

In order to establish whether the columnar defects and their mutual spatial distribution can provide a satisfactory explanation for the observed diffuse scattering effects computer simulations were carried out and diffraction patterns obtained from them for comparison with the observed patterns.

18.2.2 *Computer simulations*

A model crystal of mordenite was first constructed with standard Cmcm symmetry entirely from columnar motifs. A crystal of $256 \times 256 \times 16$ unit cells was used. Since all models used were considered to be perfect in the z-dimension there was no need to have an extensive crystal in that direction. Because each unit cell contains two columnar motifs in the x-direction and two in the y-direction (labelled A–D in Fig. 18.1(a)) an array of 512×512 random variables, $X_{i,j}$, was used to represent all the columnar motifs in the crystal. A value of $X_{i,j} = 0$ indicates that the motif labelled i, j is in its normal position, while a value of $X_{i,j} = 1$ indicates that the motif has been displaced by $\frac{1}{2}[001]$. The resulting defect fraction p is simply the probability of obtaining a value of $X_{i,j} = 1$. While observed estimates of the defect concentration are $\sim 2\%$, a higher concentration of 10% was used in the simulations to reduce the statistical noise in the scattering patterns. This modification does not alter the structure of the diffuse scattering pattern, but only its overall intensity relative to that of the Bragg reflections.

The first model considered (model 1) was one in which isolated randomly positioned column defects were present. With only this type of defect the broad areas of scattering seen in the observed patterns were reasonably well reproduced. These broad features essentially correspond to the Fourier transform of the single column of 12-tetrahedra motifs. However to obtain the narrow line features that run normal to $[110]$ and $[1\bar{1}0]$ it is necessary to have some fairly long-range correlation present. It was therefore supposed that there were in addition to the isolated defects long chains of defects which extend appreciable distances in the $[110]$ and $[1\bar{1}0]$ directions. Figure 18.3(a) shows a plot of the $X_{i,j}$ variables used to represent this. In this and other similar figures a black pixel represents one of the structural motif columns in its normal position ($X_{i,j} = 0$) while a white pixel represents a motif column which is shifted by $\frac{1}{2}[001]$. In Fig. 18.3(a) there are approximately 3.3% of isolated defects and another 3.3% of defects present in the long chains in each of the $[110]$ and $[1\bar{1}0]$ directions. In Fig. 18.3(c) the number of isolated defects has been increased to 6.6% and defects present in the two types of long chain correspondingly decreased to 1.6%. The diffraction patterns of the $(hk5)$ section calculated from these distributions are shown in Fig. 18.3(b) and (d), respectively. These both show linear streaking and the broad regions of scattering but the relative intensity of the two is different. The relative intensity of the two features in Fig. 18.3(d) is in better agreement with the observed pattern than in Fig. 18.3(b).

Although the broad diffuse features and the sharp lines are reproduced quite well by this model the star-like features at the points where two lines intersect are not reproduced at all well. In addition, the model seems unrealistic from the point of view that

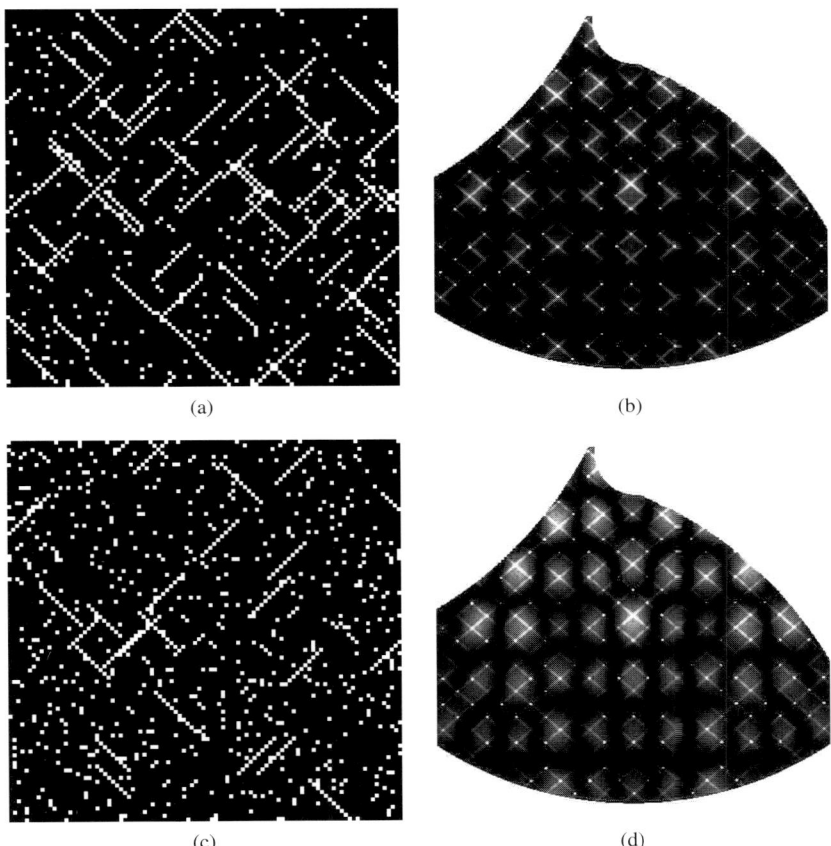

(a)

(b)

(c)

(d)

Fig. 18.3 Computer simulations of the first model of mordenite. (a) shows part of the array in which there are approximately 3.3% isolated defects and 3.3% defects in correlated rows in each of the directions $[1\,1\,0]$ and $[1\,\bar{1}\,0]$. In (c) there are approximately 6.6% isolated defects and 1.6% defects in correlated rows in each of the directions $[1\,1\,0]$ and $[1\,\bar{1}\,0]$. (b) and (d) show the calculated diffraction patterns of the $(h\,k\,5)$ section obtained using (a) and (c) respectively.

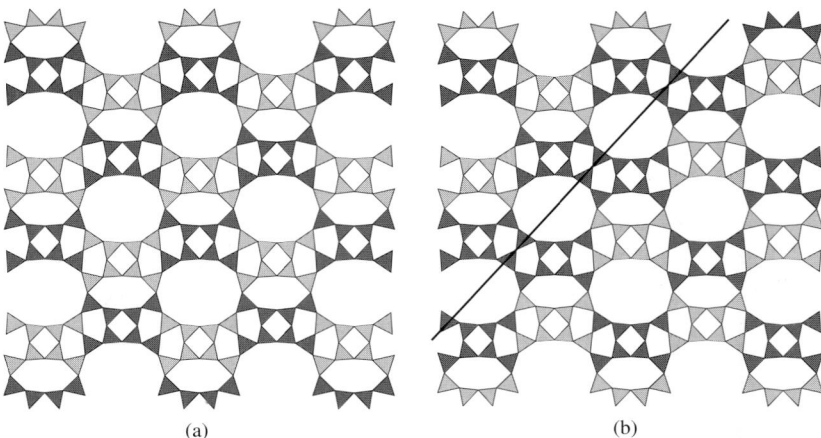

Fig. 18.4 Comparison of (a) the perfect mordenite structure with (b) the same region in which a $[1\,1\,0]$ stacking fault has been introduced. In (b) the part of the structure to the right of the dark line has undergone a $\frac{1}{2}[0\,0\,1]$ translation.

there are single defects and long rows of defects but no intermediate sized defects. If such defects were present they would have the effect of broadening the narrow diffraction lines. Consequently an alternative explanation for the sharp lines was sought.

Stacking faults are well-known to produce linear streaks in diffraction patterns and a second model (model 2) involving stacking faults in mordenite was therefore tested. Figure 18.4 shows how a stacking fault oriented in the $[1\,1\,0]$ direction can exist. Rather than a single column of motifs being displaced by $\frac{1}{2}[0\,0\,1]$, all of the material to the right of the dark line is displaced from its normal position. This displaced part of the crystal is seen to be still a region of perfect mordenite but it is in anti-phase to the region to the left of the line. Such a sharp boundary between two crystal domains produces a streak in reciprocal-space normal to its length. However unlike in the case of the linear defects discussed above the intensity in the streak due to a stacking fault is not uniform. The intensity is a maximum at the Bragg peak position and decays away from it. A model was therefore constructed in which stacking faults were introduced into an initially perfect mordenite structure. With all the variables $X_{i,j}$ initially set to 0, $[1\,1\,0]$ or $[1\,\bar{1}\,0]$ faults were placed one by one at random within the area of the crystal. At each new addition all the variables to the right of the fault line were flipped. That is, $X_{i,j} \Longleftrightarrow 1 - X_{i,j}$. After the insertion of a large number of faults the model crystal had the block mosaic appearance shown in Fig. 18.5(a), with any particular variable having been flipped a number of times. In addition to the stacking faults each of the two types of anti-phase region were also subject to the same random isolated column defects used in model 1. Figure 18.5(b) shows an enlargement of one small region of the total distribution showing how these isolated defects appear as white pixels in the dark domains and black pixels in the light domains.

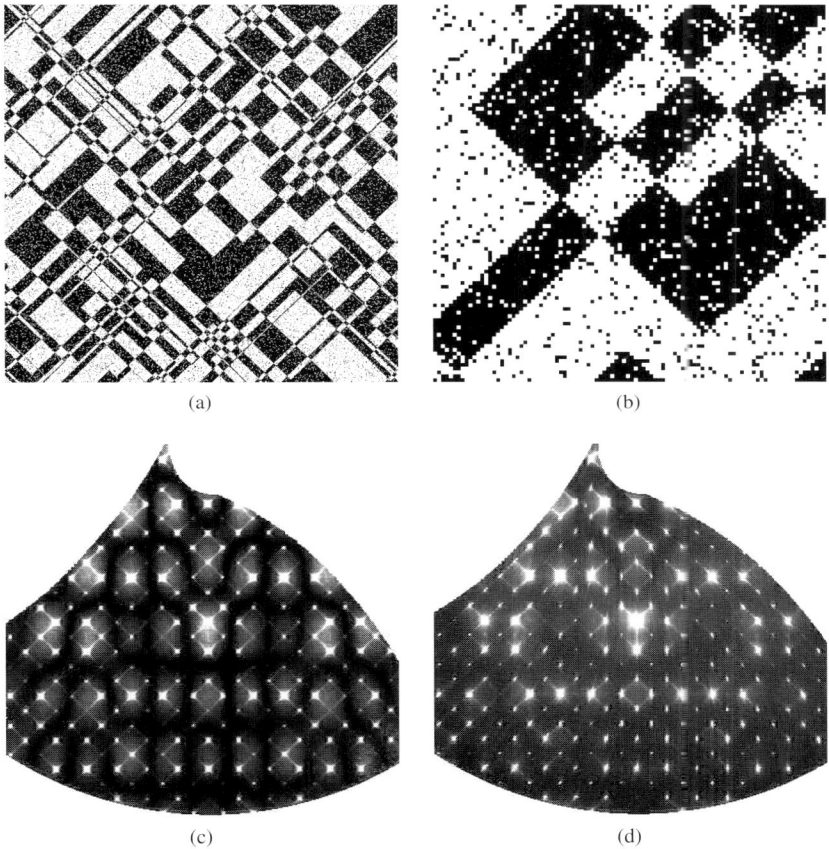

Fig. 18.5 Computer simulations of the second model of mordenite which includes stacking faults and random isolated column defects: (a) shows the whole 512×512 array of random variables, $X_{i,j}$. (b) shows an enlargement of a small region of (a). (c) shows the calculated diffraction pattern for the $(hk5)$ section calculated using the random variables in (a). (d) shows the observed diffraction pattern for comparison.

The diffraction pattern of the $(h\,k\,5)$ section obtained from this model is shown in Fig. 18.5(c) with the observed pattern reproduced in Fig. 18.5(d) for comparison. The star-shaped features on the strong Bragg peak positions are now reproduced well. As for model 1 the relative magnitudes of the streaking and broad continuous scattering features depend on the relative densities of the stacking faults and the isolated defects. Figure 18.5(b) should not be taken as indicative of the actual density of these defects but only their relative density. In order to establish estimates for the densities of either type of defect quantitative measurements on an absolute scale need to be made.

18.3   The defect structure of sodium bismuth titanate

18.3.1   *Background*

The rhombohedral perovskite $Na_{0.5}Bi_{0.5}TiO_3$ (NBT) is an example of a relaxor ferroelectric. Such materials have attracted considerable attention since the discovery of ultrahigh strain and giant piezoelectric properties in relaxor-based single crystals. NBT-based piezoelectrics offer exceptionally high strain and are thus a promising environmentally friendly alternative to the more usual lead oxides. NBT is unusual amongst perovskite materials since it appears to have an exact 1:1 ratio of the cations Na and Bi in the A sites (see Fig. 18.6(a)). In this section MC simulation experiments are described which help provide an explanation for the diffraction patterns that have been observed for NBT at ambient pressure and during a series of high pressure experiments.

Although NBT has rhombohedral symmetry the deviations from the primitive cubic perovskite model shown in Fig. 18.6(a) are small and in the description given here all directions and spacings refer to the cubic cell. Figure 18.6(b) shows the low-angle region of a diffraction pattern of a single crystal of NBT recorded at ambient temperature and pressure. This figure mainly shows the broad diffuse scattering due to the substitutional disorder of the Bi and Na atoms in the A sites. The sample used for this experiment had some surface damage and tangential streaking is evident near some Bragg peaks but the basic pattern of the broad regions of diffuse scattering is clear. This broad diffuse scattering also shows strong asymmetry with respect to the Bragg peak positions, indicative of the presence of the 'size-effect'. The intensity is stronger within the central square of reflections than outside it.

A second feature of the diffraction pattern of NBT that is barely perceptible in Fig. 18.6(b) is a set of linear streaks that emanate from the Bragg peaks along $\langle 1\,0\,0 \rangle$ directions. These are most visible in Laue (stationary crystal) diffraction patterns when the sphere of reflection is close to the Bragg reflection. This confirms that they correspond to rods of scattering in reciprocal-space rather than being the intersection of the Ewald sphere with diffuse planes of scattering. The rods extend away from the Bragg peaks only on the low angle side of the Bragg position so that they have the appearance of a letter 'L'. They have been observed at high pressure using a diamond anvil cell and Fig. 18.6(c) shows a series of images of the feature near the 320 reflection. Unlike the broad diffuse scattering, which remains virtually unchanged at high pressure, these rods of scattering are seen to disappear completely above a pressure of 2.8 GPa. The problem therefore is to obtain a model for NBT which can explain both of these effects.

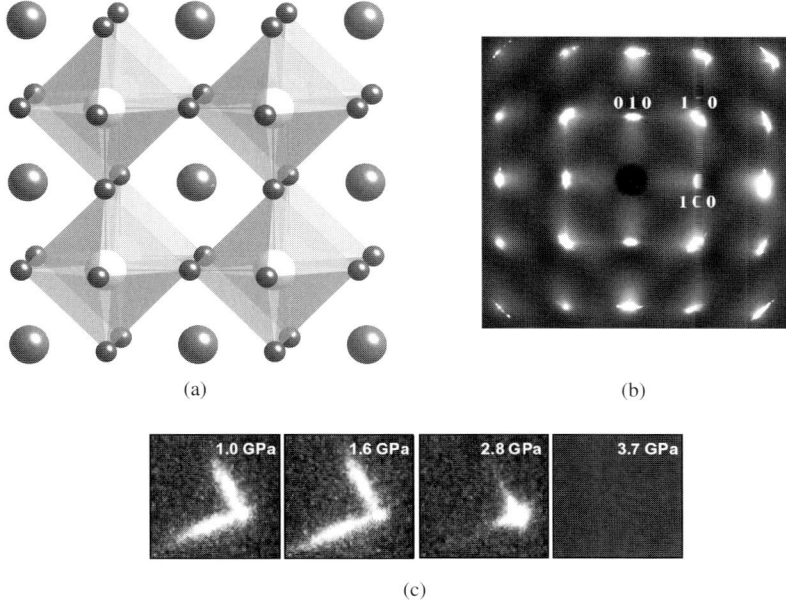

Fig. 18.6 (a) The perovskite ABO_3 structure viewed down $[0\,0\,1]$. The large dark spheres are the A cations, the small spheres are oxygen and the large light spheres within the O_6 octahedra are the B cations. (b) The $(h\,k\,0)$ diffraction pattern of NBT showing SRO diffuse scattering, 'size-effect' asymmetry and evidence of streaking between Bragg peaks. (c) Detail of the streaking near the $(3\,2\,0)$ Bragg peak as a function of pressure. (Data used in (c) are reproduced with kind permission of Dr Jens Kreisel.)

18.3.2 MC simulation—SRO

For simplicity a 2D model which tries to capture the essence of the structural disorder in NBT has been developed. Since it is only the perovskite A site that is disordered the TiO_3 part of the structure is completely ignored. The structure is therefore considered as a simple square array of A cations represented by random variables $\sigma_{i,j}$. If $\sigma_{i,j} = +1$ the site i, j is occupied by a Bi atom and if $\sigma_{i,j} = -1$ by a Na atom.

Near-neighbour correlations were introduced into the array of random variables using an Ising model with energy given by,

$$E_{\text{SRO}} = \sum_{\substack{\text{all } \langle 100 \rangle \\ \text{neighbours}}} J_1 \sigma_{i,j}\sigma_{i-n,j-m} + \sum_{\substack{\text{all } \langle 110 \rangle \\ \text{neighbours}}} J_2 \sigma_{i,j}\sigma_{i-n,j-m}. \tag{18.1}$$

Figure 18.7(a) and (b) show small portions of two example simulations generated using eqn (18.1). In these the white pixels represent Na and the grey pixels represent Bi. For Fig. 18.7(a) the $[1\,0\,0]$ nearest-neighbour correlation is 0.0 and the $[1\,1\,0]$ correlation is -0.1. For Fig. 18.7(b) the $[1\,0\,0]$ nearest-neighbour correlation is 0.1 and the $[1\,1\,0]$ correlation is -0.1. The corresponding diffraction patterns of these two distributions are shown in Fig. 18.7(c) and (d), respectively. Since all the atoms lie on the square lattice the diffuse patterns are periodic except for the fall-off in intensity at higher angles due to the atomic scattering factors.

'Size-effect' relaxation was then applied to these distributions using a second stage of MC simulation with an energy,

$$E_{Size} = \sum_{\substack{\text{all } \langle 100 \rangle \\ \text{neighbours}}} f_1(d_0 - d)^2 + \sum_{\substack{\text{all } \langle 110 \rangle \\ \text{neighbours}}} f_2(d_1 - d)^2. \tag{18.2}$$

Here d is the instantaneous length of a given inter-cation vector and f_1 and f_2 are respective force constants for the $[1\,0\,0]$ and $[1\,1\,0]$ interactions. d_0 and d_1 are the corresponding target distances for $[1\,0\,0]$ and $[1\,1\,0]$ neighbours. These depend on the size-effect parameters for particular pairs of cations. Thus,

$$d_0 = a(1 + \varepsilon_{a-b}); \qquad d_1 = a\sqrt{2}(1 + \varepsilon_{a-b}), \tag{18.3}$$

where $\varepsilon_{Bi-Bi} = 0.05$; $\varepsilon_{Bi-Na} = 0.00$; $\varepsilon_{Na-Na} = -0.5$.

Figure 18.7(e) and (f) show the diffraction patterns calculated from the resulting distributions of cations after 'size-effect' relaxation. Figure 18.7(f) shows good qualitative agreement with the observed distribution shown in Fig. 18.6(b).

18.3.3   *MC simulation—GP zones*

In order to explain the 'L-shaped' features in the observed pattern, it is necessary to assume some kind of planar defect (in 3D) reminiscent of Guinier–Preston zones (GPZs), as seen in metal alloys such as AlCu solid solutions (e.g. see Guinier, 1963, page 291). Such planar defects give rise to rods of diffuse scattering in reciprocal-space, in directions normal to the planes of the defects. Moreover the asymmetry of the intensity of these rods with respect to the Bragg peak positions can be understood in terms of the 'size-effect' that results from the deformation of the structure around the chemical segregation planes. It is thus tempting to invoke a similar explanation for the 'L-shaped' diffuse scattering in NBT. The problem is that it is not easy to see what these defects in NBT might be comprised of. If they consisted of segregated planar islands of excess Bi atoms (or Na atoms), for example, then it is difficult to imagine how their effect could be removed by the application of pressure. A clue to the possible origin of the effect is given by the fact that the Bi atoms are grossly underbonded when placed at the A-site of the average perovskite structure and can only achieve their bonding requirements by moving off centre. The requirement for Bi to improve its coordination environment is seen as the main impetus for the deviations from the average structure that gives rise to the diffuse scattering.

With this in mind the GPZs could be considered to be regions where such cation displacements are correlated from site to site, effectively producing a thin platelet of a lower-symmetry phase. Figure 18.8 shows schematically the envisaged form of such a GPZ. The rows of dark circles represent cations that are displaced in concert along

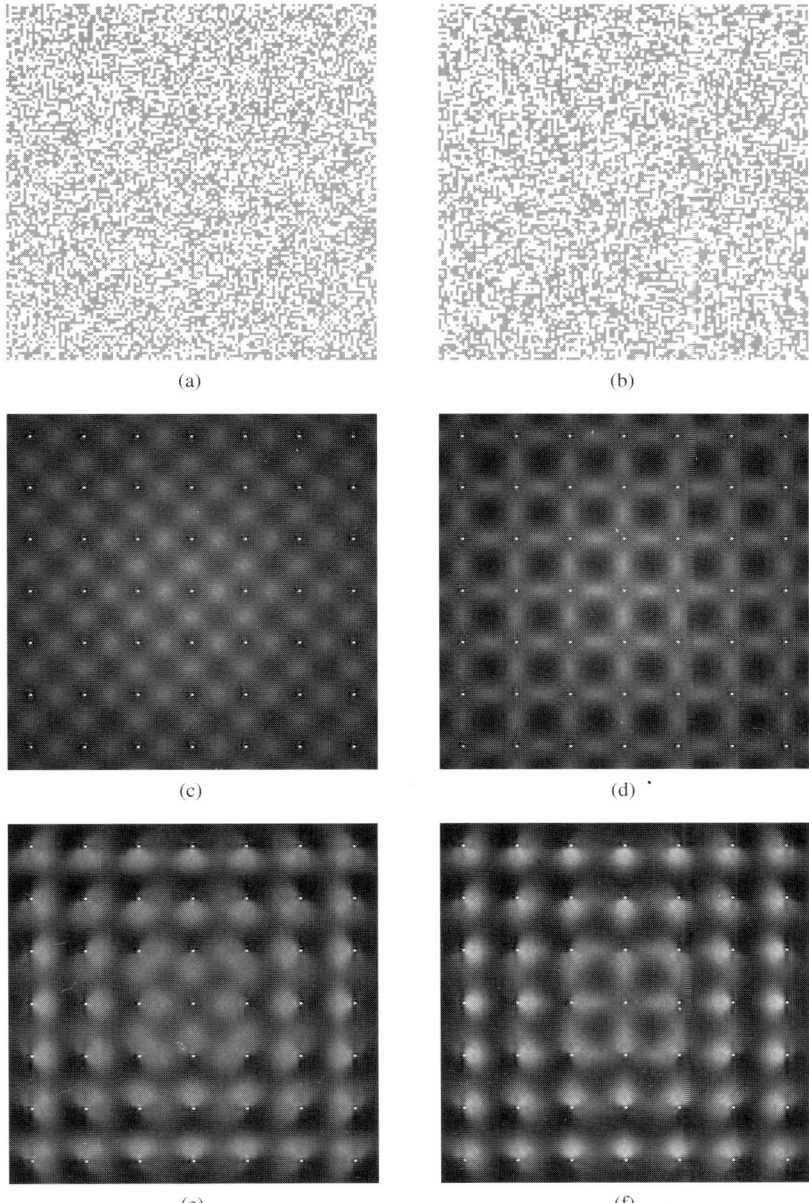

Fig. 18.7 (a)–(b) show small representative regions of the distribution of random variables representing the disorder in NBT. (a) $[1\,0\,0]$ correlation 0.0; $[1\,1\,0]$ correlation -0.1. (b) $[1\,0\,0]$ correlation 0.1; $[1\,1\,0]$ correlation -0.1. (c) and (e) show diffraction patterns calculated from (a). (d) and (f) show diffraction patterns calculated from (b). (c) and (d) without 'size-effect'. (e) and (f) with 'size-effect'.

$[1\,0\,0]$ or $[0\,1\,0]$. At the same time the structure around the defect might be expected to relax so that neighbouring rows of cations are drawn in towards the defect, as shown. If this model for the defects is correct then the application of pressure could be seen as acting to force the cations back to the average A-site positions at the centre of the unit cell. Since the magnitude of the scattering effects (both the SRO term, I_0, and the size-effect term, I_1) due to the GPZs would be expected to be proportional to $(F_{GPZ} - F_{matrix})^2$ then the intensity will clearly fall to zero if the structure factor for the GPZ becomes identical to that of the surrounding crystal matrix.

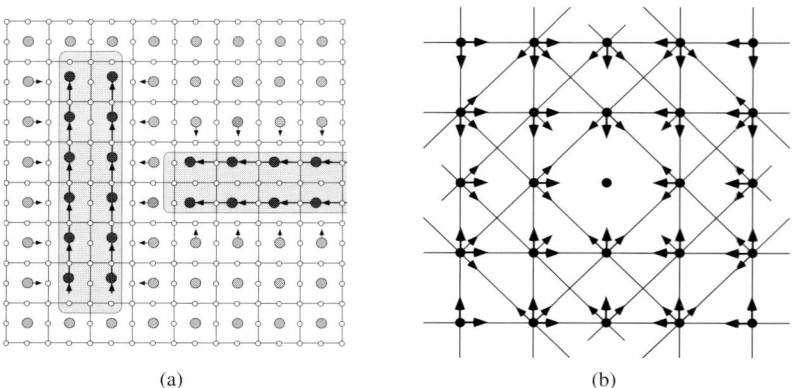

(a)   (b)

Fig. 18.8 Schematic diagram of the cation displacements in the model used for the GP zones in NBT.

In order to demonstrate the feasibility of this explanation for the 'L-shaped' features in NBT MC simulations were carried out for several different models of the defects. All of the example diffraction patterns shown in Fig. 18.9 are based on the real-space distribution part of which is shown in Fig. 18.9(a). The crystal matrix is seen to consist of the same disordered distribution of Bi and Na shown in Fig. 18.7(b). In addition, however, a small number of GPZs have now been included in two different orientations. The diffraction patterns shown in Figs. 18.9(b)–(d) were all calculated from the distribution in (a) but the exact nature of the GPZs was different for each.

For Fig. 18.9(b) the GPZs contained Na atoms at the cell centres. The parameters used to apply size-effect distortions over the crystal matrix were the same as for the examples in Fig. 18.7. In addition an extra size-effect term was introduced which tended to reduce the distance between the cations within the GPZ and those in the surrounding crystal matrix. In this case the scattering power within the GPZ, F_{Na} is less than the average scattering power from the matrix and the resulting size-effect transfer of intensity produces the 'L-shape' feature on the low-angle side of the Bragg peaks (cf. Fig. 7.2). Since for this case $(F_{GPZ} - F_{matrix})^2$ is non-zero at the origin there is a complete 'cross' of scattering there. In addition it may be seen that the intensity of the 'L-shape' near the low-angle $(1\,1\,0)$ reflection is stronger than that near the higher-angle $(3\,3\,0)$ reflection.

This is contrary to observation.

In contrast to this cation segregation model for Fig. 18.9(c) the cation displacement model described above was assumed. In this case the GPZ contains cations with the same average scattering power as the crystal matrix. Now the value of $(F_{GPZ} - F_{matrix})^2$ is zero at the origin since (F_{GPZ} is the same as F_{matrix} apart from a phase shift. In addition the value increases with distance from the origin in the direction of the cation shifts. The diffraction pattern still shows the 'L-shaped' features but now their intensity increases so that the feature near the $(3\,3\,0)$ reflection is much stronger than that near $(1\,1\,0)$. In addition there is no 'cross' at the origin. Figure 18.9(d) shows a diffraction pattern of exactly the same model except for the fact that the number of GPZ defects was doubled. The result is that the intensity of the 'L-shaped' features is enhanced relative to the SRO scattering so they may be seen more clearly.

Figure 18.9(e) was obtained from exactly the same model as (c) but in this case the cation shifts away from the centre of the cell were reduced to zero. The 'L-shaped' features have completely disappeared. Figure 18.9(f) has been included to show the form of the scattering when no size-effect relaxation is applied. The GPZs now produce a 'cross' of scattering around each Bragg peak.

18.4   'Size-effect'-like distortions in quasicrystalline structures

18.4.1   *Background*

The discovery of systems with diffraction patterns that exhibit icosahedral symmetry (Shechtman *et al.*, 1984), which is forbidden in classical crystallography, has lead to the development of a new area of study—namely the study of quasicrystals. Since that first discovery the number of systems with quasicrystalline phases possessing such non-classical symmetries as 8-, 10- or 12-fold rotation axes, has grown rapidly (Steurer, 1990) and the phenomenon can no longer be considered merely a curiosity. As a model for the explanation and understanding of quasicrystallinity there has been much interest in Penrose tiling and its generalisations (Penrose, 1974). In such tiling the *unit cell* of classical crystallography is replaced by two differently shaped *tiles*. The complete tiling pattern fills space without defects and possesses long-range orientational order, but does not possess translational symmetry.

Although much progress has been made in understanding the nature of quasicrystals and in elucidating details of the atom arrangements in them they nevertheless still present considerable challenges to the structural scientist. Most models of quasicrystals are formulated in a higher-dimensional space (e.g. 5D or 6D) and the real-space structure is obtained by projection from the higher dimension into the normal 3D laboratory space. During this process it is often difficult to remember that the structure is comprised of real atoms in 3D space which are subject to the same kinds of interatomic forces that exist in ordinary crystalline materials.

The diffraction pattern of a perfect quasicrystal contains only sharp Bragg peaks but many real quasicrystals also contain diffuse scattering of various kinds (e.g. see Hradil *et al.*, 1995). One area of current interest is in trying to understand how transitions between crystalline and quasicrystalline phases can occur. Such changes in the atomic structure are reflected in complex diffuse scattering patterns. So far only limited

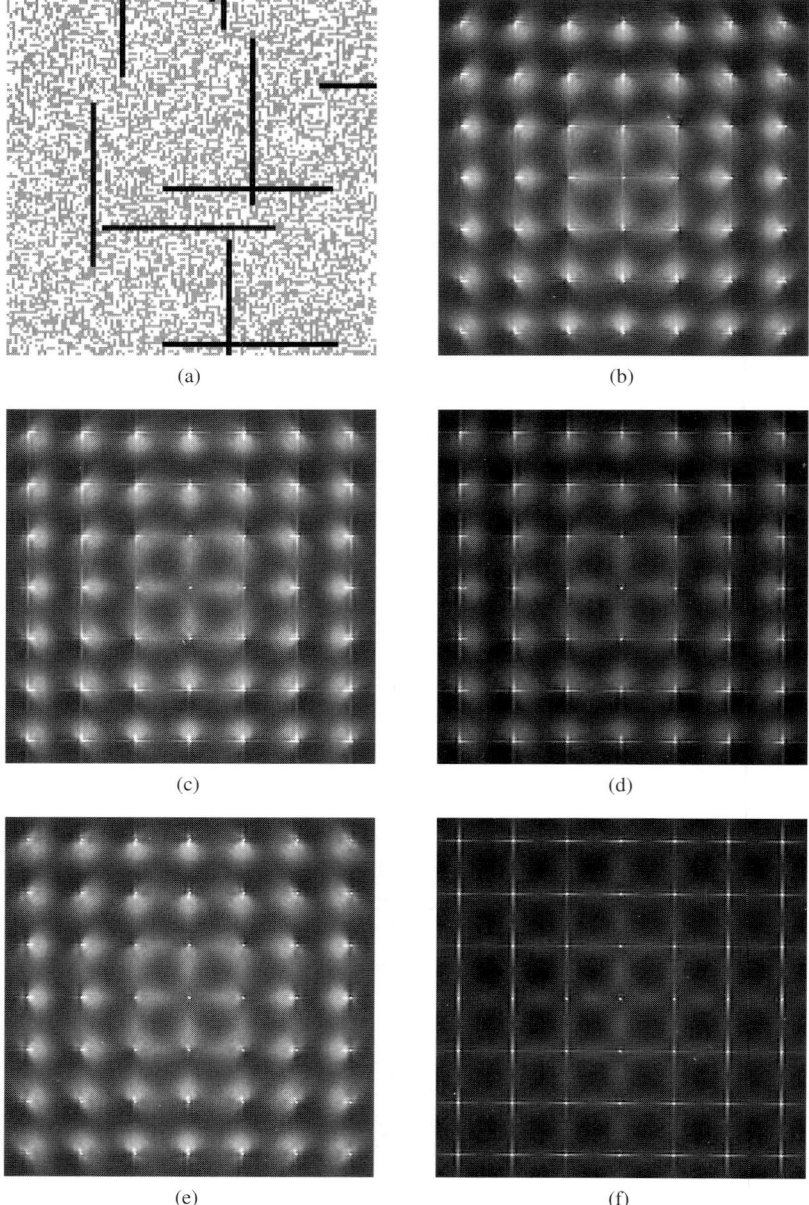

Fig. 18.9 (a) A small part of the model structure of NBT. White pixels represent unit cells containing Na, grey pixels unit cells containing Bi and black pixels represesnt GPZ's the nature of which are interpreted differently in making the calculations of the diffraction patterns in (b)–(f). See text for details.

progress has been made in understanding the details of this scattering. As for crystalline materials it is clear that diffuse scattering arises whenever there are departures from perfect quasicrystallinity but the higher dimensional description needed adds considerable extra complexity. Nevertheless it is likely that the same kinds of computer simulation methods that have proved successful for crystalline materials will be equally useful for aiding the interpretation and analysis of problems in the quasicrystal area.

Computer simulation studies have been carried out using various tiling models to investigate the relationship between quasicrystals and crystalline twin aggregates (Welberry, 1989), to investigate the possibility of inducing short-range crystalline order in quasicrystals via phason fluctuations (Welberry, 1991) and to investigate a possible mechanism for the quasicrystal-to-crystal transformation (Honal and Welberry, 2002). Although these investigations using tiling models have not at this stage dealt with real atomic structures they have nevertheless given useful insights into the different effects that can occur in quasicrystals. In this section a similar study is described which illustrates how such investigations of tiling models can be useful. This has been described in detail by Welberry and Honal (2002). This particular example is interesting as it emphasises how the same kinds of diffraction phenomena that occur in crystals can also exist in quasicrystals.

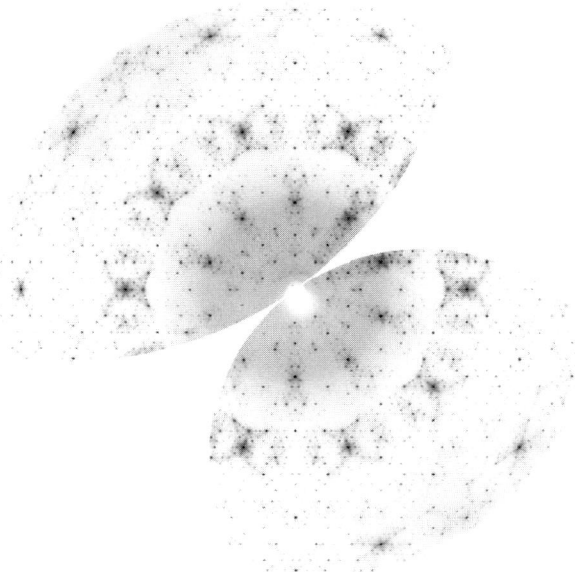

Fig. 18.10   The zero-level ($h_5 = 0$) section diffraction pattern of $Al_{71}Co_{13}Ni_{16}$ (Estermann *et al.*, 2000). Reproduced with kind permission of Prof. W. Steurer.

The study was prompted by the publication of the diffraction patterns of the decagonal quasicrystal $Al_{71}Co_{13}Ni_{16}$ (Estermann *et al.*, 2000). The ($h_5 = 0$ section of these patterns is reproduced in Fig. 18.10 for reference. [It should be noted that the diffraction patterns for this example are displayed as negative images with black representing high-intensity and white low-intensity]. The pattern clearly shows Bragg peaks and diffuse scattering but the most noticeable feature of the patterns is the clearly delineated central decagon-shaped region of lower intensity. The intensity immediately outside the decagon of both diffuse scattering and Bragg peaks is clearly much greater than that on the inside. This feature is very reminiscent of the type of 'size-effect' transfer of intensity observed in crystals, for example, in the urea inclusion compounds (see Fig. 10.3(d)). In order to test whether some kind of 'size-effect' distortion could account for this kind of intensity transfer in quasicrystals some MC simulations were carried out using the basic Penrose tiling pattern.

18.4.2 MC simulation

A small piece of perfect Penrose rhomb tiling is shown in Fig. 18.11. In addition to the two types of rhomb-shaped tiles the figure contains four different types of symbol placed at the vertices. These are a square, a triangle, a star and a circle. The different symbols in fact indicate that the vertices of the 2D tiling originate from different parts of the projection volume when the tiling pattern is obtained by projection from higher dimension (see Yamomoto and Ishihara, 1988; Ishihara and Yamomoto, 1988). Although there is no actual disorder in the pure Penrose pattern it seemed a possibility that the distinct types of vertex could be treated as different atomic species with different physical sizes, so that the rhomb edge vectors connecting two species expanded or contracted from their ideal length.

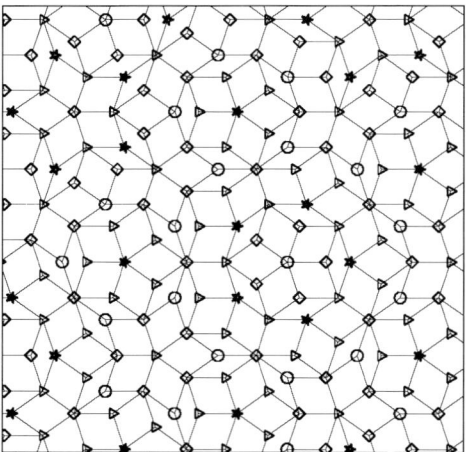

Fig. 18.11 A small portion of the Penrose rhomb-tiling showing the four different types of vertices: star, circle, triangle and square.

If the rhomb-edges are classified by the vertices they join, reference to Fig. 18.11 reveals that there are only three different types: triangle-star, square-triangle and circle-square. The frequencies with which these different edge types occur are approximately in the ratios $1:2:1$ respectively. In order that there be no net change of the average rhomb-edge length, it seemed reasonable to use MC simulation to try to reduce the length of the most common type of edge (square-triangle) and correspondingly increase the length of the less common edges (circle-square and triangle-star). This was achieved using MC simulation with the energy

$$E = \sum_{\text{vectors } i,j} k(d_{ij} - a_0(1 + \varepsilon_{ij}))^2. \tag{18.4}$$

Here d_{ij} is the instantaneous length of the rhomb-edge vector connecting vertices i and j and a_0 is the ideal rhomb-edge length. A large region of the Penrose rhomb pattern containing 10656 vertices was used. The x and y cartesian coordinates of each vertex were stored in a separate array. MC iteration was carried out by selecting a vertex at random and comparing the energy before and after a small random shift was applied to its position. The shift was accepted if the energy was lower and otherwise it was rejected (i.e. the iteration was essentially carried out at zero temperature). Simulation was carried out for 200 MC cycles where a cycle is defined as the number of individual MC steps required to visit each vertex once on average.

18.4.3   *Diffraction patterns*

Diffraction patterns were calculated from the final values of the (x, y) vertex coordinates using the program DISCUS (Proffen and Neder, 1997). A single atomic scatterer was placed at each vertex. Example diffraction patterns for different values of the 'size-effect' parameters ε_{ij} are shown in Fig. 18.12 together with small samples of the tiling patterns used to calculate them. Figure 18.12(a),(b) show the undistorted Penrose tiling and its diffraction pattern. Figure 18.12(c),(d) show the corresponding tiling and its diffraction pattern after MC simulation with $\varepsilon_{ij} = +0.15$ and Fig. 18.12(e),(f) show the corresponding tiling and its diffraction pattern for $\varepsilon_{ij} = -0.15$. It should be noted that both (c) and (e) are topologically identical to the pure Penrose pattern (a) and so are still essentially perfectly quasicrystalline. Consequently all three of the diffraction patterns show Bragg peaks and (virtually) no diffuse scattering. However, the 'size-effect' distortion has had a marked effect on the distribution of intensity in the Bragg peaks. Most remarkably the diffraction pattern Fig. 18.12(d) shows a marked decagon of lighter intensity similar to that observed in decagonal AlCoNi in Fig. 18.10.

It is interesting to look in more detail at the structure shown in Fig. 18.12(c) that has been produced by application of the size-effect. Figure 18.13(a) shows a plot of the frequency with which rhomb edge-lengths occur in this simulation. Here the vectors that have been reduced or increased in length during the MC simulation are immediately apparent. However, the figure further shows that the vectors have split into a number of different groups which have changed by differing amounts. One grouping which clusters about the original edge-length of 1.00 occurs both for vectors that have been subjected to contraction forces and for vectors that have been subject to expansion forces. These vectors must correspond to regions of the Penrose pattern which are not easily distorted. The vectors that have changed substantially comprise a cluster at around 0.94

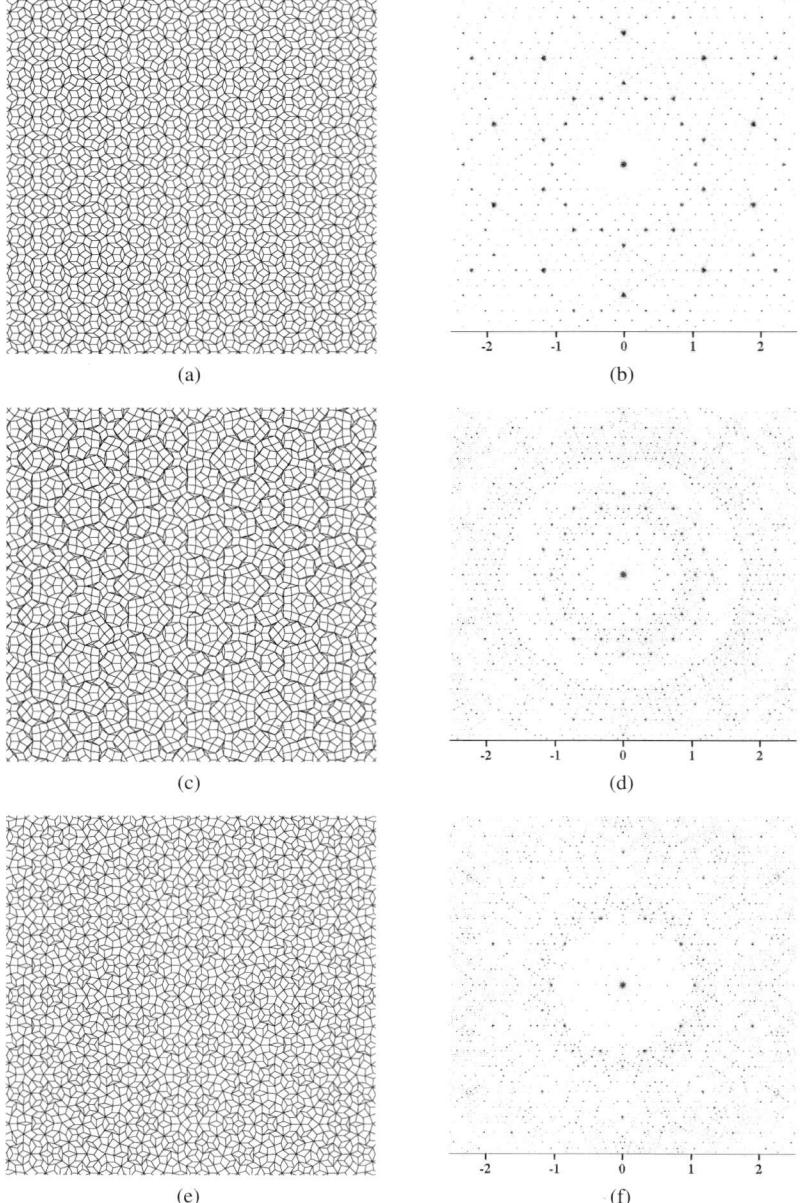

Fig. 18.12 Small representative samples of the tiling patterns and their corresponding diffraction patterns: (a), (b) the undistorted Penrose pattern; (c), (d) the same region of tiling after MC simulation with $\varepsilon_{ij} = +0.15$; (e), (f) the same region of tiling after MC simulation with $\varepsilon_{ij} = -0.15$. The horizontal scale is in reciprocal lattice units relative to a real-space scale in which the rhomb edge has unit length.

(6% contraction) and three additional clusters around 1.06, 1.12, 1.18 (6%, 12%, 18% expansion respectively). Note that the tail of very small distances (< 0.9) originates from rhomb edges near the periphery of the sample where the absence of some neighbouring tiles allows unconstrained contraction. The two different profiles for the expanded vectors, which are nevertheless quite similar, correspond to the triangle-star and circle-square vectors respectively.

In order to see where in the tiling these differently expanded/contracted vectors occur Fig. 18.13(b) shows a region of the tiling in which vectors are coloured according to whether they are shorter (black) or longer (shades of grey) than their original length. It is seen that the vectors in the centre of the pentagonal clusters have expanded the most, while vectors on the outside of the pentagonal clusters have contracted the most. It is also seen in this figure that many of the thin rhombs have become very thin while others have become almost triangular. On the other hand many of the fat rhombs have become more like squares.

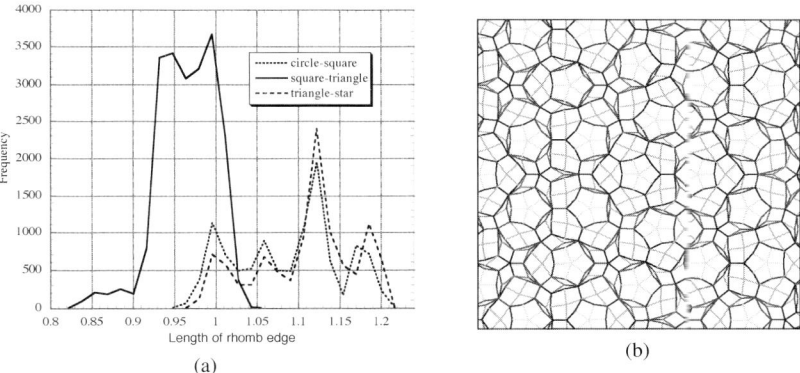

(a)   (b)

Fig. 18.13   (a) Shows a histogram of the different rhomb-edge lengths in the distorted tiling of Fig. 18.12(c) relative to the original length a_0. (b) Shows an enlarged plot of the central region of the tiling drawn to show which rhomb edges have contracted (black) and which have expanded (grey to light grey).

Figure 18.12(d) suggests quite convincingly that the 'size-effect' must play a significant role in decagonal Al—Co—Ni. However, since the atomic radii of Ni and Co differ by less than 0.5% it seems unlikely that size-effect distortions involving disordered Ni/Co sites can explain the observed diffraction effects. Aluminium has an atomic radius substantially higher ($\sim$ 14%) than either Ni or Co so could potentially be a source of strains of the required magnitude. It is perhaps a more likely possibility that size-effect-like distortions result from relaxations due to vacancies.

REFERENCES

Allpress, J. and Rossell, H. (1975). *Journal of Solid State Chemistry*, **15**, 68–78.

Amorós, J. L. and Amorós, M. (1968). *Molecular Crystals: Their Transforms and Diffuse Scattering*. Wiley: New York.

Angel, R. and Prewitt, C. (1987). *Acta Crystallographica B*, **62**, 116–126.

Bar, I. and Bernstein, J. (1982). *Journal of Physical Chemistry*, **86**, 3223–3231.

Bar, I. and Bernstein, J. (1983). *Acta Crystallographica*, **39**, 266–272

Bartlett, M. S. (1967). *Journal of the Royal Statistical Society*, **130**, 457–474.

Bartlett, M. S. (1968). *Journal of Applied Probability*, **131**, 579–580.

Bartlett, M. S. (1971). *Journal of the Royal Statistical Society*, **8**, 222–232.

Braga, D., Grepioni, F., Farrugia, L. J. and Johnson, B. F. G. (1994). *Journal of the Chemical Society. Dalton transactions.*, pp. 2911–2918.

Brese, N. E. and O'Keeffe, M. (1991). *Acta Crystallographica*, **B47**, 192–197.

Brink, F. J., Withers, R. L., Friese, K., Madariaga, G. and Norén, L. (2002). *Journal of Solid State Chemistry*, **163**, 267–274.

Butler, B. D., Haeffner, D. R., Lee, P. L. and Welberry, T. R. (2000). *Journal of Applied Crystallography*, **33**, 1046–1050.

Butler, B. D. and Hanley, H. J. M. (1999). *Journal of Sol–gel Sciience and Technology*, **15**, 161–166.

Butler, B. D. and Welberry, T. R. (1992). *Journal of Applied Crystallography*, **25**, 391–399.

Butler, B. D. and Welberry, T. R. (1994). *Journal of Applied Crystallography*, **27**, 742–754.

Cahn, J. W. (1967). *Transactions of the Metallurgical Society of AIME*, **242**, 168–180.

Campbell, B. and Cheetham, A. (2002). *Journal of Physical Chemistry*, **106**, 57–62.

Clapp, P. C. (1969). *Journal of Physics and Chemistry of Solids*, **30**, 2589–2598.

Clapp, P. C. (1971). *Physical Review*, **4**, 255–270.

Cotton, F. A. and Troup, J. M. (1975). *Journal of the American Chemical Society*, **96**, 4155–4159.

Cowley, J. M. (1950). *Physical Review*, **77**, 669–675.

Cowley, J. M. (1968). *Acta Crystallographica*, **24**, 557–563.

Cox, D. R. and Miller, H. D. (1965). *The Theory of Stochastic Processes*. Methuen: London.

de Ridder, R., van Dyck, D., van Tenderloo, D. and Amelinckx, S. (1977). *Physica Status Solidi (a)*, **40**, 669–683.

de Ridder, R., van Tenderloo, D., van Dyck, D. and Amelinckx, S. (1976). *Physica Status Solidi (a)*, **38**, 663–674.

Dobrushin, R. L. (1968). *Theory of probability and its applications*, **13**, 197–224.

Enting, I. G. (1977*a*). *Journal of Physics*, **10**, 1379–1388.

Enting, I. G. (1977*b*). *Journal of Physics*, **10**, 1023–1030.

Enting, I. G. (1977*c*). *Journal of Physics*, **10**, 1737–1743.

Enting, I. G. (1978*a*). *Journal of Physics*, **11**, 555–562.

Enting, I. G. (1978*b*). *Journal of Physics*, **11**, 2001–2013.

Estermann, M., Lemster, K., Haibach, T. and Steurer, W. (2000). *Zeitschrift f Kristallographie*, **215**, 584–596.

Forst, R., Jagodzinski, H., Boysen, H. and Frey, F. (1987). *Acta Crystallographica B*, **43**, 187–197.

Galbraith, R. F. and Walley, D. (1976). *Journal of Applied Probability*, **13**, 548–557.

Galbraith, R. F. and Walley, D. (1980). *Journal of Applied Probability*, **17**, 124–133.

Garstein, E. and Cohen, J. (1980). *Journal of Solid State Chemistry*, **33**, 271–272.

Grimmett, G. R. (1973). *Bulletin of the London Mathematical Society*, **5**, 81–84.

Guinier, A. (1963). *X-ray Diffraction in Crystals, Imperfect Crystals and Amorphous Bodies*. W. H. Freeman: San Francisco.

Haller, K. J., Rae, A. D., Heerdegen, A. P., Hockless, D. C. R. and Welberry, T. R. (1995). *Acta Crystallographica*, **51**, 198–197.

Hammersley, J. M. (1967). In *Proceedings of the fifth Berkeley Symposium on Mathematical Statistics and Probability*, vol. 3, pp. 89–118. University of California Press: Berkeley.

Honal, M. and Welberry, T. (2002). *Zeitschrift f. Kristallographie*, **217**, 109–118.

Hosemann, R. and Bagghi, S. (1962). *Direct Analysis of Diffraction by Matter*. North Holland: Amsterdam.

Hradil, K., Proffen, T., Frey, F., Kek, S., Krane, H. G. and Wroblewski, T. (1995). *Philosophical Magazine, Letters*, **71**, 199–205.

Hurst, C. A. and Green, H. S. (1960). *Journal of Chemical Physics*, **33**, 1059–63.

Ishihara, K. and Yamomoto, A. (1988). *Acta Crystallographica*, **A44**, 508–516.

Ising, E. v. (1925). *Physikalische Zeitschrift*, **31**, 253–258. In German.

Kakinoki, J. and Komura, Y. (1952). *Journal of the Physical Society of Japan*, **7**, 30–36.

Kakinoki, J. and Komura, Y. (1954*a*). *Journal of the Physical Society of Japan*, **9**, 169–176.

Kakinoki, J. and Komura, Y. (1954*b*). *Journal of the Physical Society of Japan*, **9**, 177–183.

Kakinoki, J. and Komura, Y. (1965). *Acta Crystallographica*, **19**, 137–147.

Kaufmann, B. and Onsager, L. (1949). *Physical Review*, **76**, 1244–1252.

Kitaigorodsky, A. I. (1973). *Molecular Crystals and Molecules*. Academic Press, New York.

Koch, F. and Cohen, J. (1969). *Acta Crystallographica B*, **25**, 275–287.

Maxwell, I. E., Williams, C., Muller, F. and Krutzen, B. (1992). Zeolite Catalysis— For the Fuels of Today and Tomorrow, Selected Papers Series. Tech. rep., Shell International Chemical Co. Ltd.

Mayo, S. C., Welberry, T. R., Bown, M. and Tarr, A. (1998). *Journal of Solid State Chemistry*, **141**, 437–451.

McConnell, J. and Heine, V. (1985). *Physical Review B*, **31**, 6140–6142.

Metropolis, N., Rosenbluth, A. W., Rosenbluth, M. N., Teller, A. H. and Teller, E. (1953). *Journal of Chemical Physics*, **21**, 1087–1092.

Montroll, E. W., Potts, R. B. and Ward, J. C. (1963). *Journal of Mathematical Physics*, **4**, 308–322.

Morinaga, M., Cohen, J. and Faber, J. J. (1980). *Acta Crystallographica*, **36**, 520–530.

Mortier, W. J., Pluth, J. J. and Smith, J. V. (1975). *Materials Research Bulletin*, **10**, 1319–1325.

Moussouris, J. (1974). *Journal of Statistical Physics*, **10**, 11–33.

Neder, R., Frey, F. and Schulz (1990). *Acta Crystallographica*, **46**, 792–798.

Ogita, N., Ueda, A. and Matsubara, T. (1969). *Journal of the Physical Society of Japan*, **26**, 145–149. Suppl.

Onsager, L. (1944). *Physical Review*, **44**, 117–149.

Osborn, J. C. and Welberry, T. R. (1990). *Journal of Applied Crystallography*, **23**, 476–484.

Penrose, R. (1974). *Bulletin of the Institute of Mathematics and its Applications*, **10**, 255–271.

Phillips, F. C. (1963). *An Introduction to Crystallography*. Longmans: London, 3rd edn.

Pickard, D. K. (1977). *Journal of Applied Probability*, **10**, 717–731.

Pickard, D. K. (1978). *Advances in Applied Probability*, **10 (Suppl.)**, 57–64. Suppl.

Pickard, D. K. (1980). *Advances in Applied Probability*, **12**, 655–671.

Preston, C. J. (1973). *Advances in Applied Probability*, **5**, 242–261.

Proffen, T. and Neder, R. B. (1997). *Journal of Applied Crystallography*, **30**, 171–175.

Proffen, T., Neder, R. B., Frey, F. and Assmus, W. (1993). *Acta Crystallographica*, **49**, 599–604.

Proffen, T. and Welberry, T. R. (1997). *Acta Crystallographica*, **53**, 202–216.

Reynolds, P. A. (1975). *Molecular Physics*, **29**, 519–529.

Roth, W. (1960). *Acta Crystallographica*, **13**, 140–149.

Sanders, J. V. (1985). *Zeolites*, **5**, 81–90.

Sauvage, M. and Parthé, E. (1972). *Acta Crystallographica*, **A28**, 607–616.

Schlenker, J. L., Pluth, J. J. and Smith, J. V. (1979). *Materials Research Bulletin*, **14**, 849–.

Schlössen, H. H. and Lang, A. R. (1965). *The Philosophical Magazine*, **12**, 283–296.

Schweika, W., Hoser, A., Martin, M. and Carlson, A. (1995). *Physical Review B*, **51**, 15771–15788.

Shechtman, D., Blech, I., Gratias, D. and Cahn, J. (1984). *Physical Review Letters*, **53**, 1951–1953.

Sherman, S. (1973). *Israel Journal of Mathematics*, **14**, 92–103.

Sie, S. T. (1994). In *Advanced Zeolite Science and Applications*, vol. 85 of *Studies in surface science and catalysis*, chap. 17, pp. 587–631. Elsevier: Amsterdam.

Speed, T. P. (1978). *Advances in Applied Probability*, **10**, 111–122. Suppl.

Spitzer, F. (1971). *American Mathematical Monthly*, **78**, 142–154.

Steurer, W. (1990). *Zeitschrift f. Kristallographie*, **190**, 179–234.

Verhagen, A. M. (1977). *Journal of Chemical Physics*, **67**, 5060–5065.

Voller, V. and Porte-Agel, F. (2002). *Journal of Computational Physics*, **179**, 698–703.

Weber, T. and Bürgi, H.-B. (2002). *Acta Crystallographica*, **58**, 526–540.

Welberry, T. R. (1977*a*). *Journal of Applied Crystallography*, **10**, 344–348.

Welberry, T. R. (1977*b*). *Proceedings of the Royal Society of London*, **353**, 363–376.

Welberry, T. R. (1989). *Journal of Applied Crystallography*, **22**, 308–314.

Welberry, T. R. (1991). *Journal of Applied Crystallography*, **24**, 203–211.

Welberry, T. R. (2001). *Acta Crystallographica*, **57**, 348–358.

Welberry, T. R. and Butler, B. D. (1994). *Journal of Applied Crystallography*, **27**, 205–231.

Welberry, T. R., Butler, B. D. and Heerdegen, A. P. (1993*a*). *Acta Chimica Hungarica— Models In Chemistry*, **130**, 327–345.

Welberry, T. R., Butler, B. D., Thompson, J. G. and Withers, R. L. (1993*b*). *Journal of Solid State Chemistry*, **106**, 461–475.

Welberry, T. R. and Christy, A. G. (1997). *Physics and Chemistry of Minerals*, **24**, 24–38.

Welberry, T. R. and Galbraith, R. (1973). *Journal of Applied Crystallography*, **6**, 87–96.

Welberry, T. R. and Galbraith, R. (1975). *Journal of Applied Crystallography*, **8**, 636–644.

Welberry, T. R. and Glazer, A. M. (1994). *Journal of Applied Crystallography*, **27**, 733–741.

Welberry, T. R., Goossens, D., David, W. I. F., Gutmann, M. J., Bull, M. J. and Heerdegen, A. P. (2003*a*). *Journal of Applied Crystallography*, **36**, 14401447.

Welberry, T. R., Goossens, D., Haeffner, D., Lee, P. and Almer, J. (2003*b*). *Journal of Synchrotron Radiation*, **10**, 284–286.

Welberry, T. R. and Honal, M. (2002). *Zeitschrift f. Kristallographie*, **217**, 422426.

Welberry, T. R. and Mayo, S. C. (1996). *Journal of Applied Crystallography*, **29**, 353–364.

Welberry, T. R. and Miller, G. H. (1977). *Journal of Applied Probability*, **14**, 862–868.

Welberry, T. R. and Miller, G. H. (1978). *Acta Crystallographica*, **34**, 120–123.

Welberry, T. R., Miller, G. H. and Carroll, C. E. (1980). *Acta Crystallographica*, **36**, 921–922.

Welberry, T. R. and Proffen, T. (1998). *Journal of Applied Crystallography*, **31**, 309–317.

Welberry, T. R. and Withers, R. L. (1990). *Journal of Applied Crystallography*, **23**, 303–314.

Welberry, T. R., Withers, R. L. and Mayo, S. C. (1995). *Journal of Solid State Chemistry*, **115**, 43–54.

Whittle, P. (1954). *Biometrika*, **41**, 434–449.

Wilson, A. J. C. (1962). *X-ray Optics: the Diffraction of X-rays by Finite and Imperfect Crystals*. Methuen: London, 2nd edn.

Wilson, A. J. C. (1988). *Acta Crystallographica*, **44**, 715–724.

Wilson, A. J. C. (1990). *Acta Crystallographica*, **46**, 742–754.

Withers, R. L., Welberry, T. R., Brink, F. J. and Norén, L. (2003). *Journal of Solid State Chemistry*, **170**, 211–220.

Withers, R. L., Welberry, T. R., Larsson, A.-K., Liu, Y., Norén, Rund öf, H. and Brink, F. (2004). *Journal of Solid State Chemistry*, **177**, 231–244.

Wölfel, E. R. (1983). *Journal of Applied Crystallography*, **16**, 341–348.

Wolfram, S. (1983). *Reviews of Modern Physics*, **55**, 601–644.

Wood, R. A., Welberry, T. R. and Puza, M. (1984). *Acta Crystallographica*, **40**, 1255–1260.

Yamomoto, A. and Ishihara, K. (1988). *Acta Crystallographica*, **A44** 707–714.

INDEX